THE PEPTIDES

Analysis, Synthesis, Biology

Treatise Editors

E. GROSS AND J. MEIENHOFER

Volume 1
Major Methods of Peptide Bond Formation

Volume 2
Special Methods in Peptide Synthesis, Part A

Volume 3
Protection of Functional Groups in Peptide Synthesis

The Peptides *Volume 3*

The Peptides

V3:

Analysis, Synthesis, Biology

VOLUME 3 Protection of Functional Groups in Peptide Synthesis | 1981

Edited by

ERHARD GROSS

National Institutes of Health
Bethesda, Maryland

JOHANNES MEIENHOFER

Chemical Research Department
Hoffmann-La Roche Inc.
Nutley, New Jersey

1981

ACADEMIC PRESS
A Subsidiary of Harcourt Brace Jovanovich, Publishers
New York London Toronto Sydney San Francisco

ACADEMIC PRESS, INC.
111 Fifth Avenue, New York, New York 10003

United Kingdom Edition published by
ACADEMIC PRESS, INC. (LONDON) LTD.
24/28 Oval Road, London NW1 7DX

Library of Congress Cataloging in Publication Data
Main entry under title:

The Peptides : analysis, synthesis, biology.

Includes bibliographies and indexes.
CONTENTS: v. 1. Major methods of peptide bond formation.
--v. 3. Protection of functional groups in peptide synthesis.
1. Peptides. I. Gross, Erhard. II. Meienhofer, Johannes. [DNLM: 1. Peptides. QU68 P424]
QP552.P4P47 574.19'2456 78-31958
ISBN 0-12-304203-8 (v. 3)

PRINTED IN THE UNITED STATES OF AMERICA

81 82 83 84 9 8 7 6 5 4 3 2 1

Contents

Chapter 5 *Differential Protection and Selective Deprotection in Peptide Synthesis*

Jean-Luc Fauchère and Robert Schwyzer

Chapter 6 *Peptide Synthesis with Minimal Protection of Side-Chain Functions*

John K. Inman

Chapter 7 *Dual Function Groups*

Brian J. Johnson

List of Contributors

Numbers in parentheses indicate the pages on which the authors' contributions begin.

Jean-Luc Fauchère (203), Institute of Molecular Biology and Biophysics, Swiss Federal Institute of Technology, 8093 Zürich, Switzerland

Rolf Geiger (1), Hoechst Aktiengesellschaft, D-6230 Frankfurt-Höchst, West Germany

Richard G. Hiskey (137), Department of Chemistry, The University of North Carolina, Chapel Hill, North Carolina 27514

John K. Inman (253), Laboratory of Immunology, National Institute of Allergy and Infectious Diseases, National Institutes of Health, Bethesda, Maryland 20205

Brian J. Johnson (303), Department of Microbiology, University of Alabama in Birmingham, Birmingham, Alabama 35294

Wolfgang König (1), Hoechst Aktiengesellschaft, D-6230 Frankfurt-Höchst, West Germany

Roger W. Roeske (101), Department of Biochemistry, Indiana University School of Medicine, Indianapolis, Indiana 46202

Robert Schwyzer (203), Institute of Molecular Biology and Biophysics, Swiss Federal Institute of Technology, 8093 Zürich, Switzerland

John M. Stewart (169), Department of Biochemistry, University of Colorado School of Medicine, Denver, Colorado 80262

Preface

This volume of "The Peptides" presents the state of the art in protection of functional groups in peptide synthesis. Despite the multitude of protecting groups described in the literature, the tactics of functional group protection are by no means a routine procedure and may indeed be of considerable concern in the synthesis of large peptides. Recent attempts at chemical syntheses of lysozyme, bungarotoxin, or Kazal trypsin inhibitor have been frustrated by incomplete protecting group cleavage in the final stages of these projects. The refined systems of differential or graded acidolysis for selective protective group cleavage, developed during the past 25 years, have made a major contribution to the rapid expansion of peptide synthesis both by solid-phase procedures and by conventional synthesis in solution. However, syntheses of large peptides or small proteins seem to require even more stringent or absolute selectivity of protective group cleavage. Ideally, each type of protection should be removable without affecting any other. It is clear that the development of new blocking groups and the continued exploration and refinement of existing systems remains a major challenge in peptide synthesis.

In the first chapter, the large variety of amine protecting groups is comprehensively reviewed and clearly presented by R. Geiger and W. König. The protection of carboxyl groups is described by R. W. Roeske in Chapter 2. The chemistry of sulfhydryl group protection in peptide synthesis is distinguished by the large variety of different reactions lucidly discussed in Chapter 3 by R. G. Hiskey. Protection of the hydroxyl groups of serine, threonine, tyrosine, and other hydroxyl-containing amino acids is dealt with in Chapter 4 by J. M. Stewart. Differential protection and selective deprotection in peptide synthesis is authoritatively treated by J. L. Fauchère and R. Schwyzer in Chapter 5. The opportunities and constraints of the tactics of minimal protection of side-chain functions during peptide synthesis are discussed in Chapter 6 by J. K. Inman. The closing chapter by B. J. Johnson deals with the interesting aspects of dual function groups, an area whose potential has not yet been fully utilized in synthesis.

"The Peptides" is an open-ended work providing comprehensive and critical reviews of important developments in all areas of peptide research. These reviews are intended to serve as a reference for the specialist, a guide for the novice, and a forum for investigators concerned with peptide and protein research.

The purpose of the volumes on synthesis is to present the current state of methodology. The polymeric nature of peptides and their wide range of different properties present unpredictable obstacles during synthesis. In Volume 1, the "Major Methods of Peptide Bond Formation" were described; Volume 2 dealt with "Special Methods in Peptide Synthesis." Volume 4 will be devoted to methods of analysis and characterization of peptides.

We are much indebted to the authors for their contributions and cooperation during the editorial process and to the staff of Academic Press for their efficient preparation of this volume.

Erhard Gross
Johannes Meienhofer

Nomenclature and Abbreviations

In several chapters of this volume the peptide-size nomenclature proposed by M. Bodanszky (1977) at the Fifth American Peptide Symposium has been used. Greek prefixes are retained for di- to decapeptides only. For peptides with more than 10 amino acid residues arabic numerals are used, for example, "15-peptide," "31-peptide," or corticotropin-(1–24), β-LPH-(61–91).

Peptide *segment* and *segment condensation* have been used instead of *fragment* or *fragment condensation.* Amino acids are of the L configuration unless otherwise indicated. In peptide structures, hyphens between D and the amino acid symbol are omitted, e.g., H-Gly-DPhe-Ala-OH.

The current *Chemical Abstracts* term 5(4*H*)-oxazolone is used for azlactone [viz. *Chem. Abstr.* **83**, 3633 CS(1975)], since it is unambiguous in contrast to other (earlier) terms in the literature (see Volume 1, p. 39, footnote).

Abbreviations are kept as short as possible, preferably three letters (four or more letters only if unavoidable). Many abbreviations are indeed those recommended by the IUPAC–IUB Commission on Biochemical Nomenclature, e.g., in *J. Biol. Chem.* **242**, 6489–6497 (1970); **250**, 3215–3216 (1975). However, the system proposed by the Commission in 1972 (viz. *J. Biol. Chem.* **247**, 997–982) of complex symbols frequently exceeding five spaces and including Greek letters and subscript numerals will not be used. These symbols are impractical for typesetting and are generally of infrequent use.

Symbols used in this volume are listed below (except those of the common amino acids).

Abbreviations

AA	amino acid
Ac	acetyl
Aca	acetoacetyl
AcOH	acetic acid
Acm	acetamidomethyl
ACTH	corticotropin

4Acz	4-acetoxybenzyloxycarbonyl
Adc	adamantyloxycarbonyl
Amv	2-acetyl-1-methylvinyl
Aoc	*tert*-amyloxycarbonyl
Azc	2-(4-phenylazophenyl)propyl(2)oxycarbonyl
Bam	benzamidomethyl
3Bb 4Bb	3- 4- } bromobenzyl
Bec	2-bromoethyloxycarbonyl
Bic	5-benzisoxazolylmethyleneoxycarbonyl
BME	β-mercaptoethanol
Bmv	2-benzoyl-1-methylvinyl
Boc	*tert*-butyloxycarbonyl
Bpoc	2-(4-biphenylyl)propyl(2)oxycarbonyl
Btf	2,2,2-trifluoro-1-*tert*-butyloxycarbonylaminoethyl
BTFA	boron tris(trifluoroacetate)
Bz	benzoyl
2Bz 4Bz	2- 4- } bromobenzyloxycarbonyl
Bzl	benzyl
2Cb 4Cb	2- 4- } chlorobenzyl
CCD	countercurrent distribution
Cla	chloroacetyl
4CNz	4-cyanobenzyloxycarbonyl
Coc	cholesteryloxycarbonyl
Cpc	cyclopentyloxycarbonyl
Cyc	2-cyano-*tert*-butyloxycarbonyl
2Cz 3Cz 4Cz	2- 3- } chlorobenzyloxycarbonyl 4-
Dbh	4,4′-dimethylbenzhydryl
Dbp	dibenzylphosphoryl
Dcb	2,6-dichlorobenzyl
DCC	dicyclohexylcarbodiimide
DCHA	dicyclohexylamine
2,4Dcz 3,4Dcz 2,6Dcz	2,4- 3,4- } dichlorobenzyloxycarbonyl 2,6-

Ddm	4,4′-dimethyloxybenzhydryl 4,4′-dimethyloxydiphenylmethyl
Ddz	2-(3,5-dimethyloxyphenyl)propyl(2)oxycarbonyl
Dim	5,5-dimethyl(3)oxocyclohexene(1)yl
Dit	diiodotyrosine
Dmb	3,4-dimethylbenzyl
Dmc	diisopropylmethyloxycarbonyl
DMF	dimethylformamide
Dmob	2,4-dimethyloxybenzyl
Dmoz	3,5-dimethyloxybenzyloxycarbonyl
DMSO	dimethyl sulfoxide
Dnp	2,4-dinitrophenyl
Dobz	dihydroxyborylbenzyloxycarbonyl
Doc	diphenylmethyloxycarbonyl
DOPA	*β*-(3,4-dihydroxyphenyl)alanine
Dpc	di-2-pyridylmethyloxycarbonyl
Dpm	diphenylmethyl (benzhydryl)
Dpp	diphenylphosphinyl
Dts	dithiasuccinoyl
DTT	dithiothreitol
Ec	(*S*)-ethylcarbamoyl
EDAC	*N*-ethyl-*N*′-(3-dimethylaminopropyl)carbodiimide
EDTA	ethylenediamine tetracetic acid
EEDQ	*N*-ethyloxycarbonyl-2-ethyloxy-1,2-dihydroquinoline
Et	ethyl
Et_3N	triethylamine
EtOH	ethanol
Fmoc (Fmc)	9-fluorenylmethyloxycarbonyl
Foc	furfuryloxycarbonyl
For	formyl
Har	homoarginine
Hcy	homocysteine
HFIP	hexafluoroisopropanol
HOBt	1-hydroxybenzotriazole
HONb	*N*-hydroxy-5-norbornene-2,3-dicarboximide
HOSu	*N*-hydroxysuccinimide
Hse	homoserine
Hyv	hydroxyisovaleric acid

Ibc	isobornyloxycarbonyl
Iec	2-iodoethyloxycarbonyl
IIDQ	*N*-isobutyloxycarbonyl-2-isobutyloxy-1,2-dihydroquinoline
Inc	γ-picolyloxycarbonyl (isonicotinyloxycarbonyl)
*i*Pr	isopropyl
*i*PrNEt	diisopropylethylamine
Lac	lactic acid
LHRH, LH/FSH-RH	luliberin
MA	mixed anhydride (method)
Mbs	4-methyloxybenzenesulfonyl
Mcb	1-methylcyclobutyloxycarbonyl
Mch	1-methylcyclohexyloxycarbonyl
Me	methyl
Meb	4-methylbenzyl
MeOH	methanol
Mez	4-methylbenzyloxycarbonyl
Mob	4-methyloxybenzyl
Moz	4-methyloxybenzyloxycarbonyl
Mpaz	4-methyloxyphenylazobenzyloxycarbonyl
Mpc	2-(4-methylphenyl)propyl(2)oxycarbonyl
MSA	methanesulfonic acid
Msc	2-methylsulfonylethyloxycarbonyl
MSH	melanotropin
Mtc	2-methylthioethyloxycarbonyl
N^g	guanidine nitrogen
N^{in}	indole nitrogen
N^{im}	imidazole nitrogen
N^{π}	*pros*-imidazole nitrogen
N^{τ}	*tele*-imidazole nitrogen
2Nb	2-nitrobenzyl
NBS	*N*-bromosuccinimide
NCA	*N*-carboxyanhydride
NEM	*N*-ethylmorpholine
NMM	*N*-methylmorpholine
NO_2	nitro
Nps	2-nitrophenylsulfenyl
NTA	*N*-thiocarboxyanhydride
NTBA	3-nitro-4-mercaptobenzoic acid

3Nz, 4Nz	3-, 4- nitrobenzyloxycarbonyl
OAm	9-anthrylmethyl ester
OBb	4-bromobenzyl ester
OBt	1-benzotriazolyl ester
O*t*Bu	*tert*-butyl ester
OBzl	benzyl ester
OCb	4-chlorobenzyl ester
ODnp	2,4-dinitrophenyl ester
ODpm	diphenylmethyl (benzhydryl) ester
OEt	ethyl ester
OMe	methyl ester
OMob	4-methyloxybenzyl ester
ONb	4-nitrobenzyl ester
ONo	2-nitrophenyl ester
ONp	4-nitrophenyl ester
ONpm	2,2′-dinitrodiphenylmethyl ester
OObt	3-hydroxy-4-oxo-3,4-dihydro-1,2,3-benzotriazine ester
OPa	phenacyl ester
OPcp	pentachlorophenyl ester
OPfp	pentafluorophenyl ester
OPh	phenyl ester
OPic	4-picolyl ester
OPmb	pentamethylbenzyl ester
OPon	piperonyl ester
OPse	4-phenylazobenzylsulfonylethyl ester
OPtm	phthalimidomethyl ester
OSu (ONSu)	*N*-hydroxysuccinimide
OTcp	2,4,5-trichlorophenyl ester
TFE	2,2,2-trifluoroethanol
OTmb	2,4,6-trimethylbenzyl ester
OTms	2-trimethylsilyl ester
OTrt	triphenylmethyl (trityl) ester
OTse	2-(4-toluenesulfonyl)ethyl ester
—Ⓟ	resin support
Pam	phenylacetamidomethyl
Paz	4-phenylazobenzyloxycarbonyl
Pec	2-triphenylphosphonoethyloxycarbonyl
Ph	phenyl
Pht	Phthaloyl

Pic	picolyl
Poc	2-phenylpropyl (2) oxycarbonyl
Prot	protecting group
Py	pyridine
REMA	repetitive excess mixed anhydride (method)
RNase	ribonuclease
Scm	*S*-carboxymethylsulfenyl
SEt	*S*-ethylsulfenyl
S*i*Pr	*S*-isopropylsulfenyl
SPS	solid-phase synthesis
S*t*Bu	*S*-*tert*-butylsulfenyl
Tac	4-toluenesulfonylaminocarbonyl
*t*Bu	*tert*-butyl
*t*BuONO	*tert*-butyl nitrite
Tcc	2,2,2-trichloroethyloxycarbonyl
TFA	trifluoroacetic acid
Tfa	trifluoroacetyl
Thp	tetrahydropyranyl
Tmb	2,4,6-trimethyloxybenzyl
Tms	trimethylsilyl (ether)
Tmz	α,2,4,5-tetramethoxycarbonyl
Tnp	2,4,6-trinitrophenyl
TOS	4-toluenesulfonic acid
Tos	4-toluenesulfonyl (tosyl)
Trt	triphenylmethyl (trityl)
Tsc	2-(4-toluenesulfonyl)ethyloxycarbonyl
Voc	vinyloxycarbonyl
Xan	xanthenyl
Z	benzyloxycarbonyl
Z-Cl	benzyloxycarbonyl chloride
Ztf	2,2,2-trifluoro-1-benzyloxycarbonylaminoethyl

Contents of Previous Volumes

VOLUME 1

MAJOR METHODS OF PEPTIDE BOND FORMATION

VOLUME 2

SPECIAL METHODS OF PEPTIDE SYNTHESIS, PART A

Chapter 1

Amine Protecting Groups

ROLF GEIGER and WOLFGANG KÖNIG

THE PEPTIDES, VOLUME 3

ISBN 0-12-304203-8

I. INTRODUCTION

Nitrogen in amino acids is a constituent of different functional groups and is located in either the backbone or the side chains of peptides. Except for certain peptide antibiotics and distinct cross-links in cell walls, all peptide bonds are amide connections between the α-amino and the α-carboxyl groups of different amino acids. The necessity for protection is clearly evident from Eq. (1).

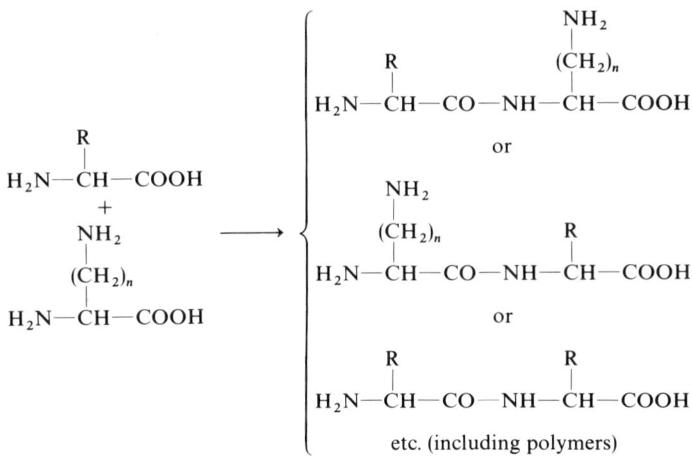

(1)

Obviously a well-defined peptide is formed only if those functional groups which should not react with each other are selectively blocked. Protecting groups have to be stable during the procedures of peptide bond formation and purification. Once a peptide has been synthesized, the protecting groups must be labile under special conditions to permit selective deblocking of α-amino and α-carboxyl functions. The permanent side-chain protecting groups that have to survive many deblocking steps during a peptide synthesis must be sensitive to specific reagents for their final cleavage.

Consequently, the choice of permanent side-chain protecting groups to be used in a synthesis depends on the stability of the target peptide under the conditions of the final deprotection, and this choice in turn governs the selection of groups for the intermediate protection of α-amino groups. Most peptides are fairly stable under acidic conditions. Therefore, N-protecting groups with graded sensitivity to cleavage by acids of different strengths play a dominant role in peptide synthesis. Some of these groups are also sensitive to catalytic hydrogenation. These two deblocking methods are used in over 80% of all syntheses. Other cleavage procedures include treatment with alkali, hydrazinolysis, an even more specialized reaction with sodium in liquid ammonia, or oxidation.

An annoying complication in peptide synthesis is racemization, which proceeds via 5(4*H*)-oxazolone formation at the activated carboxyl group under the influence of base [Eq. (2)].

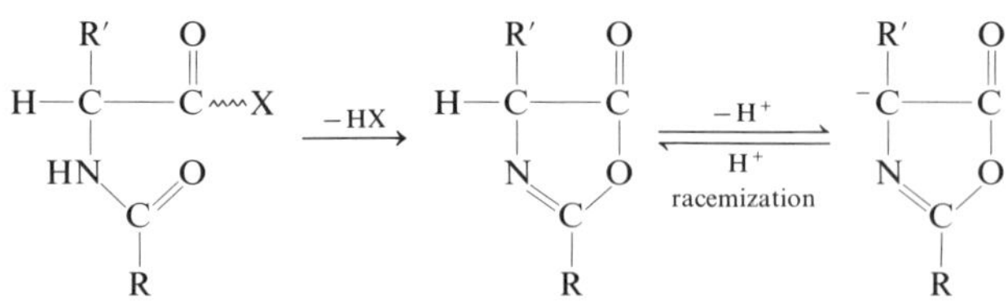

(2)

The carbonyl oxygen of any acyl residue bound to the α-amino group is prone to undergo this reaction. Fortunately, with urethane-type amine protecting groups (**1**) the tendency of racemization is largely suppressed.

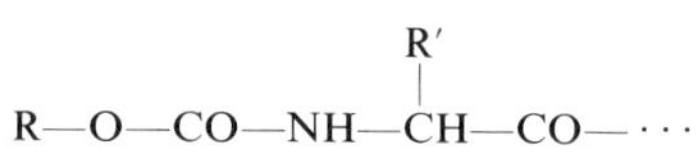

(**1**)

This favorable property and the great variety of cleavage conditions dependent on the nature of the ester moiety have rendered the urethane-type amine protection most widely used. The availability of racemization-inhibiting protective groups is a strong argument for the stepwise synthesis of peptides (Bodanszky, 1960). This strategy is also applied in solid-phase synthesis (Merrified, 1963).

The number of useful protecting groups increased rapidly in the fifties and sixties. During the last few years this development has abated somewhat, but interesting groups are still being developed. In most cases they are derived from familiar types and adapted to special requirements, e.g., to a more highly differentiated behavior under deblocking conditions. Much has been accomplished by minor changes in chemical structure or by elaborating new deblocking conditions for already known groups. A good example of the systematic application of the latter approach is the total synthesis of insulin by Sieber *et al.* (1974).

II. AMINE PROTECTION

The following sections deal with the properties of different types of N-protecting groups and the scope and limitations of their application. Since the size of this chapter does not permit a complete enumeration of all possible techniques, the reader is referred to the excellent detailed works of Schröder and Lübke (1965, 1966) and Wünsch *et al.* (1974). The latter reference includes literature citations up to 1973 and devotes nearly 400 pages to amine protection. In this review, most types of N-protecting groups and their main representatives are described synoptically. A critical evaluation, especially in the final sections, should help produce useful answers to the more complex questions on this topic. The combination of different protecting groups is only partially treated within these sections, since Chapter 5, by Fauchere and Schwyzer, deals with these aspects.

The nitrogen-containing functional groups of amino acids differ in structure and chemical properties and, consequently, in their behavior under the conditions of peptide synthesis. For some groups protection is obligatory, for others it is facultative. Table I lists these groups and includes α- and β-azaamino acids (Gante, 1966; Niedrich and Grupe, 1965; Bentley and Morley, 1966; see also Wünsch *et al.*, 1974, Part I, pp. 894–901). Hydrazides in protected form are frequently used if their formation from an ester in a later step of the synthesis is not advisable (Hofmann *et al.*, 1952; Schwyzer, 1959; Weygand and Steglich, 1959; Wang and Merrifield, 1969; see Chapter 7, this volume). The fact that several nitrogen-containing groups do not need protection in every case provides a certain freedom in the chemist's decision. Thus, strategies ranging from minimal to maximal protection have been applied. The advantages of maximal protection lie in avoiding as many side reactions as possible. On the other hand, syntheses on a technical scale which use the less expensive minimal protection have been successfully elaborated.

Table I Nitrogen-Containing Functional Groups in Amino Acids and Peptides

Structure	Group	Protection status
$NH_2(\alpha)$-	All α-amino acids	+ [a]
$NH_2(\omega)$-	Ornithine, lysine	+
HN<	Proline, hydroxyproline	+
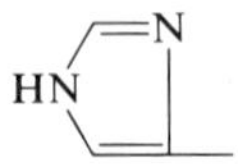	Histidine	– [b]
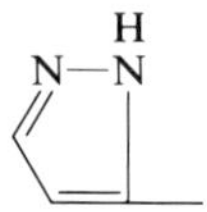	β-(Pyrazolyl-3)alanine	–
H_2H—C(=NH)—NH—	Arginine	+ [c]
—CO—NH_2	Asparagine, glutamine COOH-terminal carboxamide	–
—CO—NH—	Peptide bond	– [d]
H_2N—N(R)—	Azaamino acids	+
—CO—NH—NH_2	Amino acid and peptide hydrazides	+
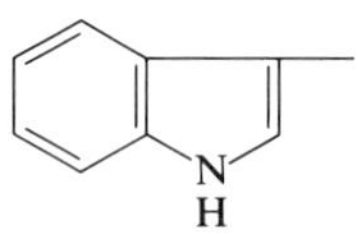	Tryptophan	– [d]

[a] + = protection obligatory.
[b] – = protection facultative.
[c] Protonation for minimal protection.
[d] Without protection in most cases.

The syntheses of ACTH peptides (Schwyzer *et al.*, 1960; Kappeler and Schwyzer, 1961; Hofmann *et al.*, 1961a; Geiger *et al.*, 1964) follow this strategy as well as the approaches of Hirschmann *et al.* (1969; Hirschmann, 1971) and Storey and Hofmann (1973) to the total syntheses of ribonucleases S and T_1, respectively.

In contrast, Wünsch (1973) resorted to maximal protection in the synthesis of gastrointestinal hormones. Even the imidazole nitrogen of histidine was blocked, whereas in most other syntheses the imidazole moiety remains free. The guanido group in arginine-containing peptides is commonly blocked by protonation. However, if the α-carboxyl group of arginine is involved in a coupling reaction, this kind of protection is less favorable. The same is true for repetitive syntheses (see Vol. 2, Chapter 6). In glutamine-containing peptides the free α-amine group reacts with its own carboxamide group to form a pyroglutamyl residue. Asparagine and glutamine undergo nitrile formation and cyclization upon activation of the α-carboxyl group (Section V). Therefore, the amide function is sometimes reversibly blocked by benzyl or benzhydryl derivatives, as described in Section V. Although the latter side reactions are not critical if special condensation techniques are applied (König and Geiger, 1970a), this kind of protection often increases the solubility of peptides. Even COOH-terminal amides have occasionally been substituted for this reason (König and Geiger, 1972; Bajusz *et al.*, 1973). Because of methodologic difficulties, peptide bonds as functional groups have only rarely been protected.

III. THE TYPES OF AMINE PROTECTING GROUPS

A. Carboxylic Acids

Acyl groups provide complete protection of nitrogen in amino, imino, and guanido groups (in the latter case by diacylation), and in imidazole against the electrophilic attack by the carbonyl C-atom of an activated carboxyl group of N-protected amino acids or peptides.

The highly stable amides of most common carboxylic acids with amine and imine groups precludes this type of protection in peptide chemistry. Only those acyl components that can be split off under mild special conditions meet the requirements described in Section I.

In addition N^{α}-acylated amino acids are subject to considerable racemization. Methods which suppress racemization partially overcome this difficulty. However, their application restricts synthetic strategy to a few procedures. This again restricts the scope of acyl groups to special cases. In addition to glycine and the nonracemizing proline, ω-amino groups (e.g., in lysine) have occasionally been protected by acylation, but even here they are more and more frequently displaced by protecting groups of the urethane type (Section III,B).

1. Formyl Group (For)

The reaction of amino acids with boiling formic acid (Fischer and Warburg, 1905), with a mixture of formic acid and acetic anhydride at 0°–20°C (Stoll and Petrzilka, 1952), with alkyl esters (Hofmann *et al.*, 1960), or with active esters of formic acid (Okawa and Hase, 1963) results in N-formylation [Eq. (3)].

$$HC(=O)X + H_2N{-}R \longrightarrow HCO{-}NH{-}R + HX \qquad (3)$$

X = OH, ONp, $OCCH_3$, OCH_3, or OC_2H_5

Formyl group cleavage is readily achieved by 0.5–1 *N* alcoholic HCl at room temperature within 12–48 hr (Sheehan and Yang, 1958). Two other methods of deprotection are oxidation with H_2O_2 (Losse and Zönnchen, 1960) or hydrazinolysis [effecting formyl group transfer to hydrazine derivatives] (Miyamoto *et al.*, 1961; Geiger and Siedel, 1968; Yajima *et al.*, 1968). These methods require conditions, however, under which not all peptides are stable.

N-Formyl peptide hydrazides, intermediates for azide couplings, have been prepared in fairly good yield (Vogler and Lanz, 1960; Hofmann *et al.*, 1961b) since hydrazinolysis of a methyl ester proceeds faster than that of the *N*-formyl group.

The *N*-formyl group is sufficiently stable to both alkali and anhydrous acids, e.g., trifluoroacetic acid. Cleavage by hydrogenation (Losse and Nadolski, 1964) is doubtful. Unsatisfactory results in the synthesis of larger peptides (Hofmann and Yajima, 1961) explain why the use of this protecting group has been largely discontinued, except perhaps for the preparation of some small peptides and intermediates.

2. *Trifluoroacetyl Group (Tfa)*

The trifluoroacetyl group is introduced into amino acids by means of trifluoroacetic acid anhydride, preferably in trifluoracetic acid as a solvent (Weygand and Geiger, 1956). Other reagents, such as methyl, thioethyl, and phenyl esters of trifluoroacetic acid, are also suitable for the trifluoroacetylation of peptides (Weygand and Geiger, 1959; Schallenberg and Calvin, 1955; Weygand and Röpsch, 1959). The trifluoroacetyl group, which is cleaved by mild alkali treatment (Weygand and Czendes, 1952) or by sodium borohydride (Weygand and Frauendorfer, 1970), gained more significance for its interesting reactions and analysis of trifluoroacetyl amino acids and peptides (Steglich *et al.*, 1967; Weygand *et al.*, 1965) than for the synthesis of higher peptides.

Alkali-labile protecting groups have always been welcomed in peptide chemistry as a supplement to the numerous protecting groups with differential sensitivity to acids. However, with the development of protecting groups that are cleaved via β-elimination under even milder alkaline conditions, the trifluoroacetyl group has lost its significance. *N*-2,2,3,3-Tetrafluoropropionylamino acids (Tomida and Kayahara, 1976) behave similarly to trifluoroacetyl amino acids, and the corresponding pseudooxazolones-(5) undergo asymmetric induction (Tomida *et al.*, 1976) as does the corresponding trifluoromethyl pseudooxazolones (Steglich *et al.*, 1967).

3. Phthaloyl Group (Pht)

The phthaloyl group is stable to both acids and hydrogenolysis, leaving hydrazinolysis as the only method to remove this diacyl protecting group. The five-membered ring is easily opened by alkali (Scheme 1) with the formation of the *o*-carboxybenzoyl compound (**3**), which can no longer be readily cleaved. Consequently, phthaloylamino acids (**2**) and peptides cannot be treated with alkali to saponify alkyl esters.

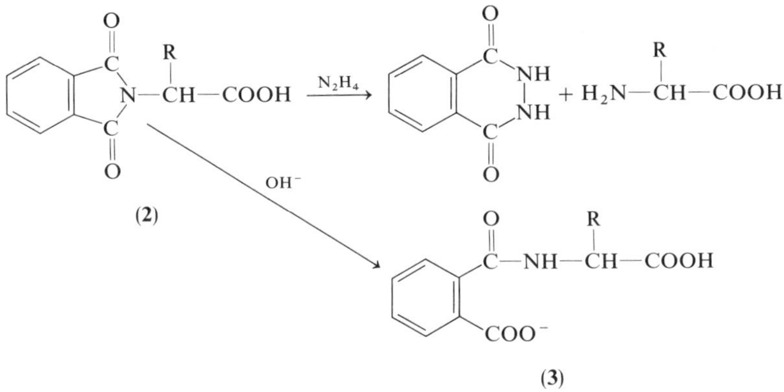

Scheme 1

The original method of introduction of the phthaloyl group by treatment of amino acids with phthalic anhydride at elevated temperature (up to 185°C, usually with racemization) has been abandoned. Nefkens *et al.* (1960) found that *N*-carbethoxyphthalimide (**4**) reacts with amino acids in weakly alkaline solution to yield phthaloylamino acids [Eq. (4)]. The conditions are mild enough to avoid reopening of the ring.

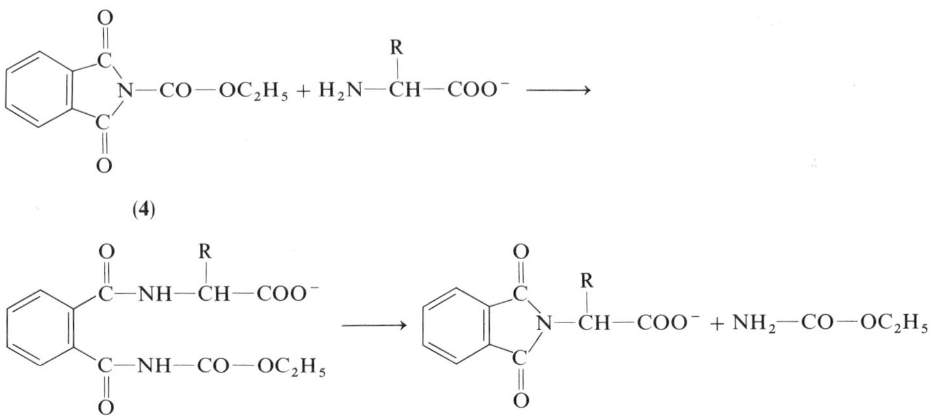

(4)

Smooth cleavage of the phthaloyl group from the ε-amine group of lysine has been achieved with hydrazine acetate in neutral solution during a synthesis of α-MSH by Schwyzer *et al.*, (1963). Even insulin, which is highly sensitive to hydrazine, could be successfully liberated from phthaloyl residues in concentrated phenolic solution (Geiger and Langner, 1975). The phthaloyl residue has been applied quite frequently for many years. With the appearance of more novel groups it has received only scant attention.

4. Acetoacetyl Group (Aca)

The acetoacetyl group is introduced by reaction of amino acid or peptide esters with ketene and subsequent saponification of the esters (D'Angeli *et al.*, 1965). Peptide synthesis can be achieved with dicylohexylcarbodiimide in the presence of *N*-hydroxysuccinimide (DCC–HOSu, see Chapter 5 in Vol. 1 of the Peptides). Treatment with dicyclohexylcarbodiimide alone leads to compound **5**, which is opened by the nucleophilic attack of an amine to form a peptide (Di Bello *et al.*, 1971).

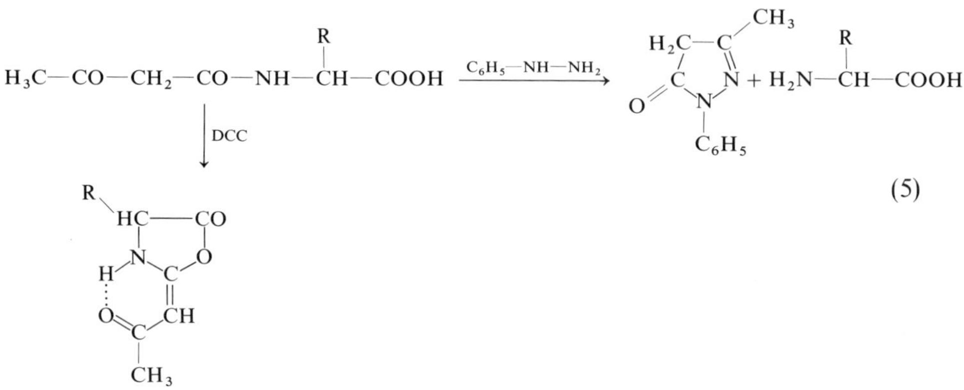

(5)

(5)

The group is stable to acids. It is cleaved by hydroxylamine, hydrazine, or phenylhydrazine [Eq. (5)] but it has found no general application in peptide synthesis.

5. Maleoyl Group

Maleoylamino acids and peptide derivatives (**6**) are not actually suitable for routine use in peptide synthesis. However, the double bond reacts rapidly and quite specifically with thiol groups of proteins. This opens the possibility for the application of such compounds in biochemistry as hapten carriers, as affinity sorbents, and for labeling or topochemical probing (Keller and Rudinger, 1975).

11. Dithiasuccinoyl Group (Dts)

A novel protecting group of the diacyl type has been described by Barany and Merrifield (1977). The dithiasuccinoyl group (**10a**) is stable toward both acids and photolysis and is selectively cleaved by reductive procedures (e.g., treatment with 0.2 *M* mercaptoethanol in dichloromethane) preferably in the presence of 0.5 *M* triethylamine. The reaction is generally complete after 5 min at 25°C.

The preparation of Dts derivatives starts from ethoxythiocarbonyl amino acid *tert*-butyl or trimethylsilyl esters. They react with chlorocarbonylsulfenyl chloride, optionally in the presence of tertiary amines, to form an intermediate which looses HCl and ethyl chloride under formation of the Dts amino acid esters [Eq. (9a)]. The esters can easily be cleaved by acids.

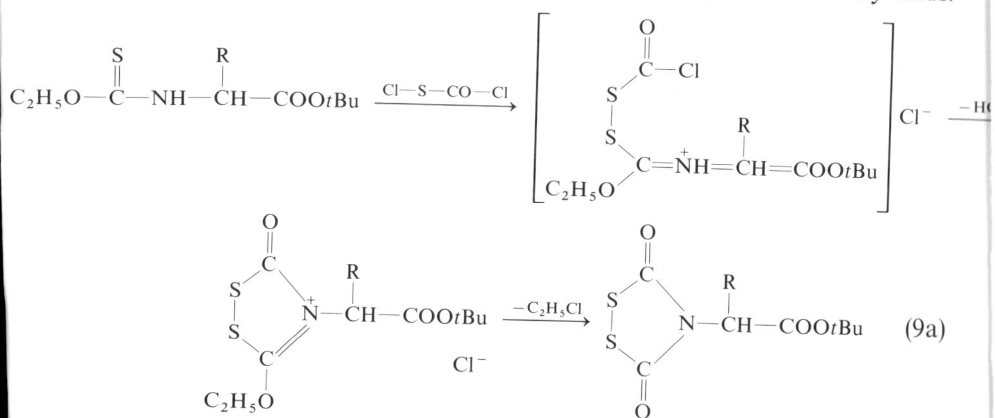

(9a)

(**10a**)

The authors recommend the dithiasuccinoyl group as a third type of protecting residues, i.e., in addition to acid-labile and photolabile groups.

12. Cysteic acid [$Cys(O_3H)$]

Cysteic acid as a constituent of peptides greatly enhances their solubility in water. Hubbuch *et al.* (1977) introduced N^{α}-protected cysteic acid as an NH_2-terminal amino acid into synthetic peptides.

The sulfo group of cysteic acid is inert toward peptide-forming reagents, especially dicyclohexylcarbodiimide. The peptide bond between cysteic acid and the following amino acid is stable to cleavage conditions for protective residues. After deblocking of the N^{α}-amino group, cysteic acid is removed by Edman degradation. The new approach may be a great help if NH_2-

A satisfactory synthesis of maleimido acids has been achieved by Keller and Rudinger (1975). They adopted the method of Nefkens *et al.* (1960) for phthaloylation [Eq. (6)]. Since saponification occurs, already at pH 8.5 the

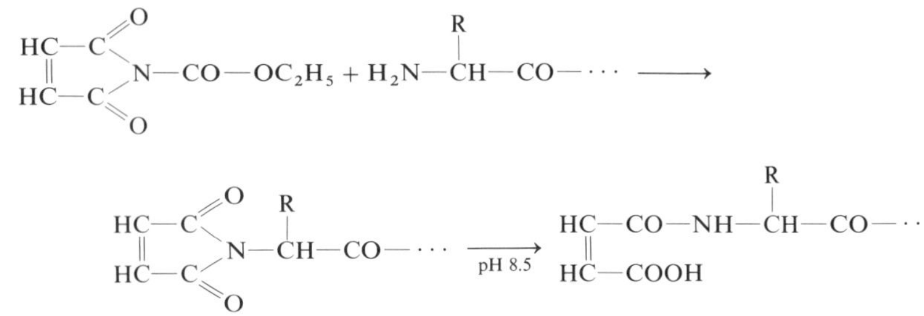

(6)

(**6**) (**7**)

reaction product is always contaminated with **7**, which can be separated by filtration over silica gel. Maleoylamino acids form peptides with amino acid or peptide derivatives upon addition of dicyclohexylcarbodiimide. The maleoyl group is quite stable toward trifluoroacetic acid but is cleaved by treatment with dilute alkali (pH 10) followed either by aqueous acid (pH 2–3) or by hydrazine.

6. Chloroacetyl Group (Cla)

This protecting group was already in use by Fischer and Otta in 1903 for the synthesis of glycine peptides by reacting chloroacetyl peptides with ammonia [Eq. (7)]. Corresponding amino acid derivatives are prepared by

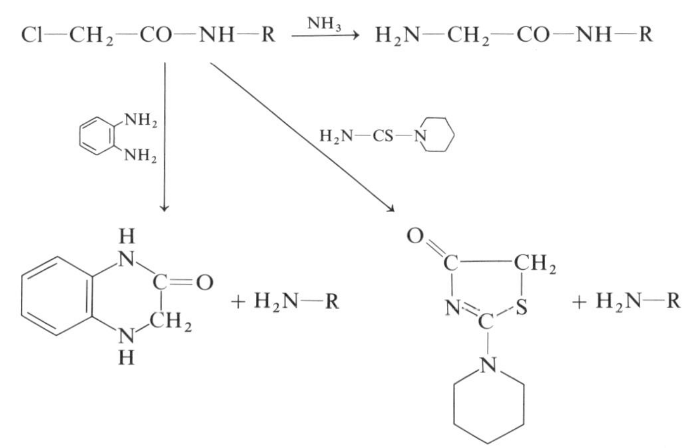

(7)

Schotten–Baumann reaction with chloroacetyl chloride. Cleavage is performed under special conditions, namely with 1,2-phenylenediamine (Holley and Holley, 1952, thiourea (Masaki *et al.*, 1968), or, better, with *N,N*-disubstituted thiourea (Steglich and Batz, 1971). However, this protecting group is of limited significance.

7. 2-Nitrobenzoyl Group

The 2-nitrobenzoyl group, introduced via the acyl chloride, is stable towards acids and cleaved under special conditions (Koul *et al.*, 1973). Reduction of the nitro group yields 2-aminobenzoyl derivatives. The amine compounds are liberated in the presence of copper(II) salts via formation of a chelate [Eq. (8)]. The protecting group has not yet been applied in practice.

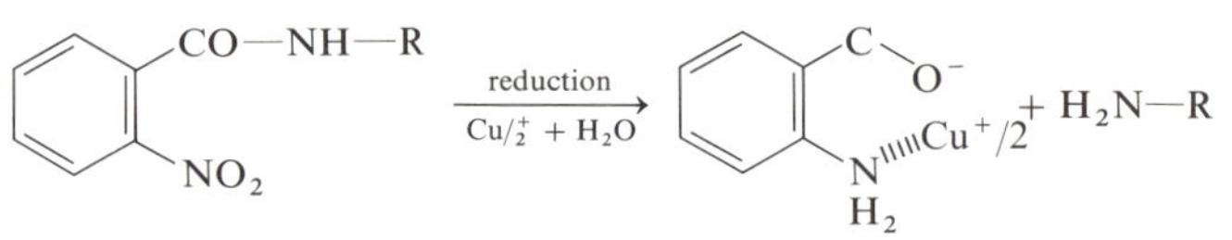

(8)

8. Acyl-Type Protecting Groups Cleavable by Neighboring Group Assistance

The first representative of these groups, the *N*-(2-nitrophenoxyacetyl) group (**8**) was proposed by Holley and Holley (1952). It is characterized by the ability to form cyclic lactams following reduction of the nitro group [Eq. (9)] thus releasing the amine moiety. The cyclization requires elevated temperatures. Cuiban (1971) studied the properties of groups in which oxygen was eliminated or replaced by CH_2, S, and NH. Entwistle (1979) used

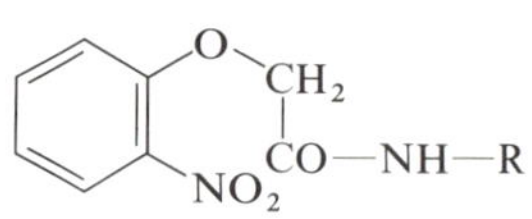

(**8**)

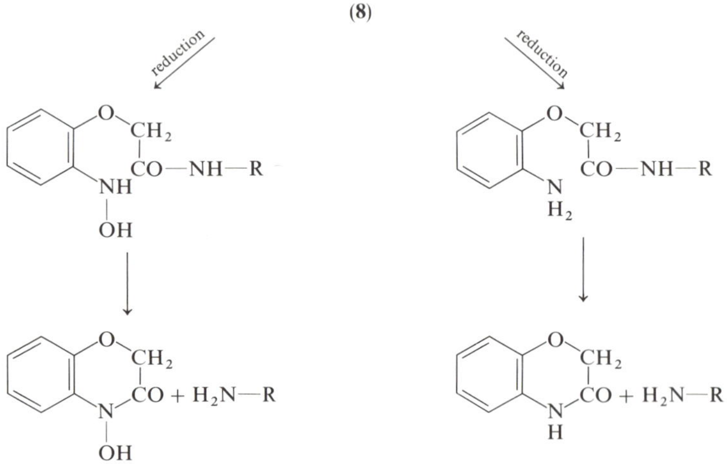

(9)

2-nitrophenylpropionyl and 2-nitrocinnamoyl residues and cyclohexene or sodium phosphinate with 10% Pd–C as catalyst to reduce the nitro group by hydrogen transfer. Panetta (1969) recommended the homologous 2,2-dimethyl compound (**9**) and Panetta and Rahman (1971) employed a phenylazo instead of the nitro substituent (**10**). The different residues show characteristic absorptions in uv and ir which may be helpful for purification of the protected peptides. The introduction of these groups via acid chlorides proceeds readily. The stability to acids distinguishes them from benzyloxycarbonyl and *tert*-butyloxycarbonyl residues.

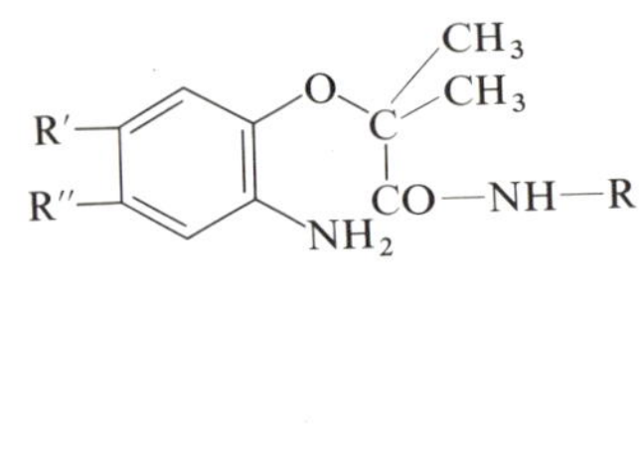

(**9**)

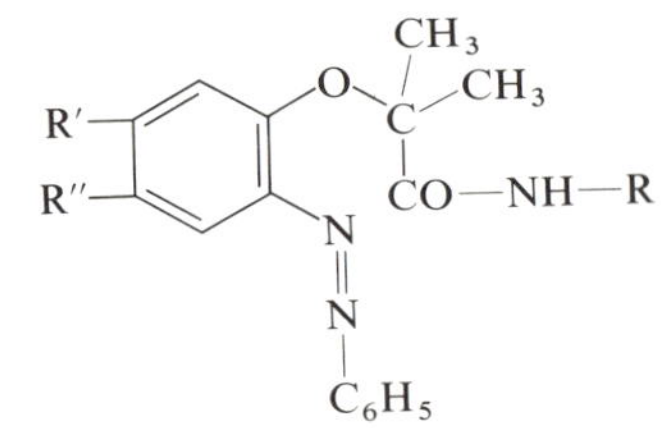

(**10**)

R′ = H or CH_3
R″ = CH_3

Despite the advantageous cleavage by the elegant cyclization at neut pH, compounds **8–10** have thus far found no practical application. practice, reduction and cyclization do not proceed completely and ea enough to render these groups competitive with the commonly emplo urethanes.

9. Enzymatically Cleavable Acyl Group

Meyers and Glass (1975) studied the tryptic cleavage of peptides NH_2-terminal benzyloxycarbonylarginine as a protecting group. Depr tion occurred with a high degree of specificity. The method has not applied to longer peptides and is limited to peptides that are devo arginine or that contain this amino acid in N^G-protected form.

10. N-Acylmethionine as Amine Protecting Group

The analytically important cleavage of methionine-containing pepti cyanogen bromide (Gross and Witkop, 1962) has been applied to eli an *N,N′*-carbonylbis(methionyl) cross-linking residue from insulin (and Obermeier, 1974/77; Busse *et al.*, 1974). Saunders and Offord elaborated methods for the application of this approach to semisy using insulin as a model.

terminal moieties of peptides are too insoluble to permit segment condensation.

B. Urethane Protecting Groups Derived from Primary Alcohols

When Bergmann and Zervas (1932) devised the benzyloxycarbonyl group for amine protection they initiated a revolution in peptide chemistry. Their idea was based on the observation of Rosenmund that benzyl esters can be cleaved by catalytic hydrogenation. The urethane protecting group embodying the benzyl ester and the amine was so fruitful that nearly all peptide syntheses have since taken advantage of it.

About 25 years after the development of the benzyloxycarbonyl group the acid-sensitive *tert*-butyl ester was adapted in the same manner for amine protection. These groups are the main pillars of support in contemporary peptide synthesis. Over the years numerous derivatives of both the benzyl- and *tert*-butyloxycarbonyl groups have been synthesized. They encompass a wide range of cleavage conditions and allow an increasing number of protecting group combinations, all derived from this type. This development is still in progress, and only recently several new substituted urethane protecting groups have been presented.

The importance of the family of urethane protecting groups deserves a more detailed discussion. Urethanes derive from primary, secondary, and tertiary alcohols. A classification in this order does not always meet with corresponding chemical properties. In spite of this difficulty we shall use this arrangement for lack of a clear order due to the overlapping properties of the groups in question.

1. Benzyloxycarbonyl Group (Z)

For detailed reviews on the benzyloxcarbonyl group see Schröder and Lübke (1965, 1966) and Wünsch *et al.* (1974). The benzyloxycarbonyl group is introduced into amino acids via the easily accessible and quite stable chloroformate, using the Schotten–Baumann procedure, mostly in an aqueous–organic solvent mixture under vigorous stirring. Amino acid esters are reacted in chloroform solution. Nearly all amino acids can be protected in this way without difficulty. The yields are good and most derivatives are crystalline. Oily benzyloxycarbonylamino acids form crystalline salts with dicyclohexylamine.

Cleavage is accomplished by several methods (see Table II). The most widely used, hydrogenation with palladium as catalyst, is clean and can even be managed if methionine is present (by the addition of triethylamine) in any but the NH_2-terminal position (Medzihradszky and Medzihradszky-Schweiger, 1965). Cysteine derivatives often inactivate the catalyst. However, with liquid ammonia as a solvent (Meienhofer and Kuromizu, 1974) N^{α}-benzyloxycarbonyl groups can be split off in the presence of *S*-benzylcysteine-containing peptides.

Catalytic hydrogenation is usually accomplished by passing hydrogen through a neutral or, preferably, slightly acidic solution of the protected peptide in the presence of the catalyst. The end of the reaction is indicated by the disappearance of CO_2 in the outcoming stream of hydrogen. More advantageously, the course of the hydrogenation can be controlled by automated titration of the liberated amine group at about pH 4, for example, with HCl or toluenesulfonic acid.

The search for additional deblocking conditions led to the cleavage by HBr in acetic acid as a standard method (Ben Ishai and Berger, 1952). Sakakibara *et al.* (1967) introduced the liquid hydrogen fluoride cleavage, which by now is generally used for the removal of all acid-sensitive protecting groups. A related reagent is pyridinium polyhydrogen fluoride (Matsuura *et al.*, 1976). A number of other acidic deblocking reagents have been described during the last few years. Weygand and Steglich (1959a) recommended boiling trifluoroacetic acid, Pless and Bauer (1973) boron tris(trifluoroacetate), and Felix (1974) boron tribromide. Sulfonic acids such as methanesulfonic acid and fluorosulfonic acid (Yajima *et al.*, 1975a) or trifluoromethylsulfonic acid (Yajima *et al.*, 1974), preferably in methylene chloride or trifluoroacetic acid, turned out to be nearly as effective as hydrogen fluoride. The benzyloxycarbonyl group is quite stable toward diluted acids, e.g., 1–2 *N* HCl, acetic acid, or cold trifluoroacetic acid. *N*-Benzyloxycarbonyl group cleavage by all of these acidolytic agents leads to the formation of benzyl cations which cause S-benzylation of methionine and C-benzylation of tyrosine and tryptophan. Addition of scavengers such as anisole or thiophenols partially suppresses this side reaction.

A third method for the removal of the benzyloxycarbonyl group is treatment with sodium in liquid ammonia (Sifferd and du Vigneaud, 1935). It is a useful technique for final deblocking of peptides. In a side reaction (see Table XVII), the X-Pro bond may be cleaved. Katsoyannis *et al.* (1971) added sodium amide to the liquid ammonia solution to suppress this adverse effect.

The application of benzyloxycarbonyl amine protection in peptide synthesis is subjected to only a few limitations. A certain sensibility towards alkali has been observed (see Schröder and Lübke, 1965, p. 25), although most saponifications proceed without severe damage of the molecule. However, benzyloxycarbonyldipeptide esters with COOH-terminal glycine give rise to formation of hydantoins and ureas under treatment with alkali or ammonia.

Therefore, saponification or amidation should not be performed with these compounds. Virtually all coupling methods are allowed; only benzyloxycarbonylamino acid chlorides tend to form *N*-carboxy anhydrides [Eq. (10)].

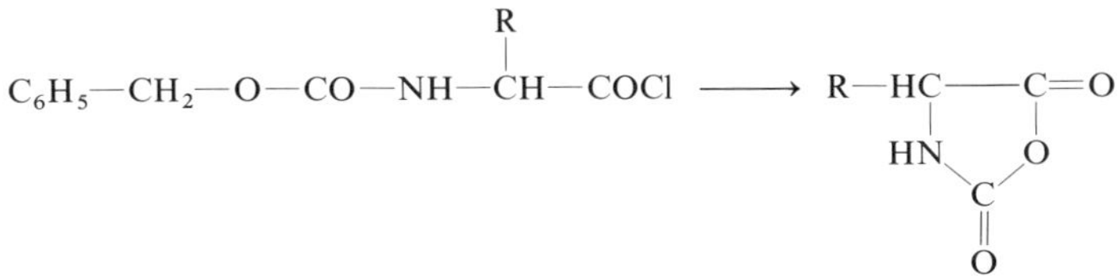

(10)

These are a source of uncontrolled polymerization under the usual conditions of peptide synthesis (see, however, Hirschmann, 1971). Despite the few reservations recorded above, the favorable stability and cleavage properties as well as the suppression of racemization render the benzyloxycarbonyl group the most widely used amine protecting group.

2. *Benzyloxycarbonyl Groups Substituted in the Aromatic Ring*

To facilitate the purification of N-protected amino acids and peptides by an increased tendency to crystallize, chloro (Boissonnas and Preitner, 1953; Kisfaludy and Dualszky, 1960), bromo (Channing *et al.*, 1951), and nitro (Carpenter and Gish, 1952) substituents have been inserted into the benzyloxycarbonyl group. In addition, the colored phenylazo residue allows one to monitor colorimetrically or visually purification by chromatography or countercurrent distribution (Schwyzer *et al.*, 1958). Tables II and III list examples of ring-substituted benzyloxycarbonyl groups. More recently, the increased stability of such derivatives toward acids has been utilized to improve the stability of side-chain protecting benzyloxycarbonyl and benzyl groups in repetitive condensation techniques, e.g., in solid-phase synthesis. The repeated treatment of peptides with acids such as trifluoroacetic acid results in a continuous loss of small amounts of unsubstituted benzyl-type side-chain protection and a concomitant increase of side-chain couplings within the growing peptide chain. The application of, for example, N^{ω}-4-chlorobenzyloxycarbonyl protection greatly reduces loss during acid treatment and improves the purity of the final products.

In contrast to electron-withdrawing groups at the benzyl moiety, methyl or, especially, methyloxy substitution render the protecting group more sensitive to acidolysis (McKay and Albertson, 1957). Thus, the 4-methyloxybenzyloxycarbonyl (Moz) group is readily cleaved by cold trifluoroacetic acid providing a clear selectivity within this class of amine protecting groups.

All of these substituted benzyloxycarbonyl groups are cleaved by catalytic hydrogenation, hydrogen fluoride, or sodium in liquid ammonia. The

Table II Benzyloxycarbonyl Protecting Groups Monosubstituted in the Aromatic Ring

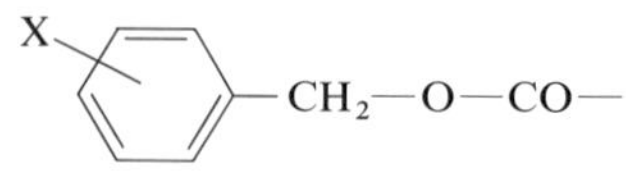

X	Abbrev	Important agents for cleavage	$t_{1/2}$(min) in 0.85 *M* HBr–AcOH at 25°C
H	Z	HBr–AcOH, HF, boiling CF_3COOH, liquid HBr, H_2(Pd), Na—NH_3, CH_3SO_3H, CF_3SO_3H, FSO_3H	66
2-Cl	2 Cz		—
4-Cl	4 Cz		120
4-Br	4 Bz		—
3-Cl	3 Cz	Analogous to Z, HF advantageous	210
4-NO_2	4 Nz		430
4-(C_6H_5—N=N)—	Paz		400
4-(CH_3O—C_6H_5—N=N)—	Mpaz	Analogous to Z	—
4-CH_3—	Mez	Analogous to Z; cold trifluoroacetic acid	7
4-CH_3O—	Moz		0.8
4-CH_3CO—O—	4 Acz	Analogous to Z; NaOH	—
4-$(OH)_2B$—	Dobz	Analogous to Z; H_2O_2	—
2-$CON(CH_3)_2$		Analogous to Z	—

Table III Benzyloxycarbonyl Protecting Groups with Multiple Substitutions in the Aromatic Ring

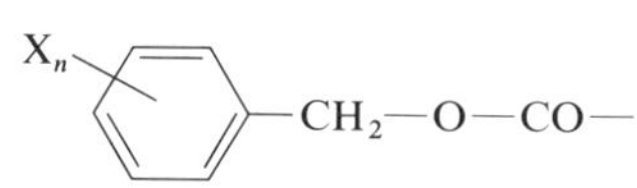

X_n	Abbrev	Important agents for cleavage	Relative cleavage rate in CF_3COOH–CH_2Cl_2 at 20°C (Z = 100)
2,4-dichloro-	2,4Dcz	H_2(Pd), Na—NH_3	1.2
3,4-dichloro-	3,4Dcz	HF	0.5
3,5-dimethyloxy	3,5Dmoz	Analogous to Z (see Table II)	—
2-nitro-4,5-dimethyloxy-		*hν*	—

methyloxy- and nitro-substituted residues offer additional advantages by their characteristic uv absorption which facilitates their identification during chromatographic procedures. Negatively substituted benzyloxycarbonyl-amino acids are prepared via the corresponding chloroformates or mixed active esters and the methyloxy derivatives via the 4-methyloxybenzyloxy-carbonyl azide (Weygand and Hunger, 1962). The 2-(*N*,*N*-dimethylcarbon-amide) substitution slightly increases the stability toward acids and improves the solubility of the protected compounds in dimethylformamide (Coyle *et al.*, 1975).

The 4-acetoxy-substituted benzyloxycarbonyl group (Wakselman and Guibé-Jampel, 1973) and related groups (Le Corre *et al.*, 1978) differ from those described above in their susceptibility to cleavage by alkali in a two-step reaction. The acetoxy group is saponified by 0.1 *N* NaOH within 10 min. The resulting 4-hydroxyl derivative is subject to β-elimination in the alkaline medium [Eq. (11)].

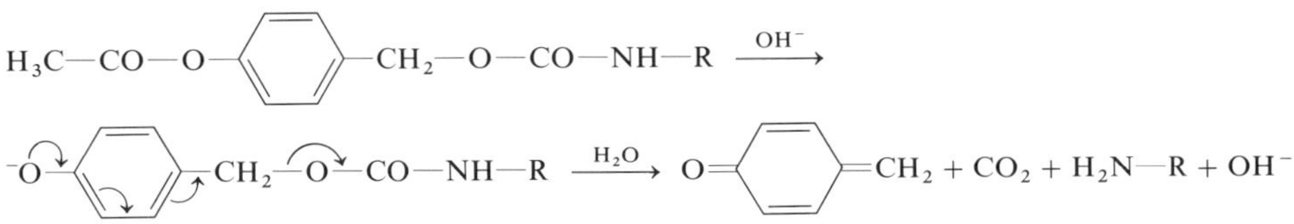

(11)

Since other alkali-labile groups are cleaved faster, this protecting group found no broad application in practice.

The *p*-dihydroxyboronatobenzyloxycarbonyl (Dobz) group (**11**) of Kemp and Roberts (1975) occupies a special place within this class.

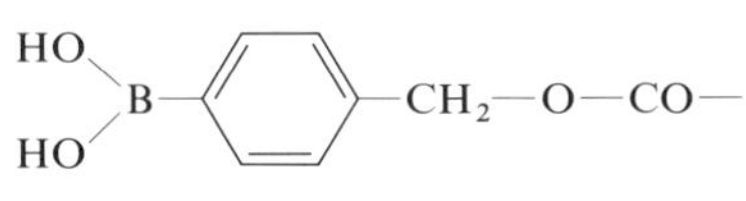

(**11**) Dobz

The boronic acid function can be masked by rapid, reversible reaction with diols to yield derivatives with more hydrophilic or hydrophobic properties, depending on the character of the applied diol. The corresponding amino acids are synthesized by means of Dobz chloride, which is protected as its catechol complex. In coupling reactions nitrophenyl esters, azides and the DCC–HOBt method (König and Geiger, 1970a) are compatible with a free dihydroboryl moiety. The efficacy of the new protecting group in combination with the above-mentioned coupling methods has been demonstrated during the synthesis of an angiotensin heptapeptide analogue. The removal of Dobz is performed by standard procedures used for benzyloxycarbonyl group cleavage. Dobz is slightly more stable to trifluoroacetic acid than the benzyloxycarbonyl group. In addition, cleavage may be smoothly accomplished by treatment with H_2O_2. Suitable reaction conditions leave methionine as well as tryptophan residues unaffected.

3. Photosensitive Benzyloxycarbonyl Derivatives

3,5-Dimethyloxybenzyloxycarbonylamino acids are prepared by reaction of the respective 4-nitrophenyl carbonate and amino acids in basic solution (Chamberlin, 1966). These compounds can be subjected to all common condensation techniques. The sensitivity to uv offers a possibility for selective removal. Irradiation in the presence of water results in the corresponding benzyl alcohol, CO_2, and peptide. One drawback is the destruction of tryptophan-containing peptides.

To avoid tryptophan decomposition during photolysis, Patchornik *et al.* (1970) recommended the 2-nitro-4,5-dimethyloxybenzyloxycarbonyl group. The acylation of amino acids is accomplished with the corresponding chloroformate. Photochemical cleavage is performed at 320 nm. Tryptophan would only be damaged at wavelengths shorter than 315 nm. The removal of the protecting group was found to proceed via a nitroso aldehyde as intermediate which may be trapped by addition of hydrazine, hydroxylamine, or semicarbazide [Eq. (12)]. However, photosensitive protecting groups are rarely used, perhaps because of uncontrolled side reactions of certain peptides.

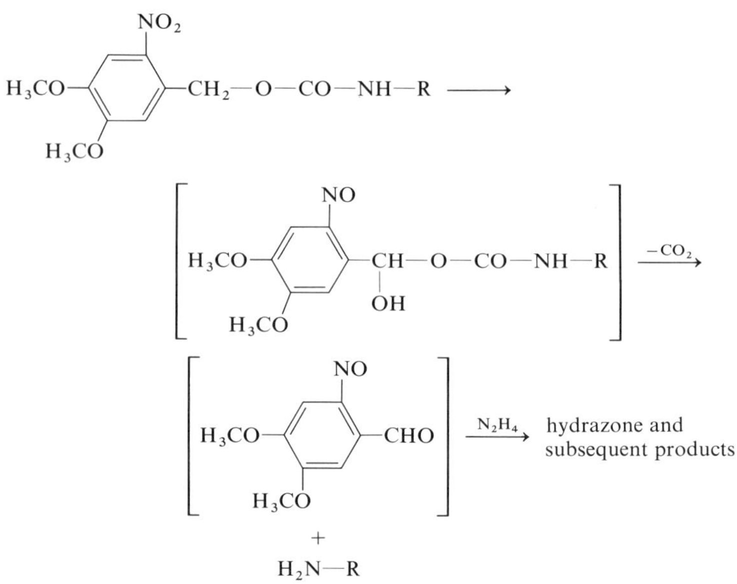

(12)

4. Alkali-Labile Urethane Protecting Groups Derived from Primary Alcohols

This family comprises a class of compounds (Table IV) which is split off with the formation of a double bond. The implication of β-elimination in the design of protecting groups was first demonstrated by Kader and Stirling (1964) with the toluenesulfonylethyloxycarbonyl residue (**12**, see Table IV). Several compounds of this series meet the requirements for β-elimination. The cleavage of the methylsulfonylethyloxycarbonyl (Msc) group (Tesser and Balvert-Geers, 1975) may illustrate the common mechanism [Eq. (13)].

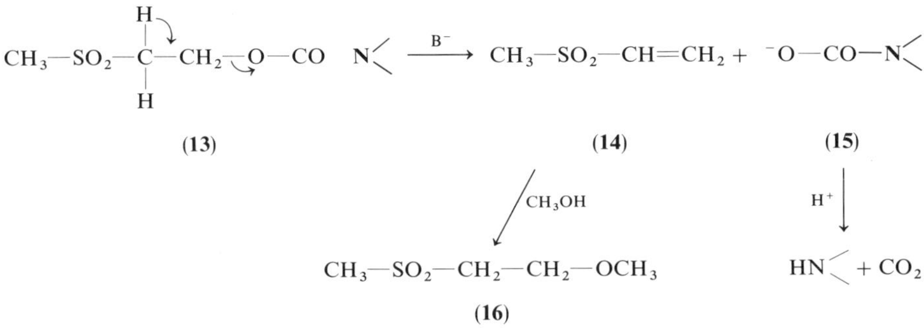

(13)

For the 5-benzisoxazolylmethyleneoxycarbonyl (Bic) group (Kemp and Hoyng, 1975), a related mechanism [Eq. (14)] can be postulated. The oxazolone ring is first opened under the influence of base; the unstable intermediate then undergoes rearrangement resulting in elimination of CO_2 and formation of the amine moiety. The Bic group is stable to medium strong acids and is consequently complementary to the acid-labile *tert*-alkyloxy-

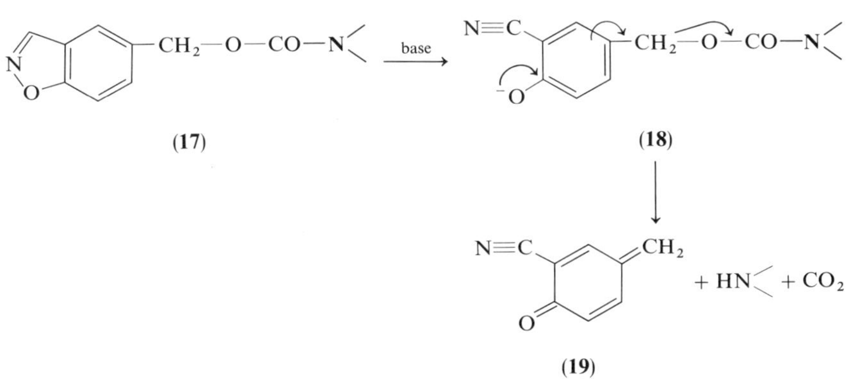

(14)

Table IV Base-labile Urethane-type Amine Protecting Groups Derived from Primary Alcohols

Protecting group	Compd no.	Structure	Abbrev
One step			
4-Toluenesulfonylethyloxycarbonyl	**(12)**	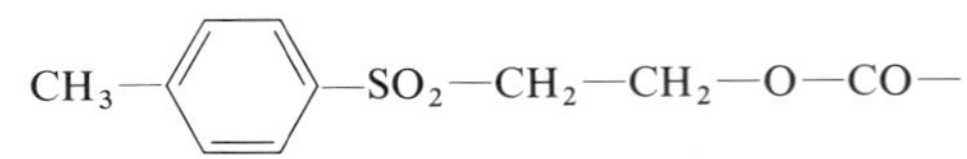	Tsc
Methylsulfonylethyloxycarbonyl	**(13)**	$CH_3—SO_2—CH_2—CH_2—O—CO—$	Msc
2-(Triphenylphosphono)ethyloxycarbonyl	**(20)**	$(C_6H_5)_3P—CH_2—CH_2—O—CO—$	Pec
9-Fluorenylmethyloxycarbonyl	**(21)**	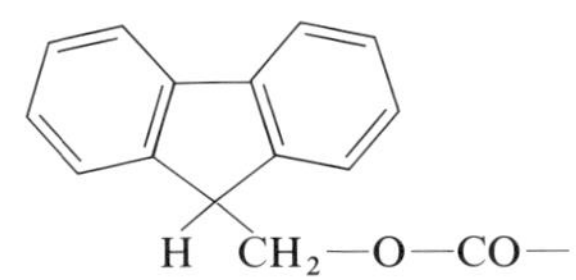	Fmoc
Two step			
Methylthioethyloxycarbonyl	**(22)**	$CH_3—S—CH_2—CH_2—O—CO—$	Mtc
Benzisoxazole-5-methyleneoxycarbonyl	**(17)**	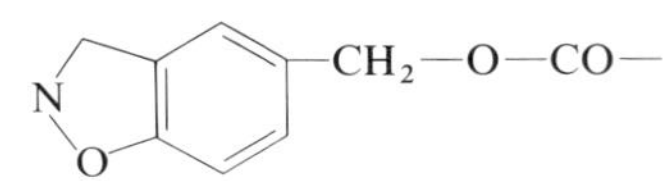	Bic

carbonyl group (Section IIID,1). On the other hand, it is cleaved by HBr–AcOH and hydrogenolysis like the benzyloxycarbonyl group. The deblocking conditions are stronger than those for the methylsulfonylethyloxycarbonyl group. Isomerization of **17** to **18** is achieved by triethylamine in acetonitrile or dimethylformamide at 25°C for 30 min, and the rearrangement of **18** occurs in good yields in trifluoroacetic acid for 90 min or in aqueous ethanolic sodium sulfite buffer at pH 7.

The Msc group (**13**) is cleaved by NaOH at pH 10–12 at 0°C within minutes, and the decomposition of the carbamate (**15**) proceeds immediately. A disadvantage of these protecting groups is the reactivity of the ensuing vinyl compound that requires to be trapped. For example, methanol reacts with **14** and forms the more stable methylsulfonylethyl methyl ether (**16**).

The 2-triphenylphosphonio (Pec) group (**20**, see Table IV) published by Kunz (1976a) shares the advantage with Msc (**13**) of being resistant to treatment with strong acids and of solubilizing the adjacent peptides, thus rendering possible peptide synthesis even in water (Kunz, 1978).

The 9-fluorenylmethyloxycarbonyl (Fmoc) group (**21** see Table IV) (Carpino and Han, 1970) can display an adverse solubilizing effect. Elimination is accomplished by piperidine at room temperature or by basic polymers (Arshady *et al.*, 1979).

The principle of the 2-methylthioethyloxycarbonyl (Mtc) group (**22**) (Kunz, 1976b) was revealed by Hardy *et al.* (1972). This residue is quite stable to acids and bases under moderate conditions, but after transformation to the sulfonium compound (**23**) or the sulfone (**13**) it easily undergoes β-elimination.

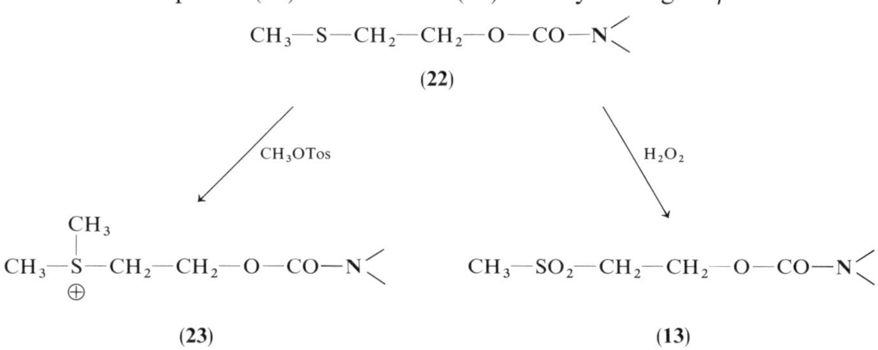

These amine-protecting groups may be introduced into amino acids by means of the corresponding stable chloroformates or active esters (carbonates). Since Peoc chloride is extremely sensitive to aqueous alkali it must be reacted with amino acid esters in chloroform. Subsequently, the esters must be saponified by aqueous acid. Benzyl or silyl esters facilitate this step, the former being susceptible to hydrogenolysis and the latter to hydrolysis (Kunz, 1976a).

Considering all known properties of the protecting groups discussed in this section, Fmoc and Msc may be most suitable for the synthesis of higher peptides. Both are stable toward acid treatment, Msc additionally to hydrogenolysis whereas Fmoc is cleaved by catalylic hydrogenation in contrast to earlier reports (König, 1978, unpublished; Atherton *et al.*, 1979). The cleavage conditions are smooth. Experience with Msc has been gained during syntheses of melanotropin (MSH) sequences (Eberle *et al.*, 1975; Smeets *et al.*, 1977) and an insulin fragment (Wolters *et al.*, 1974). Msc protection has also been successfully applied during semisynthetic investigations with a sensitive molecule like insulin (Geiger *et al.*, 1975). Only aspartic acid β-ester residues are incompatible with this group since cyclization to the aspartimide occurs during treatment with alkali. Even aspartic acid β-*tert*-butyl ester residues undergo this latter reaction (König and Volk, 1977).

In contrast to Msc, Fmoc is cleaved by piperidine without affecting aspartic acid β-*tert*-butylester residues. On the other hand, free amino acids or peptides have always been observed together with peptide formation in dimethylformamide using Fmoc for α-amino group protection (Bodanszky *et al.*, 1979). The lability of Fmoc toward bases is less dangerous in solid-phase synthesis where the spatial separation of the resin-bound amine component reduces the incidence of N^{α}-Fmoc group deblocking and subsequent double acylation. In a model study resin-bound cyclic secondary amine cleaved Fmoc to a much lesser degree than did cyclic secondary amine in solution (Carpino *et al.*, 1978). If acylation precedes rapidly with strong activation, the resin-bound free amine disappears rapidly and the extent of undesired Fmoc cleavage and subsequent side product formation is negligible.

Although the loss of Fmoc is minimized in solution, too, by coupling with DCC and excess HOBt (König, 1979, unpublished), Fmoc has been favoured for N^{α}-protection mainly in solid-phase synthesis. It permits the use of *tert*-butyl-type side-chain blocking and peptide-to-resin linkage cleavable by mild acidolysis. Thus, strong deblocking conditions, which cause severe damage of more complicated peptides, are avoided. The novel combination technique may considerably improve solid-phase synthesis (Chang and Meienhofer, 1978; Atherton *et al.*, 1978).

A derivative of Fmoc, 9-(2-sulfo)fluorenylmethyloxycarbonyl, has been recommended for the purification of synthetic peptides (Merrifield and Bach, 1978).

5. Urethane Protecting Groups Derived from Various Primary Alcohols

a. Furfuryloxycarbonyl Group (Foc). The furfuryloxycarbonyl (Foc) group (**25**, see Table V) can be introduced by reacting furfurol with isocyanatocarbonic acid esters (**24**) (Losse *et al.*, 1964) [Eq. (15)].

Table V Urethane Protecting Groups Derived from Various Primary Alcohols

Protecting group	Compd no.	Structure	Abbrev	Cleavage
Furfuryloxycarbonyl	(**25**)	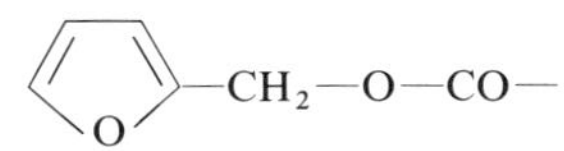	Foc	H^+; H_2(Pd)
γ-Picolyloxycarbonyl (Isonicotinyloxycarbonyl)	(**26**)	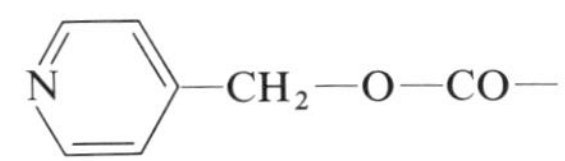	Inc	Zn–50% AcOH; H_2(Pd)
2-Bromoethyloxycarbonyl	(**27**)	Br—CH_2—CH_2—O—CO—	Bec	$AgClO_4$
2-Iodoethyloxycarbonyl	(**28**)	I—CH_2—CH_2—O—CO—	Iec	e^-
2,2,2-Trichloroethyloxycarbonyl	(**29**)	Cl_3C—CH_2—O—CO—	Tcc	Zn in methanol or dilute AcOH

(24)

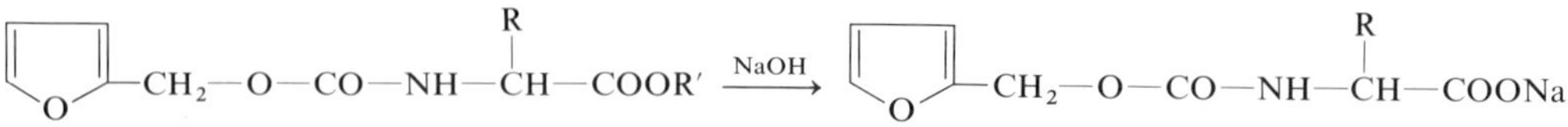

(15)

This general method can be applied if chlorocarbonates are unstable and if the resulting urethanes are not affected by the subsequent saponification of the ester. The fluorocarbonate, however, is sufficiently stable under Schotten–Baumann conditions (Schnabel *et al.*, 1968), thus enabling the classical preparation of N-protected amino acids. Furfuryloxycarbonyl-amino acids and peptides can be deprotected by treatment with trifluoroacetic acid or by catalytic hydrogenation. The Foc group offers no advantage over the 4-methyloxybenzyloxycarbonyl (Moz) group.

b. Pyridine-4-methyloxycarbonyl Group (Inc). The favorable properties of picolyl esters (see Chapter 2, this volume) encouraged Veber *et al.* (1972) to employ this alcohol as a constituent of an urethane N-protecting group that is also described as the isonicotinyloxycarbonyl group (**26**).

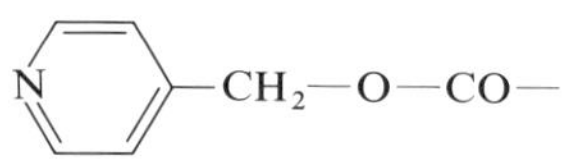

(**26**)

It is introduced via its hydroxysuccinimidoyl or nitrophenyl carbonate and is cleaved by hydrogenation or Zn in 50% acetic acid. Because of its stability to acid and its solubilizing effect, it may also provide good protection for the ε-amino group of lysine (see also Veber *et al.*, 1977).

c. 2-Halide-Substituted-Ethyloxycarbonyl Groups. This type of amine protection has been applied when smooth deprotection in neutral solution was desirable. The chlorocarbonates of these compounds are stable reagents for the acylation of amine groups. The deprotection of 2-bromoethyloxycarbonyl (Bec) amine compounds (**27**, see Table V) (Brugger, 1965; Büchi, 1965) by $AgClO_4$ in acqueous acetone is so mild that protecting groups that are sensitive to acids, bases, or hydrogenation are not affected. Therefore, the main application of these residues lies in the preparation of intermediates bearing labile protection. Scheme 2 shows the mechanism of cleavage of the Bec group (see also Wünsch *et al.*, 1974).

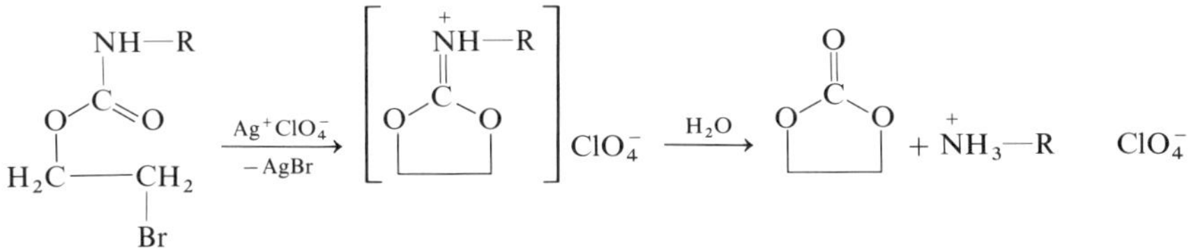

Scheme 2

The corresponding iodo compound (**28**, Table V) (Grimshaw, 1965) is easily cleaved by Zn in methanol or by electrolytic reduction (Kasafirek, 1972). Recently, Eckert and Ugi (1976) removed such protecting groups by means of the supernucleophilic cobalt(I) phthalocyanine. Treatment of **28a** with alkali should be avoided because of side reactions resulting in the formation of oxazolidones [Eq. (16)]. These protecting groups have not found any general application.

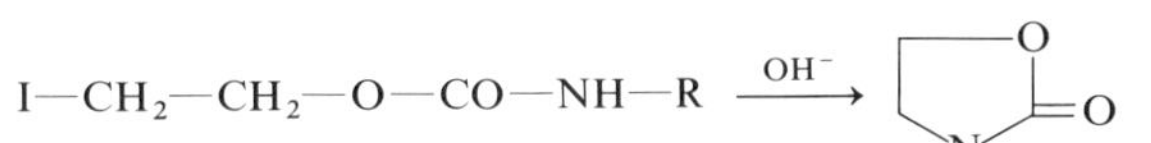

(**28a**) (16)

The 2,2,2-trichloroethyloxycarbonyl residue (**29**, Table V) has been applied in the total synthesis of cephalosporin by Woodward *et al.* (1966). The reaction of amines with 2,2,2-trichloroethyl chloroformate is carried out in the presence of sodium carbonate (Windholz and Johnston, 1967; Yajima *et al.*, 1971). The protecting group is sufficiently resistant against acid treatment. Even saponification or hydrazinolysis of ester groups may be carried out in its presence. Reports on the sensitivity to hydrogenolysis are ambiguous. The group is cleaved by means of zinc in 90% acetic acid within 2.5 hr at 0°C or by short heating with zinc in methanol. The mechanism for the cleavage of the trichloro- and iodoethyloxycarbonyl groups seems to be similar. Reaction starts with electron transfer from zinc to a halide [Eq. (17)].

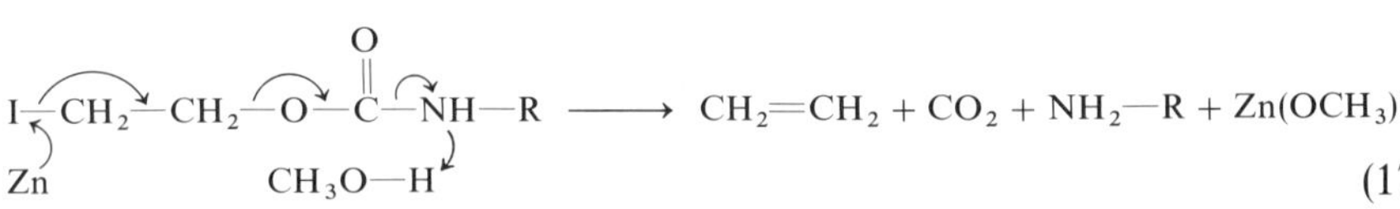

(17)

(**28a**)

d. Vinyloxycarbonyl Group (Voc). Olofson *et al.* (1977) studied the interesting vinyloxycarbonyl amine protecting group that is introduced via the known vinyl chloroformate and cleaved, for example, by an acid-induced reaction sequence [Eq. (17a)].

$$H_2C{=}CH{-}O{-}CO{-}N\Big< \xrightarrow[\text{dioxane}]{\text{HCl}} H_3C{-}\overset{\displaystyle Cl}{\overset{|}{C}}H{-}O{-}CO{-}N\Big< \xrightarrow[50^{\circ}\text{C}, 30\ \text{min}]{\text{EtOH}}$$

$$H_3C{-}\overset{\displaystyle OEt}{\overset{|}{C}}H{-}O{-}CO{-}N\Big< \longrightarrow HCl\cdot HN\Big< + CO_2 + H_3C{-}CH(OEt)_2 \qquad (17a)$$

Because of the exceptional reactivity of the vinyl moiety toward electrophilic reagents, vinyloxycarbonyl peptide derivatives could gain more attention for subsequent modification than for routine peptide synthesis.

C. Urethane Protecting Groups Derived from Secondary Alcohols

Representative amine-protective urethane groups derived from secondary alcohols are listed in Table VI. All groups are introduced via the corresponding chlorocarbonates. The cleavage conditions reach from high lability to complete stability to acids.

The diisopropylmethyloxycarbonyl (Dmc) group (**30**, Table VI) was first described by McKay and Albertson (1957). It is stable under all conditions commonly used for the cleavage of N-protecting groups. Thus, it appeared to be useless for peptide synthesis until Sakakibara *et al.* (1970) found that it can be cleaved by liquid HF at 20°C within 60 min in the presence of anisole. In spite of its excellent selectivity of cleavage this protecting group received only scant attention. The same is true for the cholesteryloxycarbonyl group (**34**) (Staab and Benz, 1961), which is cleavable by HBr–AcOH and most likely also by liquid HF. It belongs to the solubilizing protecting groups and in this respect deserves attention for side-chain protection.

The acid stability of the cyclopentyl- (**31**) (McKay and Albertson, 1957) and *cis*-2-methylcyclohexyloxycarbonyl (**32**) groups (Blaha and Rudinger, 1965) is similar to that of the benzyloxycarbonyl group; however, both residues are resistant to hydrogenolysis. The methylcyclohexyl derivative exhibits a configuration-dependent solvolysis. The rate of cleavage of the cis-compound is much faster than that of the trans-isomer. The same dependence was observed with the acid-labile isobornyloxycarbonyl (Ibc) group (**33**) (Jäger and Geiger, 1973a,b; Fujino *et al.*, 1972), whereas the corresponding bornyl derivative is stable in acids (Carpino, 1966). The Ibc group is optically active and has been applied in the L, D, and DL forms. A great number of isobornyloxycarbonylamino acids have been prepared by means of the fairly stable chlorocarbonate (see Wünsch *et al.*, 1974). This very active reagent renders the diacylation of the guanidine group in arginine possible (Jäger and Geiger, 1973c) (see Section VII,C,2). The disadvantage that diastereoisomers are formed with optically active amino

Table VI Urethane Protecting Groups Derived from Secondary Alcohols

Protecting group	Compd no.	Structure	Alcohol component	Abbrev	Cleavage
Diisopropylmethyloxycarbonyl	(**30**)	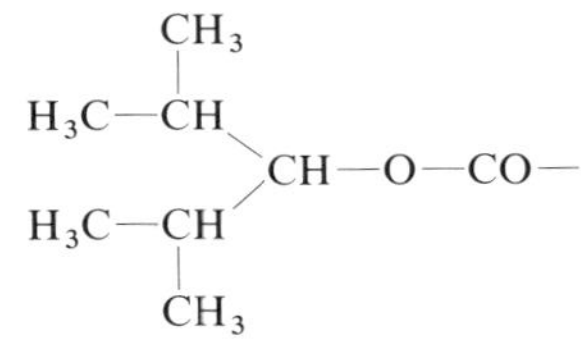	2,4-Dimethyl-3-pentanol (Diisopropylcarbinol)	Dmc	HF
Cyclopentyloxycarbonyl	(**31**)	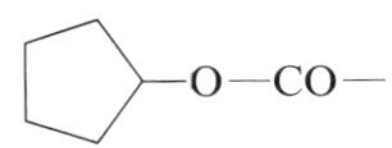	Cyclopentanol	Cpc	HBr–AcOH
cis-2-Methylcyclohexyloxycarbonyl	(**32**)	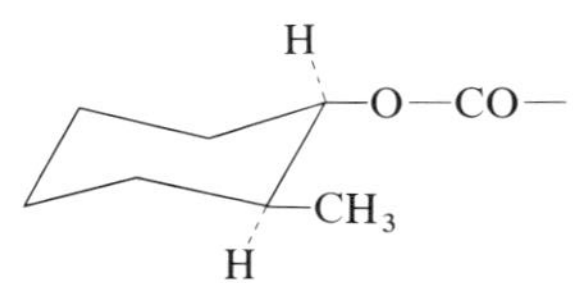	*cis*-2-Methylcyclohexanol		HBr–AcOH
Isobornyloxycarbonyl	(**33**)	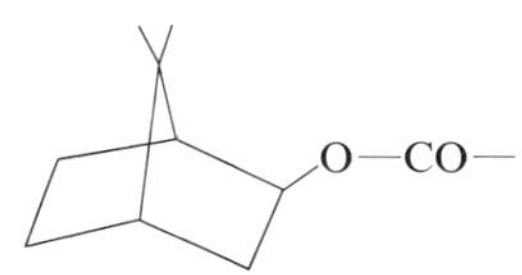	Isobornyl alcohol (isoborneol)	Ibc	CF_3COOH

(continued)

Table VI (*Continued*)

Protecting group	Compd no.	Structure	Alcohol component	Abbrev	Cleavage
Cholesteryloxycarbonyl	(**34**)	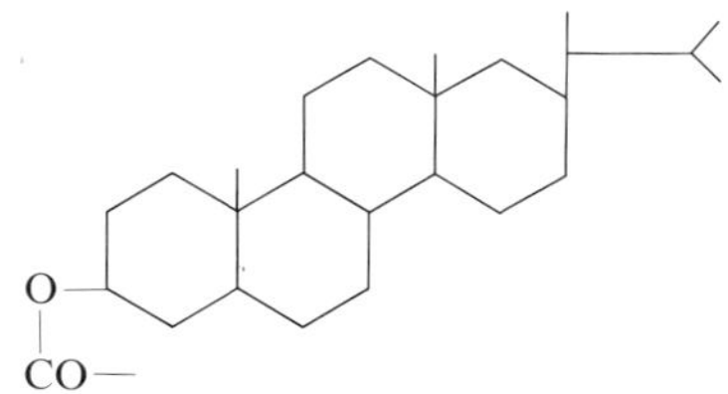	Cholesterol	Coc	HBr–AcOH HF(?)
Diphenylmethyloxycarbonyl (Benzhydryloxycarbonyl)	(**35**)	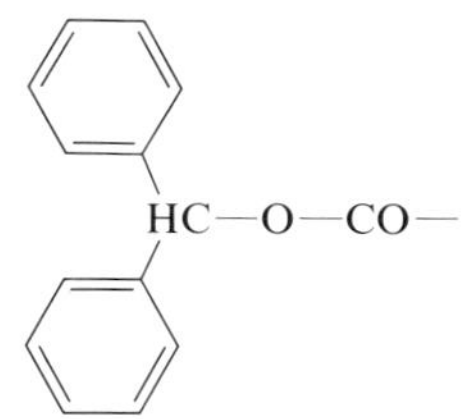	Benzhydrol	Doc	CF_3COOH H_2–Pd
Di-2-pyridylmethyloxycarbonyl	(**36**)	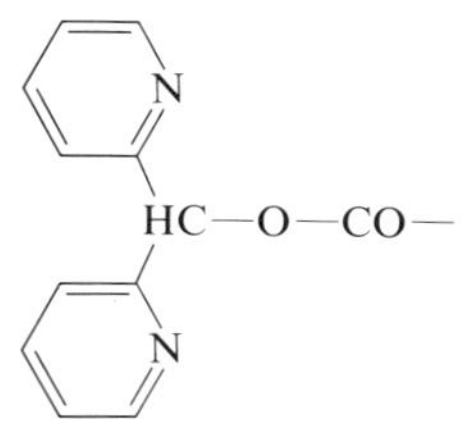	Di-2-pyridylcarbinol	Dpc	Zn–AcOH e^-

acids and peptides must be weighed against the advantage that the solubility of such "impure" peptides is often increased.

The Ibc protecting group has been applied successfully to syntheses of bradykinin (Fujino and Shinagawa, 1972) and gonadoliberin (LHRH) analogues (Fujino *et al.*, 1973). The acid stability is comparable to that of the *tert*-butyloxycarbonyl group.

Another type of urethanes derived from secondary alcohol is represented by the benzhydryloxycarbonyl group (**35**) (Staab and Benz, 1961) and the pyridine analogue (**36**, see Table VI). The former has been introduced into amino groups via the azide by Hiskey and Adams (1965) and applied in the synthesis of cystine peptides. It can be eliminated by, for example, 1.7 *N* HCl in tetrahydrofuran at 25°C within 5–60 min. During such treatment, benzhydryl esters are not affected and rearrangement of asymmetric disulfides has not been observed. In contrast to groups **30**–**34**, the benzhydryloxycarbonyl group is cleaved by catalytic hydrogenation.

The di-2-pyridylmethyloxycarbonyl group (**36**) (Coyle *et al.*, 1975) behaves quite differently. It resists acid treatment but can be eliminated by zinc in acetic acid, by electrolytic reduction, or by hydrogenation.

D. Urethane Protecting Groups Derived from Tertiary Alcohols

1. tert-Butyloxycarbonyl Group (Boc)

The *tert*-butyloxycarbonyl group (**37**) (McKay and Albertson, 1957; Anderson and McGregor, 1957) is the prototype of an acid-labile protecting group. Similar to the hydrogenolytically cleavable benzyl ester, the acid-

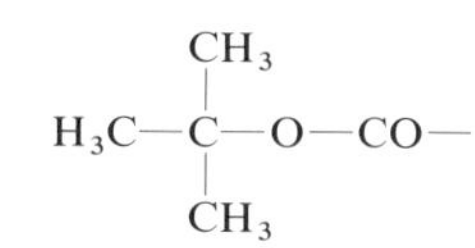

(**37**) Boc

sensitive *tert*-butyl ester was adapted to give an N-protecting group via urethane formation. The residue is resistant to catalytic hydrogenation, alkali, and sodium in liquid ammonia. Thus, it offers an excellent complement to the benzyloxycarbonyl residue, and together they have become the most frequently used ones (see Schröder and Lübke, 1965, 1966; Wünsch *et al.*, 1974). Many peptides have been synthesized using both types combined for main-chain and side-chain protection. Since the chlorocarbonate of the *tert*-butyloxycarbonyl group is only fairly stable above −30°C, active esters have been prepared (Anderson and McGregor, 1957; Frankel *et al.*, 1966;

Broadbent *et al.*, 1967; Fujino and Hatanaka, 1967). Schwyzer *et al.* (1959) recommended the azide. Schnabel *et al.* (1968) proposed the sufficiently stable and very active fluorocarbonate for the introduction of the *tert*-butyloxycarbonyl group. The azide was the most widely used reagent for many years, being conveniently prepared from the *tert*-butyloxycarbonyl-hydrazide. Excellent yields of *tert*-butyloxycarbonylamino acids are obtained, especially if the synthesis is performed in the presence of NaOH using the pH stat at optimal pH (Schnabel, 1967). Recently, the highly active and stable pyrocarbonate (Tarbell *et al.*, 1972) and 2-*tert*-butyloxycarbonyl-oximino-2-phenylacetonitrile (Boc-ON; **37a**; Itoh *et al.*, 1974) became commercially available. No hazards have to be ascribed to these reagents

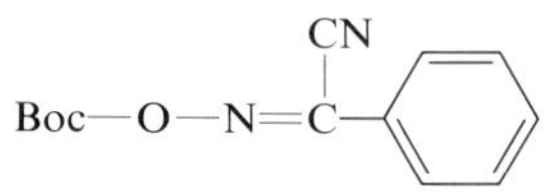

(**37a**)

since the explosive $BocN_3$ and HN_3, which have caused occasionally severe accidents, are omitted. Further oxime derivatives suitable for the introduction of *tert*-butyloxycarbonyl and other protective groups have also been described by Itoh *et al.* (1977).

tert-Butyloxycarbonylamino acids are compatible with all kinds of carboxyl activation. However, attention is drawn to the observation made by Bodanszky *et al.* (1975) that the activating procedure may cleave the *tert*-butyloxycarbonyl group to a small degree. The *tert*-butyloxycarbonyl group is rapidly cleaved by cold trifluoroacetic acid, hot formic acid, HCl, and TosOH in alcohol or acetic acid, etc. An annoying side reaction caused by *tert*-butyl cations is *tert*-butylation of tryptophan and methionine (Löw *et al.*, 1978a,b, investigated this reaction). It is diminished or avoided by deblocking with HCl in formic acid, acetic acid, or methanol, and by adding suitable scavengers during the cleavage reaction (Lundt *et al.*, 1978; see also Suzuki *et al.*, 1978; Chino *et al.*, 1978). The *tert*-butyloxycarbonyl group is also slightly thermolabile especially in aqueous solution (Bailey and Griffith, 1964). For example, recrystallization of Boc-Ser-Tyr-Ser-Met-N_2H_3 from hot water resulted in a partial loss of the Boc group (L. Kisfaludy and R. Geiger, unpublished observations).

2. *tert-Amyloxycarbonyl Group (Aoc)*

tert-Amyl chlorocarbonate is more stable than the corresponding derivative of *tert*-butanol. Therefore, Sakakibara *et al.* (1965) recommended this protecting group (**38**, Table VII) instead of the Boc group. However, there are no further advantages to the *tert*-amyl analogue.

Table VII Urethane Protecting Groups Derived from Tertiary Alcohols

Protecting group	Compd no.	Structure	Abbrev
tert-Butyloxycarbonyl	(**37**)	$H_3C-C(CH_3)_2-O-CO-$	Boc
tert-Amyloxycarbonyl	(**38**)	$CH_3-CH_2-C(CH_3)_2-O-CO-$	Aoc
Adamantyloxycarbonyl	(**39**)	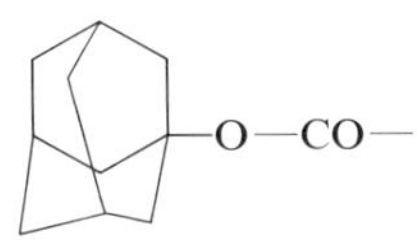	Adc
1-Methylcyclobutyloxycarbonyl	(**40**)	$C_2H-C(-O-CO-)$; C_2H-CH_2 (ring)	Mcb (McBoc)
1-Methylcyclohexyloxycarbonyl	(**41**)	$H_2C(CH_2-CH_2)_2C(CH_3)-O-CO-$	Mch

3. *Adamantyloxycarbonyl (Adc) and 1-(1-Adamantyl)-1-methylethyloxycarbonyl Groups*

Adamantyl chlorocarbonate displays a further increase of stability as compared to *tert*-amyl chlorocarbonate. Haas *et al.* (1966) used this reagent to protect (**39**, see Table VII) amine groups and the imidazole nitrogen of histidine. The reagent is stable enough to allow the introduction of two Adc groups into the guanido group of arginine, employing strongly alkaline conditions (Jäger and Geiger, 1970). Recently, the more stable fluorocarbonate was proposed by Moroder *et al.* (1976). It may displace the chlorocarbonate in the future, since it is readily prepared from fluorophosgene. The adamantyloxycarbonyl residue is an excellent protecting group, stable toward both alkali and hydrogenation. Similar to the *tert*-butyloxycarbonyl and the *tert*-amyloxycarbonyl groups, it is cleaved by trifluoroacetic acid. It is more resistant to nucleophilic attack than the *tert*-butyloxycarbonyl group. The lipophilic character improves the solubility of hydrophilic peptides in organic solvents. However, application is curtailed because of the relatively high price of adamantanol.

The 1-(1-adamantyl)-1-methylethoxycarbonyl group (Kalbacher and Voelter, 1978) is acidolytically removed 10^3 times faster than the adamantyloxycarbonyl and the *tert*-butyloxycarbonyl residues, thus allowing selective cleavage by 3% trifluoroacetic acid in dichloromethane at 0°C, leaving the latter residues unaffected.

4. 1-Methylcyclobutyl- (Mcb) and 1-Methylcyclohexyloxycarbonyl (Mch) Groups

The acidolysis rates of **37** and **38** (Table VII) are similar. The adamantyloxycarbonyl group is slightly more stable but precise values are not available. The 1-methylcyclobutyloxycarbonyl (**40**, Table VII) and 1-methylcyclohexyloxcarbonyl (**41**, Table VII) groups have been published by Veber *et al.* (1972) and Brady *et al.* (1977). Compound **40** is more and **41** less stable than Boc: ($t_{1/2}$ of **40**:**37**:**41** = 4:2:1, respectively, in trifluoroacetic acid and formic acid). The increased stability of **40** is demonstrated by the fact that Mcb-Phe-OH results in a more than 99% recovery from 50% acetic acid after 48 hr at room temperature, whereas 10–15% Boc-Phe-OH has been cleaved under the same conditions. Thus, chromatographic purification of Mcb-peptides can be performed safely using acetic acid-containing solvents. Complete removal of the Mcb group was achieved by treatment with trifluoroacetic acid at 25°C for 30 min. Thus far, the Mcb group has only been applied to the synthesis of a simple model peptide.

5. Urethanes Derived from 2-Arylisopropanols

The excellent and promising properties of urethan protecting groups derived from tertiary alcohols has inspired chemists to look for derivatives exhibiting an even higher acid lability. This could be accomplished by substitution of a methyl group of *tert*-butyl by an aromatic residue. Consequently, these groups can also be considered as di- α-methyl-substituted benzyl derivatives. Such structures (**42**–**45**) are listed in Table VIII.

The most commonly used substituted N-protecting group of this series is the 2-(4-biphenylyl)propyl(2)oxycarbonyl (Bpoc) group (**43**) proposed by Sieber and Iselin (1968). Birr's 2-(3,5-dimethyloxyphenyl)propyl(2)oxycarbonyl (Ddz) group (**44**) (Birr *et al.*, 1972) also turned out to be valuable. The protons of the methyloxy groups are easily detected by means of nmr. This is helpful if quantitative deprotection, especially of large peptides, can no longer be analyzed accurately by tlc. In solid-phase peptide syntheses, reactions can be monitored by uv, where Ddz-amino acids show absorption at 224–229 (ε = 7220), 276, and 282 nm (ε = 2320).

Table VIII 2-Aryl-isopropyloxycarbonyl Amine Protecting Groups

Protecting group	Compd no.	Structure	Abbrev	Relative cleavage rate in 80% acetic acid at 22°–25°C (Boc = 1)
2-Phenylpropyl(2)oxycarbonyl	**(42)**	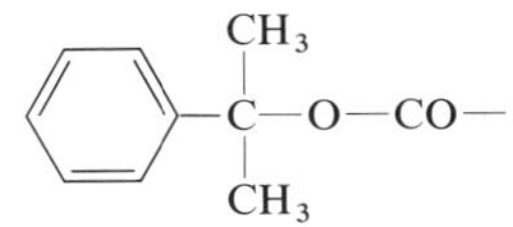	Poc	700
2-(4-Biphenylyl)propyl(2)-oxycarbonyl	**(43)**	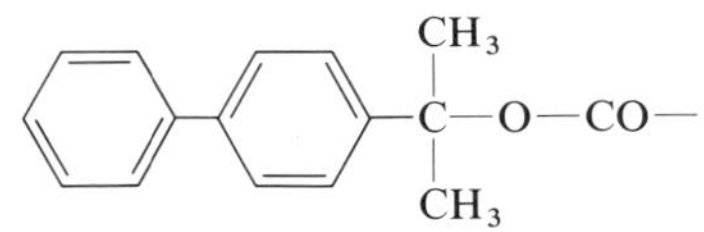	Bpoc	3,000
2-(3,5-Dimethyloxyphenyl)propyl(2)-oxycarbonyl	**(44)**	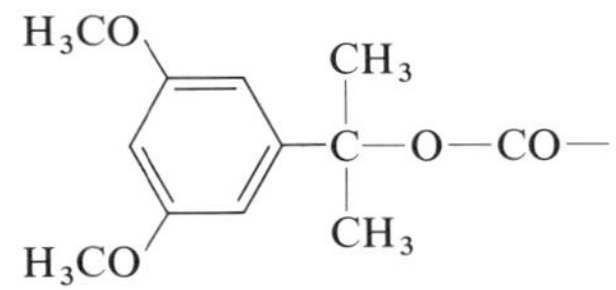	Ddz	1,400
2-(4-Methylphenyl)propyl(2)-oxycarbonyl	**(45)**	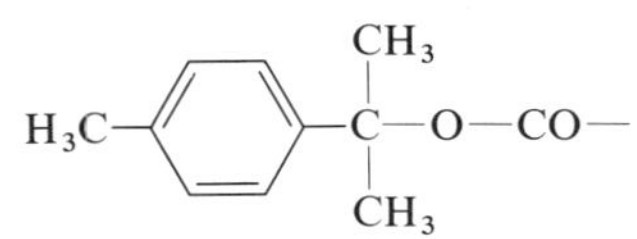	Mpc	11,000
For comparison:				
Triphenylmethyl (trityl)	**(55)**	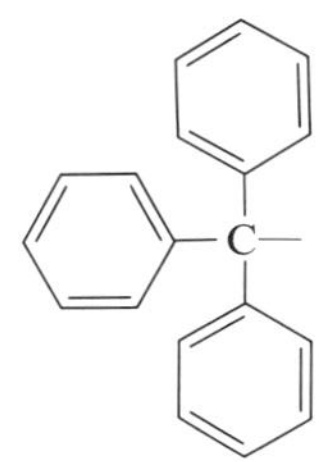	Trt	21,000

Bpoc and Ddz are introduced into amino acids via their azides or activated esters (see Wünsch *et al.*, 1974) and are usually isolated as dicyclohexylammonium salts. Ddz azide is crystalline and stable.

The relative cleavage rates given in Table VIII are not constant but depend on the solvent and the adjacent peptides. The Ddz residue, for example, is usually removed by trifluoroacetic acid in methylene chloride, either 5% for 8–20 min, 2% for 15–40 min or 1% for 30–60 min (C. Birr, personal communication). We have observed that with large peptide chains the cleavage time has to be extended even more (König *et al.*, 1979). The Bpoc group may be more advisable in such cases. The Ddz group is also cleaved by uv irradiation (Birr *et al.*, 1972) but, like Adc is stable to catalytic hydrogenation (König and Kernebeck, 1979), whereas Bpoc is cleaved by hydrogenolysis.

These novel groups filled the gap between the *tert*-butyloxycarbonyl and the trityl groups, thus providing opportunities to use selective protection. The gradually decreasing acid lability permits constant or temporary protection. This principle has found its most admirable realization in a synthesis of insulin (Sieber *et al.*, 1974).

6. *Nitrogen-Containing Substituted Urethanes Derived from Tertiary Alcohols*

When nitrogen is located at a suitable distance to the tertiary C-atom, it generally increases the resistance of the protecting groups to acids markedly. However, the influence also depends on the kind and arrangement of the substituents. Thus, structures **46** and **47** (Table IX) are stable to trifluoroacetic acid. Compound **46** is even stable to 2 *N* HBr in acetic acid for 48 hr at room temperature. It is cleaved by zinc in acetic acid, electrolytic reduction, and catalytic hydrogenation. Compound **47** is removed by 45% HBr in acetic acid at room temperature within 1 hr (Coyle *et al.*, 1975). In group **48**, the nitrogen effect is counterbalanced by two phenyl substituents so that the protecting group behaves as *tert*-butyloxycarbonyl. Two methyl groups and the involvement of the nitrogen in an amide bond results in a similar effect with compound **49**. Both groups are cleaved by trifluoroacetic acid within 1 hr at room temperature (Coyle *et al.*, 1975). An additional solubilizing effect is brought about by the basic groups in **46**, **47**, and **48** (see Table IX). In **50** the phenylazophenyl substituent renders this protecting group easily cleavable by acids (Tun-Kyi and Schwyzer, 1976). The rate of cleavage is similar to those of the Bpoc (**43**) and Ddz (**44**) groups (see Table VIII).

Table IX Nitrogen-Containing Substituted Urethane Protecting Groups Derived from *tert*-Alcohols

Protecting group	Compd no.	Structure	Abbrev	Introduction	Cleavage	Stability to
2-(4-Pyridyl)propyl(2)oxycarbonyl	(**46**)	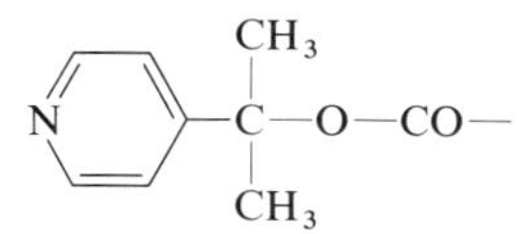	—	Active ester	H_2–Pd Zn–AcOH e^-	2 *N* HBr–AcOH 48 hr, 20°C
1,4-Dimethylpiperidyl(4)oxycarbonyl	(**47**)	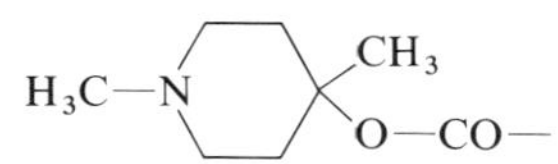	—	Active ester	45% HBr–AcOH 1 hr, 20°C	CF_3COOH 2 *N* HCl in THF 1 hr, 20°C
N,N-Dimethylaminoethyldiphenylmethyloxycarbonyl	(**48**)	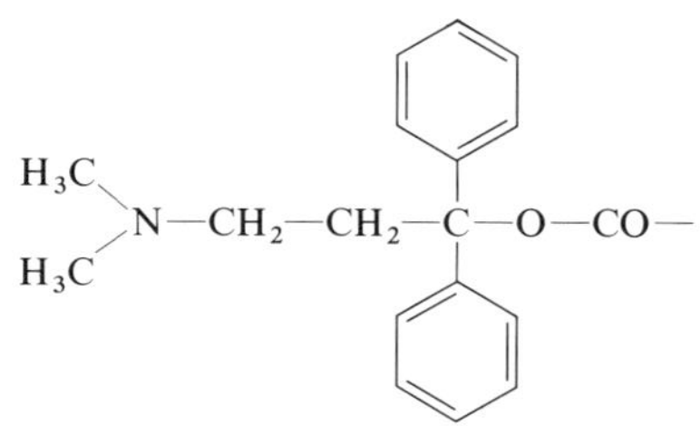	—	Active ester	H_2–Pd CF_3COOH, 1 hr, 20°C	

(*continued*)

Table IX (*Continued*)

Protecting group	Compd no.	Structure	Abbrev	Introduction	Cleavage	Stability to
2-(*N*,*N*-Dimethylcarbamoylethyl)-propyl(2)oxycarbonyl	(**49**)	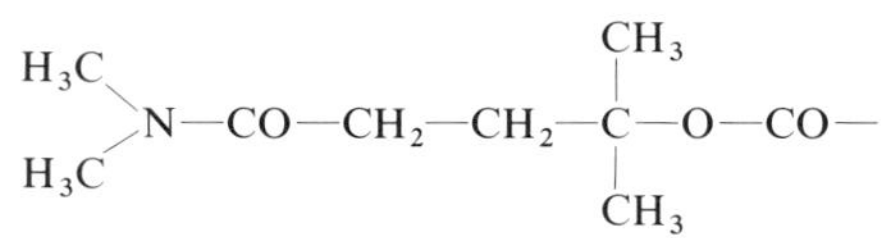	—	Active ester	H_2–Pd CF_3COOH, 1 hr, 20°C	
2-Azobenzenepropyl(2)oxycarbonyl	(**50**)	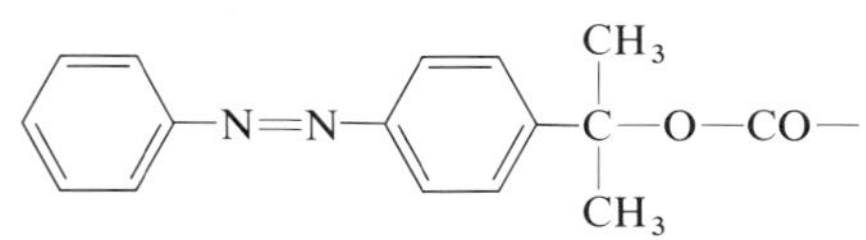	Azc	Azide	1.5% CF_3COOH in CH_2Cl_2, 5 min	
2-Cyano-*tert*-butyloxycarbonyl	(**51**)	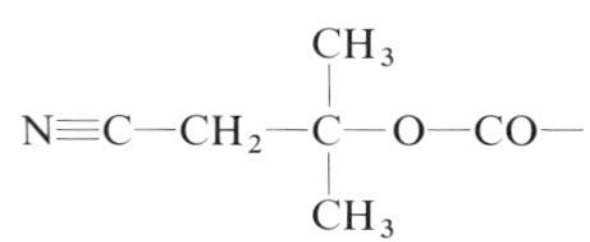	Cyc	Chloride	OH^-	

The 2-cyano-*tert*-butyloxycarbonyl group (**51**) resembles the 2-pyridyl-propyl(2)oxycarbonyl group (**46**) in its stability to acids, but it is unique in this series since it is cleaved by bases under β-elimination (Wünsch and Spangenberg, 1971). The sensitivity to hydrogenation limits its application compared to protecting groups selected from Table IV.

7. N-Dialkyl-substituted N-Oxycarbonyl Protecting Groups

Stevenson and Young (1967) extended the work on 1-piperidyl esters and prepared piperidyloxycarbonylamino acids (**51a**) and the corresponding *N*-dimethyl-substituted compounds (**51b**) (Stevenson and Young, 1969).

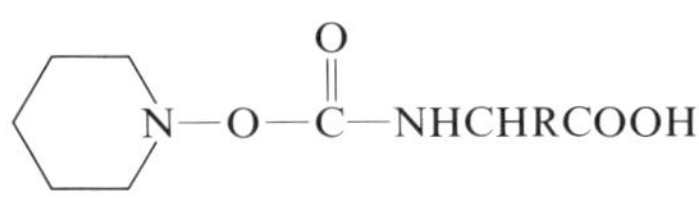

(**51a**)

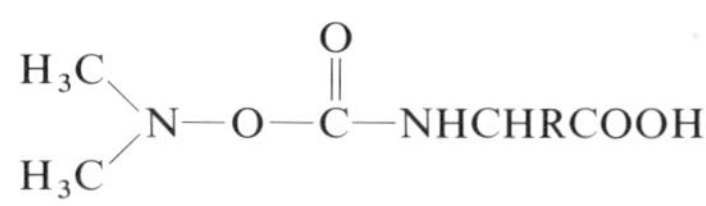

(**51b**)

Both protecting groups behave similarly. They are introduced into amino groups by means of active esters. Another route is the reaction of the *N*-hydroxy compounds with isocyanatocarboxylic acid esters. Subsequent saponification yields N-protected amino acids which are stored as dicyclohexylammonium salts.

These protecting groups are stable to cold acids, even to 30% HBr–AcOH. However, they are cleaved by boiling 50% acetic acid within 5 min. Purer products are obtained with zinc dust in aqueous acetic acid (10 min, 25°C), sodium dithionite (5 min, 25°C), electrolytic reduction, or catalytic hydrogenation. In view of their specific conditions of cleavage, these groups may be favorable for side chain protection of lysine or ornithine.

8. Amine-Protection by Carbamate Formation

Künzi and Studer (1975) prepared tosylaminocarbonyl (Tac) amino acids by reaction of the cheap tosylisocyanate with amino acids in boiling benzene [Eq. (18)]. The Tac group can be readily removed by boiling alcohols, but

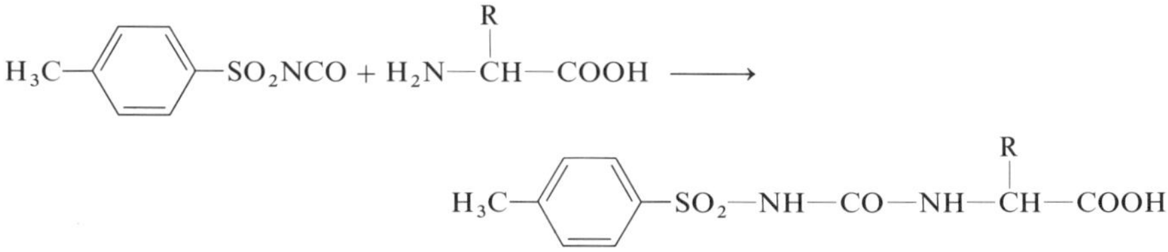

(18)

the reaction proceeds to completeness only with difficulty. Moreover, the activation of Tac-amino acids results in the formation of tosylhydantoin. Therefore, this protecting group might not find general application.

E. Derivatives of Sulfur and Phosphorus as Amine Protecting Groups

In spite of many useful novel amine-protecting groups offering a wide range of cleavage conditions, the typical sulfur-containing 2-nitrophenylsulfenyl (Nps) and 4-toluenesulfonyl (Tos) groups have by no means been displaced. Compared to these quite frequently used groups of high cleavage selectivity, phosphorus-containing derivatives have never played a great role.

1. 2-Nitrophenylsulfenyl Group (Nps)

The application of the 2-nitrophenylsulfenyl group has its origin in the work of Goerdeler and Holst (1959) who synthesized the first Nps-amino acids. Zervas *et al.* (1963) prepared the stable dicyclohexylammonium salts and successfully attempted peptide syntheses with these compounds.

The readily available 2-nitrophenylsulfenyl chloride reacts with amino acids and peptides at pH 7–8 under formation of Nps-amino acids and peptides [Eq. (19)] (see Schröder and Lübke, 1965, 1966; Wünsch *et al.*, 1974). The sulfenamide provides a complete protection of the amine group. On the other hand, it is easily cleaved in weakly acidic medium, leaving *tert*-butyl residues unaffected. This opened an early possibility to complement these important protecting groups, since the extremely acid-labile trityl group (Section III,F) was only reluctantly used at that time, due to steric hindrance of the carboxyl group of trityl amino acids.

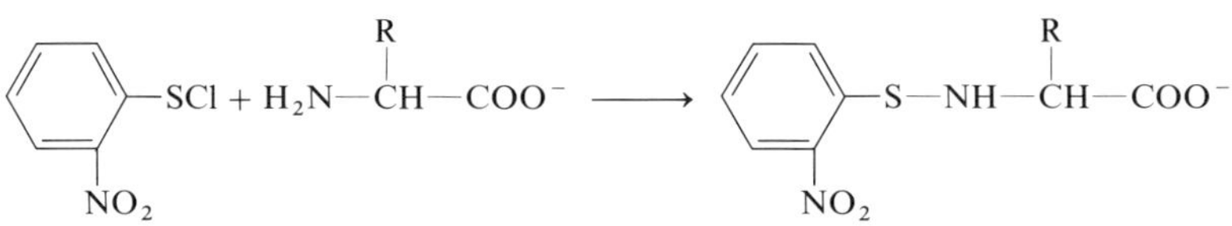

(19)

The 2-nitrophenylsulfenyl group has found wide application in peptide synthesis. Wünsch (1967) used it, for example, as a third amine protecting group besides the benzyl and *tert*-butyl-derived groups for the synthesis of glucagon. Kessler and Iselin (1966) investigated a great number of deblocking procedures for the Nps group. Cleavage starts with an attack of a proton at the sulfur [Eq. (20)]. Intermediates, formed by reaction of the sulfenium ion with a nucleophile, are often active enough to transfer the Nps residue onto another molecule; in weakly acidic media it may then be reattached to an amine group. Scavengers must be added to bring the reaction to completeness under nearly neutral conditions.

Wünsch and Drees (1967) and Fontana *et al.* (1966) observed that activated Nps derivatives react with tryptophan to yield 2-(2-nitrophenylthio)-

tryptophan. Since the tryptophyl moiety of glucagon underwent this reaction too, Wünsch and Drees (1967) added a tenfold molar excess of indole (better 2- or 3-methylindole) and were able to suppress the side reaction.

This technique was the basis for a mild deblocking procedure: Nps-peptides are treated in weakly acidic solution with rhodanide as a strong nucleophile. The initially formed sulfenylrhodanide irreversibly transfers the Nps residue onto methylindole (Wünsch and Spangenberg, 1972) [Eq. (20)].

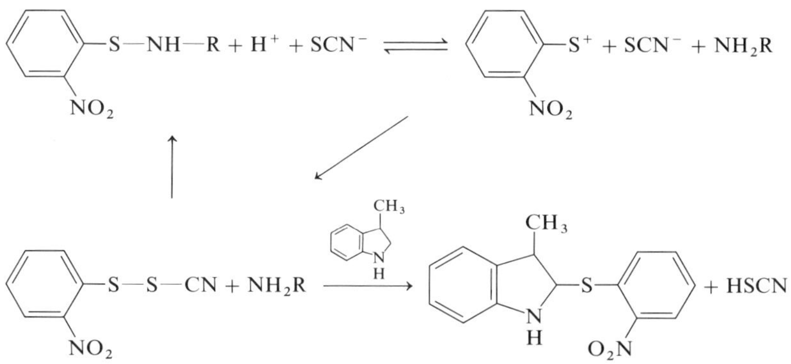

(20)

Thiolysis is another valuable method of Nps group cleavage [Eq. (21)] (Fontana *et al.*, 1966). Tun-Kyi (1978) successfully used 2-mercaptopyridine in a corresponding reaction sequence without affecting tryptophane. How-

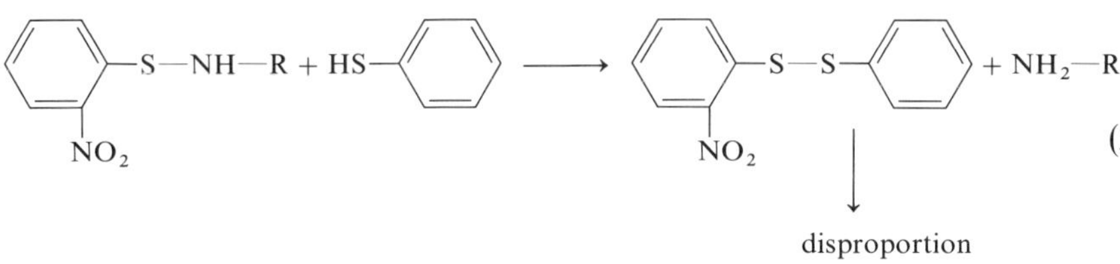

(21)

ever peptide syntheses with Nps-protected intermediates suffer from some obstacles too. Basic conditions should be avoided, and highly effective condensation methods such as dicyclohexylcarbodiimide in the presence of 1-hydroxybenzotriazole (DCC–HOBt; König and Geiger, 1970a) can interfere with the protecting group since HOBt is also a strong nucleophile. Ito *et al.* (1978) recommended the 2,4-dinitrophenylsulfenyl residue for amine protection because of its enhanced stability toward acid.

2. 4-Toluenesulfonyl Group (Tosyl, Tos)

The 4-toluenesulfonyl or tosyl group is one of the earliest among frequently employed N-protecting groups (Rundinger, 1973). The cleavage by HI and PH_4I in acetic acid at 60°C has been already described by Fischer (1915). In contemporary practice the protecting group is almost exclusively split off by sodium in liquid ammonia (du Vigneaud and Behrens, 1937). Sulfonamides are acidic and the ionized sulfamide residue is a strong nucleophile. It can be alkylated yielding *N*-alkylamino compounds after cleavage of the sulfonamide group. The acylation proceeds easily too. Thus, *N*-tosylglutamic acid chloride forms *N*-tosylpyroglutamic acid. The tosyllactam in turn behaves as an activated carboxyl group and undergoes saponification and amidation, thus yielding *N*-tosylglutamic acid γ-amides [Eq. (22)]. Tosylation of amine groups is performed by means of 4-toluene-

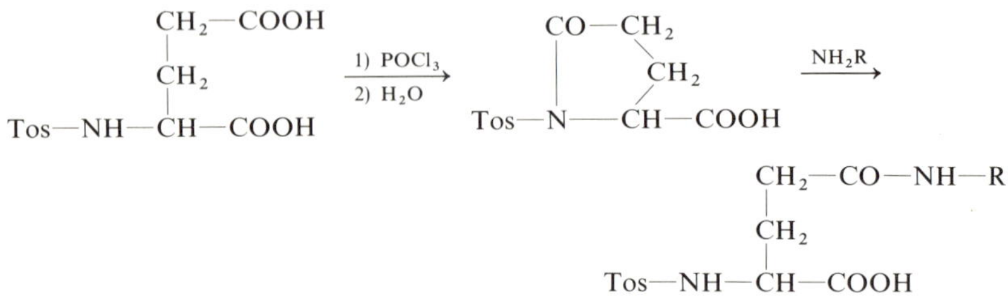

(22)

sulfonyl chloride. Unfortunately, activated *N*-tosylamino acids display a tendency to side reactions and the cleavage of the protecting group by sodium in liquid ammonia is accompanied by partial decomposition of amino acids such as arginine, cysteine, hydroxylamino acids, and tryptophan. The X-Pro bond has been shown to be partially cleaved and racemization has been observed (Table XVII). Under carefully controlled conditions most of these side reactions are suppressed and thus can be tolerated. For instance, the addition of sodium amide to the reaction mixture avoids the cleavage of the X-Pro bond (Katsoyannis *et al.*, 1971).

The tosylamine bond is extremely stable. It is not attacked by HBr–AcOH (except at elevated temperature), alkali, or hydrogen fluoride. However, electrolytic reduction has been described (Horner and Neumann, 1965; Okumura *et al.*, 1971). In spite of some of the aforementioned problems the tosyl group is still used because of its stability as side-chain protection for ornithine, lysine, arginine, and histidine. N^G-Tosylarginine has become an especially important derivative in solid-phase peptide synthesis since the tosylguanidine bond is cleavable by liquid hydrogen fluoride (see Section VII,B) which is commonly used as the final synthetic step to remove peptides from the polymer support. An excellent review on this protecting group has been published by Rudinger (1973).

3. *N-Dibenzylphosphoryl Group (Dbp)*

Zervas and Dilaris (1955, 1956) found that dibenzylphosphoric acid chloride reacts with amino acid esters in various solvents to yield *N*-dibenzylphosphorylamino acids upon saponification of the ester [Eq. (23)].

3. Amine Protection by Dimedone (Dim)

In contrast to the aforementioned groups, the 5,5-dimethyl-3-oxocyclohexen-1-yl (Dim) group is stable to both acids and hydrogenation. (Halpern and James, 1964a,b). The N-protected derivatives are prepared by reaction of dimedone (**52**) with amino acid esters and subsequent cleavage of the ester [Eq. (27)] (Halpern, 1965).

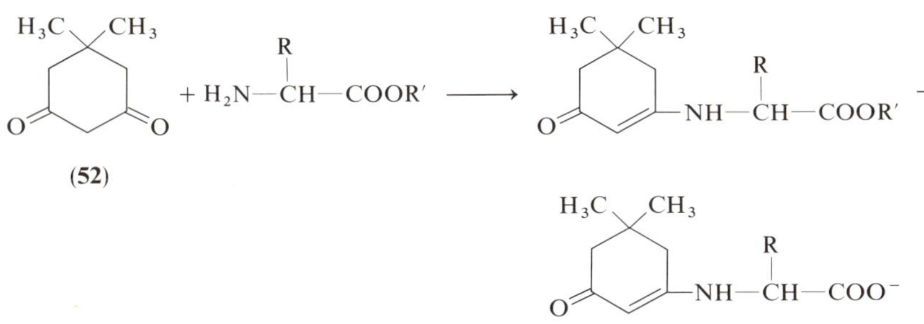

(27)

The Dim group protects activated amino acids against racemization. Cleavage is accomplished by treatment of protected compounds in solution or suspension with bromine water or nitrous acid in acetic acid (Halpern and James, 1964a; Halpern and Cross, 1965) [Eq. (28)]. The application of bromine is less feasible, especially in the presence of tyrosine. Experience

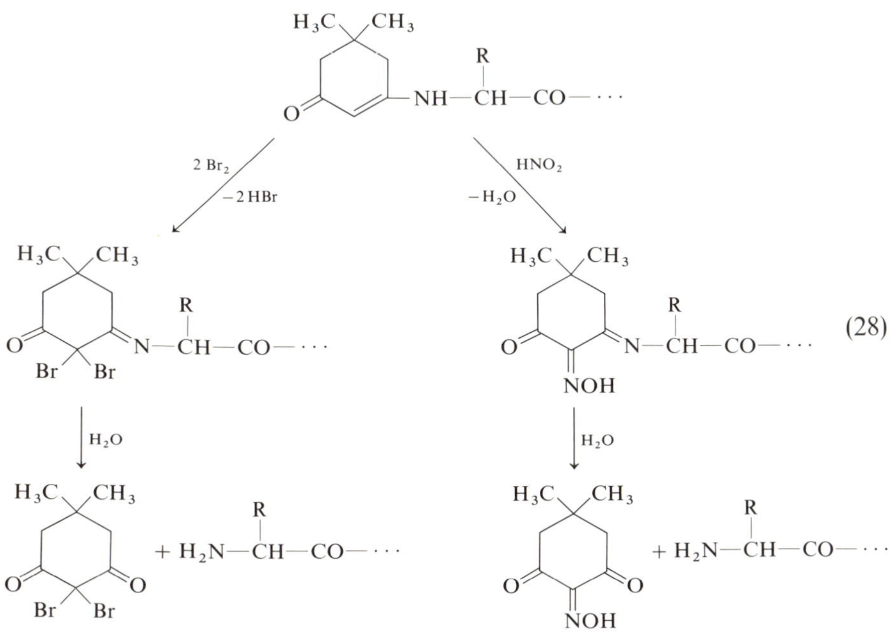

(28)

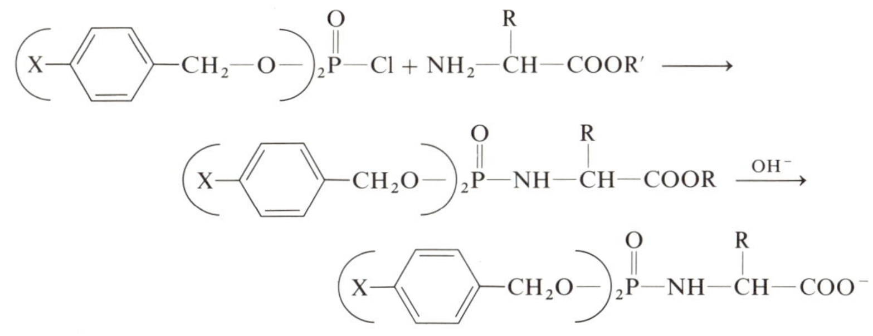

(23)

$X = H, NO_2, Br, I$

Only the dicyclohexylcarbodiimide method has been applied to synthesize *N*-dibenzylphosphoryl peptides. The mixed anhydride method failed as well as did the azide couplings of single amino acids, since hydrazinolysis of Dbp-amino acid esters turned out to be impossible (with the exception of glycine). Peptide esters, however, yielded hydrazides. The cleavage conditions are mild: hydrogenation at pH 4 or HBr in chloroform.

4. Diphenylphosphine Group (Dpp)

Kenner *et al.* (1976) described novel N-protected amino acids that are accessible by reaction of the readily available phosphinic acid chloride with amino acid esters [Eq. (24)]. Saponification of the corresponding esters or hydrogenation of benzyl esters yield Dpp-amino acids.

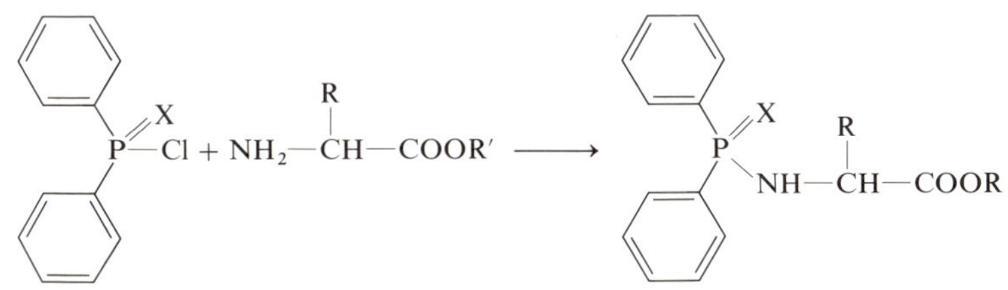

(24)

$X = O$

The Dpp group allows the formation of peptides, without racemization, by all known condensation methods (Jackson *et al.*, 1976). It resists mild alkaline hydrolysis, catalytic hydrogenation, and hydrazinolysis. Investigation of the cleavage conditions revealed this protecting group to be slightly more acid-labile than the *tert*-butyloxycarbonyl group. This type of N-protecting group avoids formation of carbonium ions which produce those, previously mentioned, severe problems of acidolytic cleavage of urethane protecting groups. Work is in progress to develop a range of variously substituted phosphinamide protecting groups.

5. Diphenylthiophosphine Group (Ppt)

Ueki *et al.* (1976) investigated a series of related phosphinyl and phosphinothioyl protecting groups. They proposed diphenylphosphinothioyl [diphenylthiophosphine; X = S in Eq. (24)] as the most useful residue. It is introduced and cleaved like the above mentioned oxygen derivative. Diphenylthiophosphinyl chloride reacts with amino and hydroxy groups as well. The Ppt—O bond is quite stable toward trifluoroacetic acid in methylene chloride but susceptible to alkali, in contrast to the Ppt—N bond (Ikeda *et al.*, 1979).

F. Alkyl Protecting Groups

Triphenylmethyl Group (Trityl, Trt)

The triphenylmethyl group (**55**) provides complete protection for the amine group and is easily introduced via trityl chloride (Helferich *et al.*, 1925; Hillmann-Elies *et al.*, 1953). It is the only alkyl group of importance for N^{α}-protection in peptide synthesis. For the introduction of the trityl group, amino acids are alkylated as diethylammonium salts or esters (Zervas and Theodoropoulos, 1956; Amiard *et al.*, 1955a,b). The latter have to be saponified by strong alkali. Racemization has not been found even when 20% KOH was used.

The carboxyl group of *N*-tritylamino acids is sterically hindered except for glycine. However, condensation with dicyclohexylcarbodiimide, DCC–HOSu, or DCC–HOBt results in the formation of peptides in good yields (see Schröder and Lübke, 1965, 1966; Wünsch *et al.*, 1974).

The trityl group is cleaved by hydrogenation, but the preferred procedure is treatment with weak acids, e.g., aqueous acetic acid. Recently, Riniker *et al.* (1975) employed 90% aqueous trifluoroethanol as solvent and succeeded in selective cleavage of the trityl group by pH-controlled titration with HCl even in the presence of the highly acid-labile Bpoc and 2-phenylpropyl(2)oxycarbonyl (Poc) groups. Thus, the *N*-trityl group provides for an additional selectivity in the series Z, Boc, Bpoc/Poc/Ddz, Trt.

G. Derivatives of Oxo Compounds as Amine Protecting Groups

1. Benzylidene Derivatives

Benzaldehyde reacts with salts of amino acids to form benzylidene derivatives (Bergmann *et al.*, 1925; Wieland and Schäfer, 1952) [Eq. (25)]. They

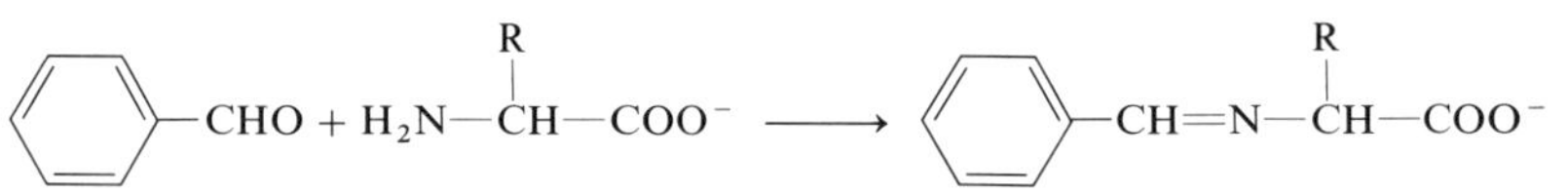
(25)

play no role in peptide synthesis except for a transient protection of ω-amino groups in the preparation of N^{α}-protected diamino acids (Section IV). Further stabilization is required for more efficient protection. Chelate formation with 2-hydroxybenzaldehyde increases the stability of benzylidene amino acids and it is additionally enhanced in 2-hydroxy-5-chlorobenzylidene compounds (McIntire, 1947) (Scheme 3).

Scheme 3

The cleavage of these protecting groups is easily accomplished by 1 *N* HCl. Transamination, observed with this type as well as with benzaldehyde itself, prohibits wide application.

Corresponding ketimine derivatives

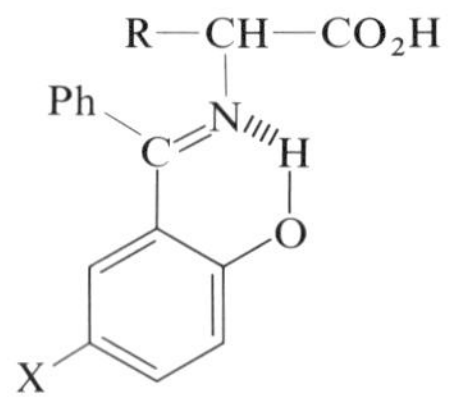

X = Cl, Me
Ph = phenyl

have also found no broad application thus far (Hope and Halpern, 1976; Abdipranoto *et al.*, 1977).

2. N-(1-Methyl-2-acylvinyl) Groups (Amv and Bmv)

Acetylacetone, benzoylacetone, and acetoacetic ester and amide easily react with alkali salts of amino acids in alcohol solution to yield the *N*-(1-methyl-2-acylvinyl)amino acid salts (Dane *et al.*, 1962; Dane and Dockner 1965) [Eq. (26)]. Although certain difficulties arose in the synthesis of d

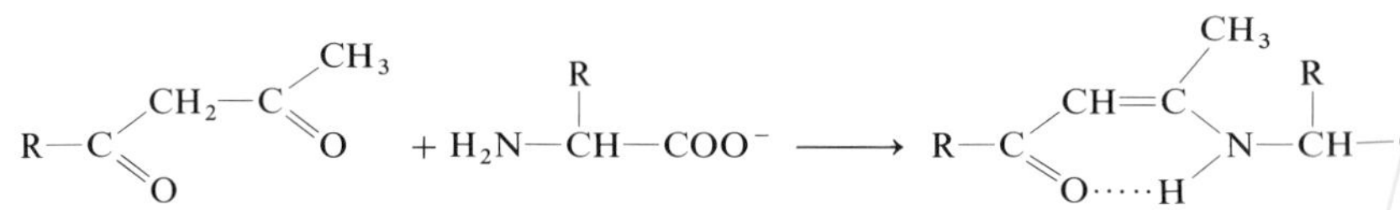

R = CH_3(Amv), C_6H_5(Bmv), OC_2H_5, NH_2

peptides, such protecting groups have been used in special cases, e.g partial synthesis of N^6-acylated penicillins. (Dane and Dockner, 19 conditions of cleavage are very mild. Short treatment, e.g., of Bmv-ṗ compounds, with dilute acetic acid readily releases the amino gr sparingly soluble benzoylacetone precipitates and is filtered off.

with protected higher peptides, which are often sparingly soluble, has not been reported.

IV. SELECTIVE PROTECTION OF α,ω-DIAMINOCARBOXYLIC ACIDS

In principle, protection of the ω-amine groups of α,ω-diaminocarboxylic acids is not different from that of the α-amino group. The decision as to which kind of group should serve as N^{ω}-protection depends on the strategy of the synthesis. N^{ω}-Protecting groups have to be stable under all the various conditions of α-amine deprotection and the problem consists of selecting the proper combination of suitable N^{α}- and N^{ω}-protecting groups. Table X outlines some proven examples of such combinations. Of course, the mode of protection of functionalities other than side-chain amino groups must also be considered in the selection of the desirable protecting group combination (see Chapter 5, this volume).

Table X Differential Protection and Selective Deprotection of α- and ω-Amine Groups

Suitable residues for N^{α}-protection	Cleavage conditions for α-amine groups	Suitable residues for N^{ω}-protection
Z and related groups (Table II and III)	Hydrogenation	Tos, Msc, Boc and related groups; Tfa, Pht, (For)
Z, 4 Cz and related groups	HBr–AcOH, HF	Tos, Msc, and related groups; Tfa, Pht
Boc and related groups, Moz, Ibc, Doc, Dpp (Table VII)	Cold CF_3COOH	Z, 4 Cz, Msc, Tos, Pht and related groups
Ddz, Bpoc, Adpc, Poc, Trt Azoc, Amv, Bmv	80% AcOH; 1–5% CF_3COOH in CH_2Cl_2	Boc, Z, Msc, Tos, Pht
Trt	Moderate acidic conditions: HCl in CF_3CH_2OH at pH 4	As above, Bpoc, Poc, Ddz, Dpp
Nps	Special conditions: weak acid + nucleophile + indole	As above
Msc and related groups (Table IV)	Alkali (β-elimination)	Boc, Z and related groups; Bpoc, Poc, Ddz

A. Protection of the ω-Amino Group

A specific introduction of appropriate residues into the ω-amino group is performed using copper salts which complex with the α-amine and carboxyl groups and block the reaction of the α-amine (Kurtz, 1941). The complexes have to be decomposed afterward with H_2S, H_2S-releasing compounds, or complexing agents (Zahn and Umlauf, 1954; Zaoral, 1965; Kuwata and Watanabe, 1965; Ledger and Stewart, 1965; Yamashiro and Li, 1972; Erickson and Merrifield, 1972; Eckstein *et al.*, 1973). Acid-stable residues allow decomposition with HCl (Schwyzer and Sieber, 1960; Bodanszky *et al.*, 1964).

A second method takes advantage of the different nucleophilicity of α- and ω-amino groups. Thus, lysine and ornithine react with formyl- (Okawa and Hase, 1963) or *tert*-butyloxycarbonyl-4-nitrobenzoate (Schwyzer and Rittel, 1961), benzyloxycarbonyl phenolate (Zahn and Falkenburg, 1960), or thioethyl trifluoroacetate (Schallenberg and Calvin, 1955) to yield the corresponding ω-protected diaminocarbonic acid. These compounds are mostly less soluble than the α-compounds that are formed in this reaction to about 10%.

B. Protection of the α-Amino Group

Specific reaction of the α-amino group occurs in acidic medium. With trifluoroacetic acid anhydride in trifluoroacetic acid N^{α}-trifluoroacetyl derivatives are exclusively formed (Weygand and Geiger, 1956). Acetic acid anhydride in formic acid yields the α-formyl derivative (Hofmann *et al.*, 1960). Another route to α-protection proceeds via the ω-benzylidene compounds, which can be isolated without difficulty (Bezas and Zervas, 1961; Wünsch and Zwick, 1964). After protection of the α-amino group, the benzylidene residue is removed by treatment with acid. The liberated ω-amino group is then derivatized with the second protecting group. In this way, for example, Z-Lys(Boc)-OH became readily accessible (Sturm *et al.*, 1963). Transamination of benzylidene cannot be completely excluded. A suitable purification is therefore required. In most cases the differing solubilities of α- and ω-protected diaminocarboxylic acids offer a satisfactory way.

C. General Aspects

Some specific reactions have to be considered to which α,γ-diaminocarboxylic acids can be subjected. When the carboxyl groups in ornithine or α,γ-diaminobutyric acid are activated, lactam formation occurs in considerable amounts [Eq. (29)].

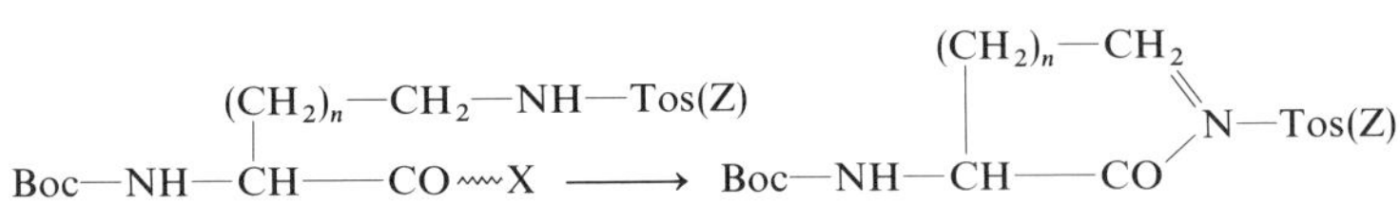

(29)

$n = 1, 2$

This cyclization is most pronounced with ω-tosyl derivatives. Monoacylated α,γ-diaminobutyric acid and α,β-diaminopropionic acid undergo an acyl exchange (Scheme 4).

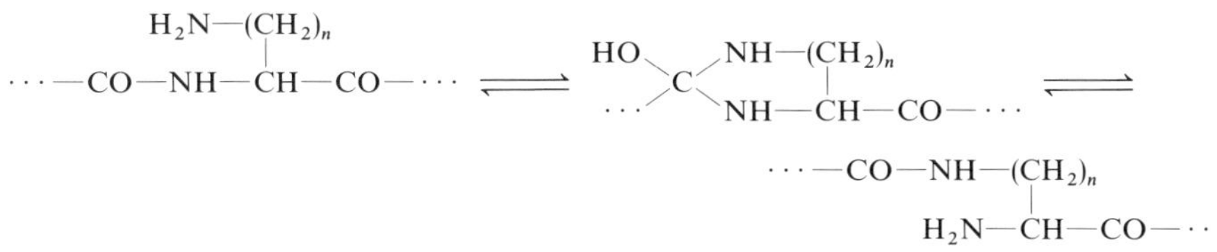

$n = 1, 2$

Scheme 4

Lysine derivatives are quite stable and offer no serious problems in the synthesis of peptides. Preferred combinations are benzyloxycarbonyl for α-amino groups and *tert*-butyloxycarbonyl for N^{ω}-protection (Schwyzer Strategy). Catalytic hydrogenation for repetitive N^{α}-deprotection proceeds smoothly and the final cleavage of the Boc (and *tert*-butyl) groups of completed target peptides occurs mostly without serious damage to the end product.

In repetitive methods such as solid- or liquid-phase peptide syntheses (see Volume 2, Chapters 1 and 2, respectively) or the REMA procedure (see Volume 2, chapter 6), easily cleavable residues for α-protection such as the *tert*-butyloxycarbonyl group are preferred, whereas the ω-amino groups are blocked by benzyloxycarbonyl or better 4- or 2-chlorobenzyloxycarbonyl to increase the difference in acid lability between both groups. Fukuda and Fujino (1978) investigated the *p*-methylbenzylsulfonyl group (Pms) which is more stable toward trifluoroacetic acid and is more easily cleaved by treatment with hydrogen fluoride (for 60 min at $-20°C$) than the 2-chlorobenzyloxycarbonyl residue. Combinations of Bpoc/Boc and Ddz/Boc have also been applied, both being cleaved under mildly acidic conditions. Other combinations are conceivable, and these will be discussed in Chapter 5 of this volume.

V. AMIDE PROTECTING GROUPS

In naturally occurring pepides three different kinds of amide groups are found: the α-amide occasionally present as the COOH-terminus of a peptide, the β-amide in asparagine, and the γ-amide in glutamine. Furthermore, every peptide bond is itself a primary or secondary amide. Despite the known stability of amides they may give rise to a number of side reactions. Consequently, protection of amides is advantageous in some cases.

A. Side Reactions of Amide Groups

1. Side Reactions of the α-Amide Group

α-Carboxamides are relatively stable and are not dehydrated by peptide coupling reagents, as for instance dicyclohexylcarbodiimide. Only several arylsulfonyl halides are capable of converting a variety of primary carboxamides to nitriles (Stephens *et al.*, 1955). Dipeptide amides, especially with proline at the NH_2-terminus may give rise to diketopiperazine formation [Eq. (30)].

$$\text{H-Pro-Gly-NH}_2 \xrightarrow{-NH_3}$$

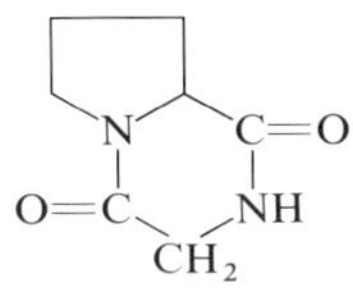

(30)

2. Side Reactions of the β-Amide Group

Asparagine is very susceptible to nitrile formation when the α-carboxylic group is activated by anhydride-forming reagents such as tetraethylpyrophosphite or dicyclohexylcarbodiimide (Gish *et al.*, 1964). To a much lower extent, mixed anhydride products have also been shown to be contaminated with nitrile (Paul and Kende, 1964).

However, Z-Asn-OMe and Asn within the peptide chain could not be dehydrated by dicyclohexylcarbodiimide (Liberek, 1962). These findings suggested that the carbonyl group of asparagine might be participating in the dehydration of the carboxamide group (Kashelikar and Ressler, 1964; Paul and Kende, 1964).

Interestingly, dehydration of asparagine occurs under conditions corresponding to those for racemization of activated peptides. For instance, Z-Asn-ONp once formed is stable to further dehydration (Bodanszky and du Vigneaud, 1959). Similarly, racemization of dipeptide *p*-nitrophenyl esters is less than 2% (Weygand *et al.*, 1966b; König and Geiger, 1973). Another example is prevention of racemization, as well as nitrile formation, of Z-Asn-OH using DCC–HOBt (König and Geiger, 1970a) or DCC–HOSu (Weygand *et al.*, 1966c; Kawasaki *et al.*, 1973).

Indeed, the proposed isoimide intermediate (**53**) (Stammer, 1961) is formed similarly to the racemizing oxazolone by the attack of an amide oxygen on the carbonyl C-atom of an activated carboxyl group. Probably, the highly nucleophilic 1-hydroxybenzotriazole successfully competes with these reactions (Scheme 5).

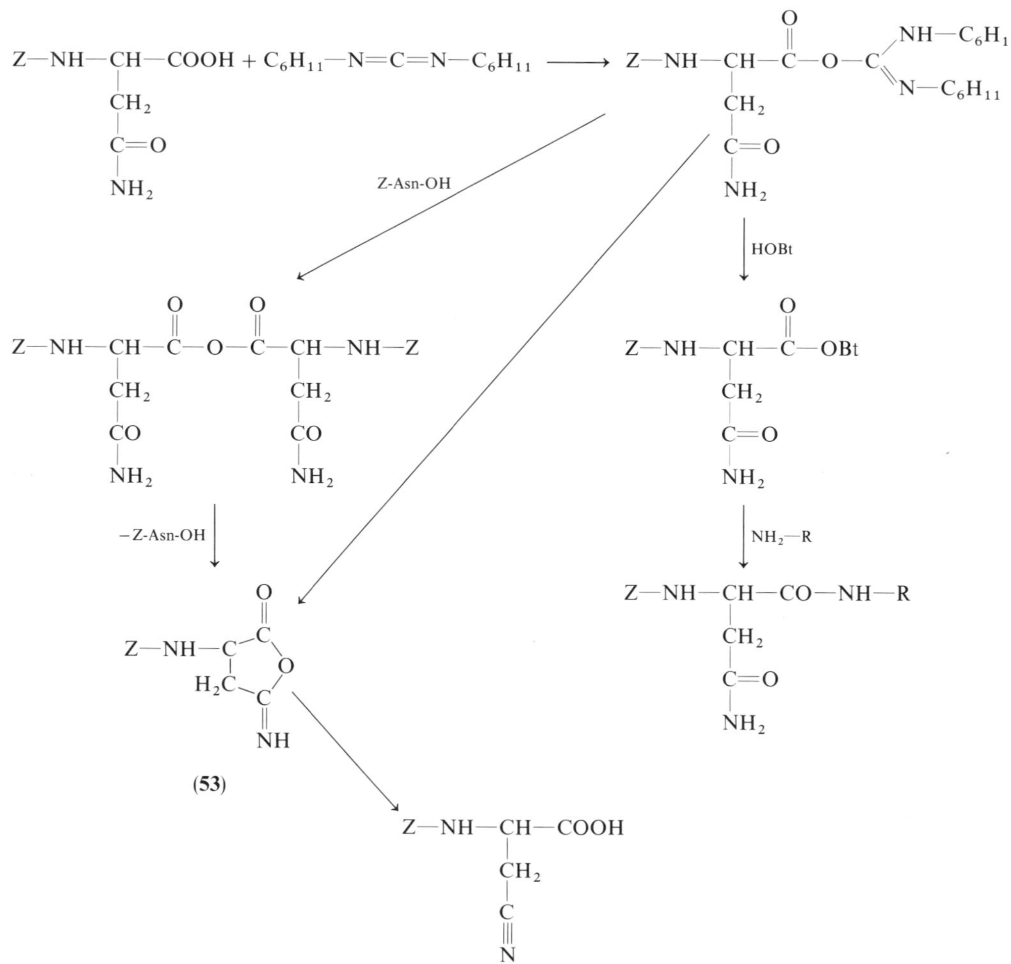

Scheme 5

Another side reaction of asparagine peptides is the formation of aspartimides under alkaline conditions. The aspartimides are rather stable and can be isolated. Further treatment with alkali forms a mixture of α- and β-aspartyl peptides (Sondheimer and Holley, 1954; Battersby and Robinson, 1955; Roeske, 1963; König and Volk, 1977).

3. *Side Reactions of the γ-Amide Group*

Similar to asparagine but to a smaller extent, the γ-amide group of glutamine is also subject to dehydration. The γ-nitrile is formed if the α-carbonyl group is activated by peptide-coupling reagents such as dicyclohexylcarbodiimide (Gish *et al.*, 1956; Ressler and Ratzkin, 1961).

Cyclization to the six-membered glutarimide was also observed during the preparations of Z-Gln-OSu (Zahn and Fölsche, 1969) or Nps-Gln-OSu (Meyers *et al.*, 1969). These side reactions can be prevented by the application of the DCC–HOBt method (König and Geiger, 1970a) as in the case of asparagine. In alkaline to weakly acidic solution, peptides with an NH_2-terminal glutamine are frequently cyclized to yield the pyroglutamyl peptides (Rudinger and Pravda, 1958; Chillemi *et al.*, 1964). Trifluoroacetic acid converts also glutamine or glutaminyl peptides to pyroglutamic acid and pyroglutamyl peptides, respectively (König and Geiger, 1972).

B. Protection of Amide Groups

To avoid or to diminish the aforementioned side reactions, a series of amide protecting groups have been recommended. Amide-protecting groups can generally be classified by their method of introduction: 1). insertion of the protecting group into an amide already present; 2). synthesis of the substituted amide from the carboxyl group and a primary or secondary amine.

1. *Introduction of Protecting Groups into a Carboxamide Function*

There are only a few substituents that can be introduced without difficulty into the carboxamide group. They are listed in Table XI. The well-known reaction of xanthydrol with amides in acetic acid yielding the corresponding xanthenylamides (Fosse and Haller, 1907) inspired Akabori *et al.* (1961) and Shimonishi *et al.* (1962) to synthesize Z-Gln(Xan)-OH and the corresponding asparagine derivative. However, the results of their investigations were discouraging because of the low solubility of the xanthenyl derivatives in organic solvents. Dorman *et al.* (1972) synthesized Boc-Gln(Xan)-OH and used it in a solid-phase synthesis, but deprotection could not be performed selectively.

Similarly to xanthydrol, 4,4′-dimethoxybenzhydrol and 4,4′-dimethylbenzhydrol react with Z-Gln-OH, Z-Asn-OH, and other carboxamide-group

Table XI Protecting Groups Being Inserted Directly into Carboxamides

Structure	Residue
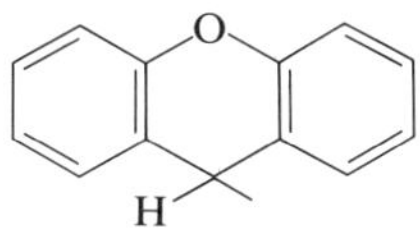	Xanthenyl (Xan)
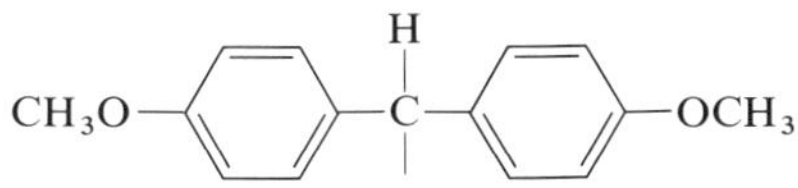	4,4′-Dimethoxybenzhydryl (Ddm)
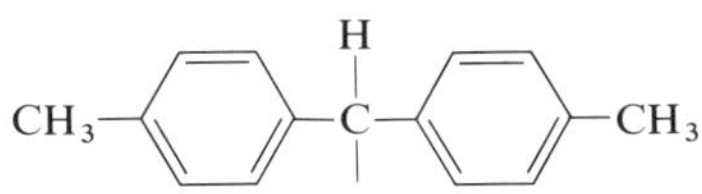	4,4′-Dimethylbenzhydryl (Dbh)
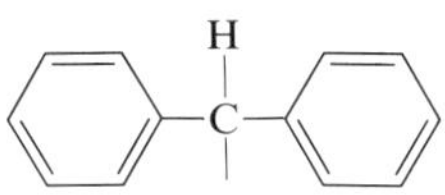	Benzhydryl (Dpm)
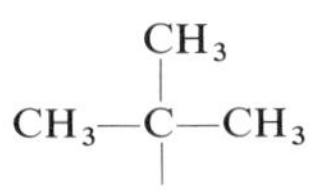	*tert*-Butyl (*t*Bu)

containing compounds in acetic acid to yield the corresponding benzhydrylamides. Traces of concentrated H_2SO_4, *p*-toluenesulfonic acid, or BF_3 are required as catalyst (König and Geiger, 1970b). 4,4′-Dimethylbenzhydrylamides attained no importance because of their troublesome preparation and difficult cleavage conditions, whereas 4,4′-dimethoxybenzhydryl-protection is frequently used. In contrast to xanthenylamides, 4,4′-dimethoxybenzhydryl derivatives are easily soluble in dimethylformamide.

The unsubstituted benzhydryl group has been introduced by reacting the amide with benzhydryl trifluoroacetate or benzhydryl chloride in trifluoroacetic acid at room temperature (Weygand *et al.*, 1968a). Protection of amide by the *t*Bu group which may be accomplished by *tert*-butyl acetate–perchloric acid (Schnabel and Schüssler, 1965) or isobutylene in dioxane–sulfuric acid (König and Kernebeck, 1979), is not useful because of irreversibility. Cleavage conditions of these protecting groups are summarized in Table XII.

Among these carboxamide protecting groups the 4,4′-dimethoxy-benzhydryl residue seems the most recommendable. It is easily introduced into amide groups and may be removed without jeopardizing the peptide backbone. However, side products derived from the 4,4′-dimethoxybenzhydryl cation can be formed if peptides contain tryptophan, tyrosine, methionine,

Table XII Cleavage Conditions for Amide Protecting Groups (I)

Protecting group	H_2–Pd	CF_3COOH–anisole (room temp)	CF_3COOH–anisole (72°C)	HBr–acetic acid (room temp)	HBr–acetic acid–indole (room temp)	3 *N* HCl–dioxane	HF (0°C)	Alkali
Xanthenyl	Stable[a]		30 min[b]	1–2 hr		Can be deprotected[f]		Glutamine derivatives seem to be stable: asparagine derivatives are not[b]
4,4′-Dimethoxybenzhydryl	Stable[b]	2–3 hr[b]	5 min[b]	1–2 hr[b]	5–10 min[b]			
4,4′-Dimethylbenzhydryl	Stable[b]		30 min[b]					
Benzhydryl	Stable	Without anisole $t_{1/2} = 9$ hr[d] stable[e]					30–60 min[e,g]	
tert-Butyl	Stable[c]	Stable	Without anisole $t_{1/2} = 2$ hr[d]	Stable[c]				

[a] Shimonishi *et al.* (1962).
[b] König and Geiger (1970b).
[c] Callahan *et al.* (1963).
[d] Weygand *et al.* (1968a).
[e] Hruby *et al.* (1972).
[f] Dorman *et al.* (1972).
[g] Sakakibara *et al.* (1967).

or cysteine. Cyclization of NH_2-terminal glutamine to the pyroglutamic acid moiety occurs in trifluoroacetic acid as deblocking agent (König and Geiger, 1972) The benzyloxycarbonyl group can be removed selectively by catalytic hydrogenation, and the 2-nitrophenylsulfenyl group may be cleaved under mild conditions (see Section III,E,1) without affecting the 4,4′-dimethoxybenzhydryl group. However, selective removal of *tert*-butyloxycarbonyl groups cannot be achieved.

2. Introduction of Amide Protecting Groups by Amidation of Carboxylic Acids

In comparative studies Weygand *et al.* (1966a, 1968a) found the 2,4-dimethoxybenzyl(Dmob) and 2,4,6-trimethoxybenzyl(Tmb) residues convenient for amide protection, whereas the 4-methoxybenzyl residue is ill-suited because of the strong conditions required for deprotection. These results were confirmed by Pietta *et al.* (1976). Weygand's first contribution to this field (Weygand *et al.*, 1966a) aimed at the protection of the peptide bond (Scheme 6). The aim of this work was to improve the solubility of peptides

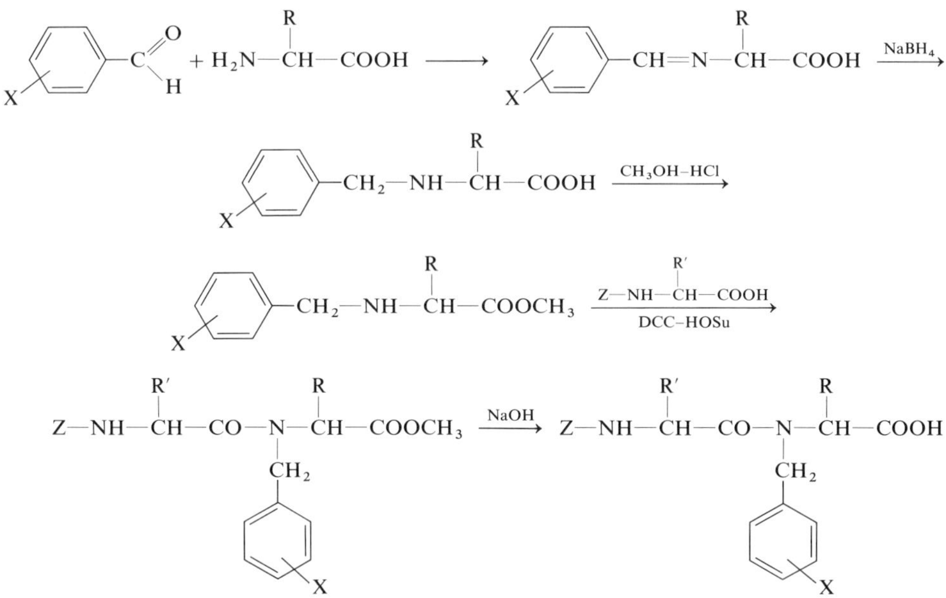

X = H, 4-OCH_3, 2,4-$(OCH_3)_2$

Scheme 6

in organic solvents. Thus for instance, Moz-Gly-DmobGly-Dmob-Gly-DmobGly-DmobGly-OH becomes readily soluble in ethyl acetate.

However, racemization should be increased using Dmob-peptides, since Bovarnick and Clarke (1938) and McDermott *et al.* (1973) observed pronounced racemization with N-alkylated amino acids and peptides.

The protection of the γ-amide group of glutamine and the β-amide group of asparagine proved to be of more importance than peptide bond protection. Weygand *et al.* (1966a, 1968b) synthesized bis(2,4-dimethoxybenzyl)- and 2,4,6-trimethoxybenzyl-protected compounds whereas Pietta *et al.* (1971, 1976) investigated in this connection the use of 2,4-dimethoxybenzyl and 4-methoxybenzyl groups (Scheme 7).

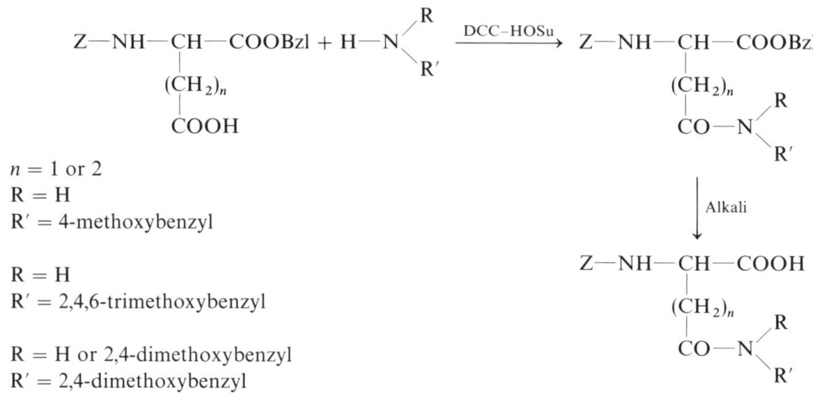

Scheme 7

Amide protection by 2,6-dimethoxybenzyl-, 3,4-dimethoxybenzyl-, 3,4,5-trimethoxybenzyl, 2,4,6-trimethylbenzyl (Weygand *et al.*, 1968a), or 2,4-dimethylbenzyl groups (Hruby *et al.*, 1972) is of little avail because conditions of deprotection are too difficult. The stability of the above-mentioned benzyl-type protecting groups are summarized in Table XIII. By increasing the acid-lability of the 4,4′-dimethoxybenzhydryl group, S. Bajusz (Personal communication, 1977) synthesized the 2,2′,4,4′-tetramethoxybenzhydrylamine. This compound was used for preparing Asn, Gln, or amino acid amide derivatives as outlined above.

The most recommendable benzyl-type amide protection seems to involve the 2,4-dimethoxybenzyl group. The cleavage in trifluoroacetic acid or HF is indeed delayed, but it is stable to hydrogenation (selective removal of Z-group and benzyl ester), to 1 *N* methanolic HCl (selective removal of Moz group), and even to sodium in liquid ammonia. Whereas the glutamine derivatives are stable to alkali, the correspondingly protected asparagine compounds seem to be sensitive under the same conditions (Pietta *et al.*, 1971).

These amide protecting groups gained further important application in solid-phase peptide synthesis preparing peptide amides (Pietta and Marshall,

Table XIII Cleavage Conditions of Amide Protecting Groups (II)

Protecting group	H_2–Pd	1 *N* methanol–HCl (room temp)	CF_3COOH (72°C)	CF_3COOH (room temp)	HF (room temp)	1 *N* HCl–acetic acid (room temp)	CF_3COOH–CH_2Cl_2 (room temp) (50/50)	Alkali on Asn derivatives	Alkali on Gln derivatives
Bis(2,4-dimethoxybenzyl	Stable[a]	Stable[a]	1 hr[b]	30 hr[b]				Stable[b] unstable[c]	Stable[b]
2,4-Dimethoxybenzyl	Stable[a]	Stable[a]	1 hr[b]	30 hr[b] 72 hr[d]	3 hr[c]	After 72 hr about 40% deprotect[d]	72 hr[d]	Unstable[c]	Stable[c]
2,4,6-Trimethoxybenzyl	Stable	Stable[a]		$+C_6H_5CH_2$-SH 4 hr[a]				Stable[b]	Stable[b]
4-Methoxybenzyl		Stable		After 72 hr about 30% deprotect[d]	After 75 min 30–90% deprotect[e]	After 72 hr about 5% deprotect[d]			
2,2′,4,4′-Tetramethoxybenzhydryl	Stable	3 *N* Aqueous HCl stable[f]		CF_3COOH–H_2O (8:2) 15 min[f]			CF_3COOH–$CHCl_3$ (8:2) 5 min[f]		

[a] Weygand *et al.* (1968a).
[b] Weygand *et al.* (1968b).
[c] Pietta *et al.* (1971).
[d] Pietta *et al.* (1976).
[e] Hruby *et al.* (1972).
[f] Bajusz (personal communication 1977).

1970; Pietta *et al.*, 1973), as for instance thyroliberin (TRH) (Pietta *et al.*, 1974) or gonadoliberin (Rivaille *et al.*, 1971). In most cases polymer-bound benzhydrylamine is reacted with *tert*-butyloxycarbonyl amino acids. After the usual deprotection and coupling procedures the peptide amide is cleaved from the solid support by liquid HF, adding anisol or mercaptoethanol as scavengers.

Recently, the more easily cleavable 4-methoxybenzhydrylamine resin has been investigated by Orlowski and Walter (1976). Another recently developed method is based on the use of 3-nitro-4-aminomethylbenzoylamide polystyrene resin (Rich and Gurwara, 1975). The peptide amides are released from this resin by irradiation at 350 nm in methanol with the exclusion of oxygen [Eq. (31)].

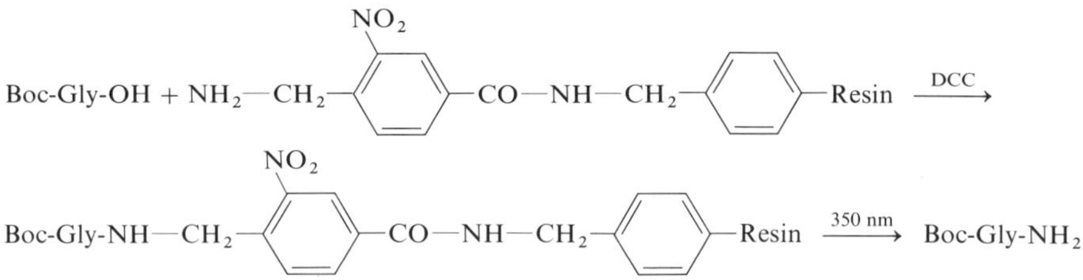

(31)

In conclusion, the protection of the β- and γ-amides of asparagine and glutamine can be favorable in many cases if the solubility of hydrophilic peptides in organic solvents is improved (König and Geiger, 1972; Bajusz and Fauszt, 1972). Furthermore, a simple purification by crystallizable derivatives may be attained and the amides are stabilized against side reactions (labile amides can be stored in N-protected form over a longer period of time). During the treatment with alkali, however, protection is limited to glutamine. Asparagine side reaction cannot be completely prevented by amide protection. An additional risk is the formation of carbonium ions under deblocking conditions which may attack tyrosine, methionine, and tryptophan; but this is a problem which also arises with many other well-established protecting groups.

As a consequence of the facts and considerations presented in this section we cannot recommend any particular amide protection in routine synthesis unless some of the aforementioned advantages justify special cases. The application of DCC–HOBt (König and Geiger, 1970a) for the coupling reaction at the carboxyl groups of asparagine and glutamine avoids nitrile and imide formation. Treatment with alkali can be circumvented by a suitable selection of protecting groups. Cyclization of glutamine to pyroglutamic acid depends on the sequence. It is suppressed in cases of rapid peptide bond formation.

Table XIV Combinations of N^{α}- and Hydrazide Protection

N^{α}-Protection	Abbrev	Hydrazine protection	Abbrev	References
Phthaloyl	Pht	Benzyloxycarbonyl	Z	Wünsch and Wendlberger (1964)
tert-Butyloxycarbonyl	Boc	Benzyloxycarbonyl	Z	Wünsch and Wendlberger (1964); Hofmann *et al.* (1965)
Formyl	For	Benzyloxycarbonyl	Z	Storey *et al.* (1972)
Benzyloxycarbonyl	Z	*tert*-Butyloxycarbonyl	Boc	Wünsch and Wendlberger (1964)
2-Nitrophenylsulfenyl	Nps	Trifluoroacetyl	Tfa	Preston and Weinstein (1967)
2-(4-Biphenyl)propyl(2)oxycarbonyl	Bpoc	Aoc-polymer	—	Wang and Merrifield (1969)
Trifluoroacetyl	Tfa	Trityl	Trt	Weygand and Steglich (1959b)
Benzyloxycarbonyl	Z	Trityl	Trt	Moroder *et al.* (1974)

VI. HYDRAZIDE PROTECTION

Peptide hydrazides are usually prepared from the corresponding methyl esters. Occasionally, peptides are unstable to hydrazinolysis. In order to render the azide method possible in such cases also, hydrazides can be temporarily N′-protected and deblocked before transforming the hydrazide into the azide (Hofmann *et al.*, 1950; for details, see Chapter 4, Volume I).

N′-Protected amino acid hydrazides are prepared either by reaction of an N^{α}-substituted amino acid with a monosubstituted hydrazine [Eq. (32)] or by reacting the N^{α}-substituted amino acid hydrazide with a protecting agent [Eq. (33)]. Some combinations of protecting groups described in the

$$\text{Z—NH—}\underset{}{\overset{\text{R}}{\overset{|}{\text{CH}}}}\text{—COOH} + \text{H}_2\text{N—NH—Boc} \xrightarrow[-\text{H}_2\text{O}]{} \text{Z—NH—}\overset{\text{R}}{\overset{|}{\text{CH}}}\text{—CO—NH—NH—Boc} \tag{32}$$

$$\text{Boc—NH—}\overset{\text{R}}{\overset{|}{\text{CH}}}\text{—CO—NH—NH}_2 + \text{Z—Cl} \xrightarrow[-\text{HCl}]{} \text{Boc—NH—}\overset{\text{R}}{\overset{|}{\text{CH}}}\text{—CO—NH—NH—Z} \tag{33}$$

literature are outlined in Table XIV. The method has not frequently been applied but its administration is advisable in the aforementioned special cases. It is restricted by the need of three types of groups for N^{α}-, side-chain, and hydrazide protection. Thus, the stability of the first two protecting groups during the deblocking of the hydrazide has to be assured.

VII. GUANIDINE PROTECTING GROUPS

The strongly basic guanidine group of arginine ($pK = 12.5$) can be protected by protonation (Gish and Carpenter, 1953; Anderson, 1953). However, the low solubility of positively charged arginine derivatives in organic solvents and the formation of lactams during activation of the carboxyl group of amine-protected arginine often requires other types of guanidine protection. Thus, for instance, it was not possible to synthesize Z-Arg-Pro-O*t*Bu from Z-Arg-OH and Pro-O*t*Bu·HCl by DCC–HOBt. Supposedly because of the basicity of Pro-O*t*Bu, the protonation of the guanidine group is incomplete and the six-membered lactam ring is formed as the main product (König and Geiger, 1975).

Furthermore, acylation of the guanidine group during peptide coupling (Zervas and Bergmann, 1928; Photaki and Yiotakis, 1976) may occur. This side reaction leads to incorporation of ornithine instead of arginine into the peptide chain [Eq. (34)].

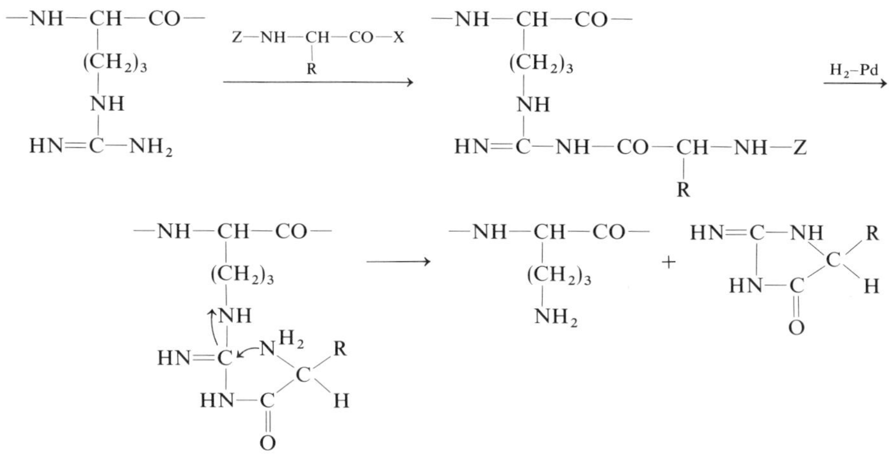

(34)

Five types of guanidine protecting groups have been applied: 1). the nitro group, 2). arylsulfonyl and aralkylsulfonyl groups, 3). acyl groups, 4). the trityl group, and 5). reaction products with bifunctional aldehydes and ketones. It is frequently observed that ω-protected arginine still gives rise to side products. In several attempts to prepare Z-Arg(NO_2)-ONp from Z-Arg(NO_2)-OH, Bodanszky and Sheehan (1960) obtained the δ-lactam instead of the desired ester [Eq. (35)]. This

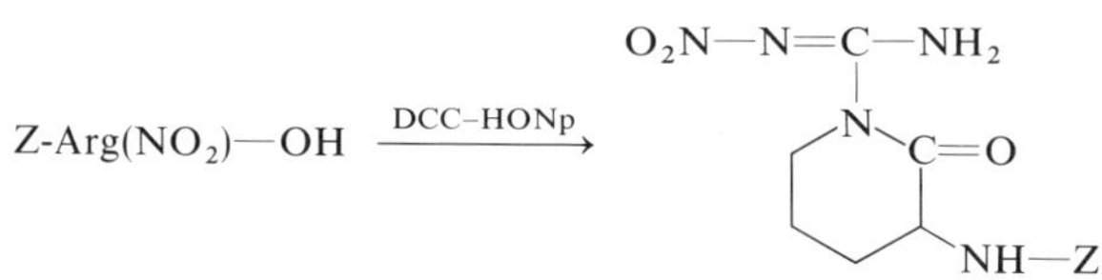

(35)

observation has been confirmed by Paul *et al.* (1961). Zervas *et al.* (1959) obtained the corresponding δ-lactam of Z-Arg(Z)-OH with dicyclohexylcarbodiimide and free amino acid esters. Only with amino acid ester hydrochloride were peptides formed. Furthermore, δ-lactam formation was observed with N^{ω}-nitrobenzyloxycarbonyl (Guttmann and Pless, 1965), N^{ω}-isopropoxycarbonyl, N^{ω}-tetrachlorobenzoyl (Guttmann and Pless, 1965), N^{ω}-*tert*-butyloxycarbonyl (Arold and Reissmann, 1968), and N^{ω}-tosyl protection (Li *et al.*, 1961). In the presence of a tertiary base, Z-Arg(NO_2)-OPcp forms the lactam about 20 times faster than Z-Arg(Tos)-OPcp. The lactams, which are in equilibrium with an active ester, racemize more quickly than the pentachlorophenyl esters. Here again the NO_2 compound racemizes about 3.5 times faster than the tosyl-protected arginine derivative (Kovacs *et al.*, 1979). These results show that all α-acylated N^{ω}-protected arginine derivatives have a tendency to δ-lactam formation which can be minimized by protonation. N^{δ},N^{ω}-Guanido protection should overcome this side reaction (Section VII,C,2).

Paul *et al.* (1961) found an additional side reaction while attempting to prepare Boc-Arg(NO_2)-OH from H-Arg(NO_2)-OH and *tert*-butyl-*p*-nitrophenyl carbonate (Boc-ONp) in aqueous sodium hydrogencarbonate. This side product is also formed without Boc-ONp, and its structure has been determined to be 2-nitrimino-4-carboxy-1,3-diazacycloheptane [Eq. (36)].

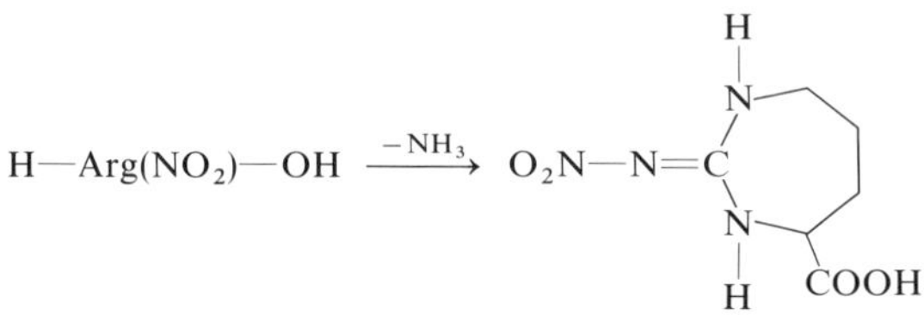

(36)

This formation of 1,3-diazacycloheptane can be observed even with N^{δ},N^{ω}-protected arginine derivatives. H-Arg($Adoc_2$)-X rearranges in acetic acid, and also to some extent in methanol, to the seven-membered ring [Eq. (37)] (Jäger and Geiger, 1970). Also

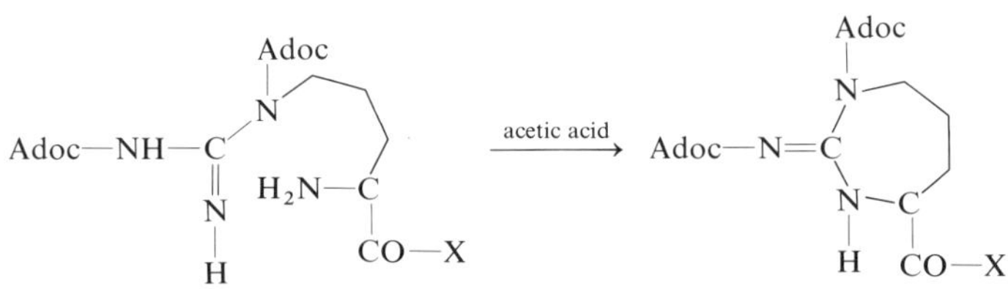

(37)

X = OH, $N(CH_3)_2$, Gly—NH_2

H-Arg(Z_2)-OH decomposes in methanol into several fragments, the structures of which have not yet been investigated. Despite guanidine protection, side products cannot be excluded during the synthesis of arginine-containing peptides; but manipulation of arginine derivatives does become more feasible if the guanidine group is protected.

A. Nitro Group

The nitro group, introduced as protection by Bergmann *et al.* (1934), is the oldest guanidine protecting group. It has been the most important guanidine protection for many years, but frequent use has also revealed difficulties during deprotection. In most cases it has been cleaved by reductive procedures such as catalytic hydrogenation (Bergmann *et al.*, 1934), electrolytic reduction (Clubb *et al.*, 1960; Scopes *et al.*, 1965), and reduction with Zn (Pless and Guttmann, 1967; Turán *et al.*, 1975) or $SnCl_2$ (Hayakawa *et al.*, 1967). Recently also 1,4-cyclohexadiene, an effective hydrogen donor, was used to remove the NO_2 group. But the efficient palladium black catalyst was required for cleavage (Felix *et al.*, 1978). Removal of the nitro group from the arginine residue by use of sodium in liquid ammonia was not satisfactory (Tritsch and Woolley, 1960). These reductive methods are not always very practicable and many side reactions have been reported. They are listed in Table XV.

The catalytic hydrogenation of nitroarginine-containing peptides is sequence-dependent. Thus, -Arg(NO_2)-Arg(NO_2)- within a peptide chain always yields impure products. Berse *et al.* (1960), for instance, have not been able to deprotect H-Arg(NO_2)-Lys-OH, whereas H-Arg(NO_2)-Asn-OH was smoothly reduced to H-Arg-Asn-OH. Long-lasting hydrogenation of larger peptides, which is often required, may give rise to hydrogenation of aromatic amino acids. The preferred solvent for the elimination of the nitro group is 80–90% acetic acid. Besides this, methanol–BF_3–diethyl etherate (Yajima *et al.*, 1968), and liquid ammonia (Meienhofer and Kuromizu, 1974) have been recommended. The latter offers several additional advantages concerning other protecting groups.

For several years proton solvolysis has been used extensively for deprotection. Sakakibara *et al.* (1967) introduced liquid hydrogen fluoride in the presence of anisole as a deprotecting agent in peptide chemistry. The nitro

Table XV Side Products Found to Occur During Reductive Cleavage of Nitroarginine

Agent	Result and side product	References
Na in liquid ammonia	Unsatisfactory removal	Tritsch and Woolley (1960)
Electrolytic reduction	N^{ω}-Aminoarginine	Gros *et al.* (1961)
Catalytic hydrogenation	N^{ω}-Aminoarginine	Gros *et al.* (1961)
	Nitrosoarginine N^{ω}-cyanoornithine and citrulline	Iselin (1966)
	Hydrogenation of aromatic amino acids	Schafer and Young (1971); Bajusz *et al.* (1973)
Zn in 85% acetic acid	60% N^{ω}-Acetylaminoarginine and 0.5% ornithine	Turán *et al.* (1975)
Zn in 2 *N* H_3PO_4	33% N^{ω}-Aminoarginine and 5.7% ornithine	
Zn in 95% formic acid	6.5% N^{ω}-Formylaminoarginine and 1.8% ornithine	
Zn in 2 *N* trifluoroacetic acid	2.4% N^{ω}-Aminoarginine and 1.7% ornithine	

Table XVI Side Reactions During the Cleavage of Protecting Groups by Liquid Hydrogen Fluoride in the Presence of Anisole

Amino acid or peptide	Side reaction	References
N^G-Nitroarginine	Ornithine	Yamashiro *et al.* (1972)
Serine- and Threonine-containing peptides	N → O acyl migration	Sakakibara *et al.* (1962); Lenard and Hess (1964)
Methionine-containing peptides	Cleavage	Lenard *et al.* (1964)
Tryptophan	Slow decomposition	Lenard and Hess (1964)
Aspartyl peptides	α,β-Rearrangement	Baba *et al.* (1973)
Glutamyl peptides	Pyrrolidones	Feinberg and Merrifield (1975)
Aspartic, glutamic acid	Oxo-compounds with anisole	Sano and Kawanishi (1975); Feinberg and Merrifield (1975)

group is eliminated within less than 30 min at 0°C. However, even this elegant method is able to damage peptides. Some side reactions that have been observed are listed in Table XVI. Thus, insulin loses two-thirds of its biological activity after treatment with HF, and gonadoliberin is destroyed to an extent of about 30% (König and Geiger, 1975).

Milder agents than HF may be used to remove the nitro group, even though they are not commonly applied. Pless and Bauer (1973) used boron tris(trifluoroacetate) in trifluoroacetic acid at 0°C. Boron tribromide (1 *M*) in methylene chloride caused only partial deprotection to arginine and, moreover, cleavage to give ornithine (Felix, 1974). Pyridinium polyhydrogen fluoride in the presence of anisole at room temperature (Matsuura *et al.*, 1976) as well as trifluoromethanesulfonic acid and anisole at 40°C (Yajima *et al.*, 1974) yielded only about 35% arginine from nitroarginine.

Methansulfonic acid in the presence of anisole (30 min at 20°C) liberates arginine from nitroarginine in 58% yield (Yajima *et al.*, 1975a). Fluorosulfonic acid and anisole, which deprotects Arg(NO_2) quantitatively in 15 min at 20°C, proved to be more favorable. The nitro group, stable to trifluoroacetic acid at room temperature, is also stable to HBr–AcOH, so that selective removal of the benzyloxycarbonyl group is possible. Ammonolysis of nitroarginine-containing peptides leads to the formation of ornithine-containing side products. This conversion is accompanied by the deprotection of nitroarginine. On the other hand, N^G-tosyl or N^G-dicarbobenzoxyarginine peptides are not decomposed to ornithine by peptide ammonolysis (Künzi *et al.*, 1974). Whereas hydrazine reacts with the nitroguanido group (Fruton, 1949; McKay, 1952; Van Orden and Smith, 1954) thus impeding ester hydrazinolysis, alkaline saponification of nitroarginine-containing peptides can be performed under normal conditions.

B. Arylsulfonyl Groups

Thus far, four different sulfonyl groups have been applied to guanidine protection. Schwyzer and Li (1958) introduced the *p*-toluenesulfonyl group (tosyl), which has since been widely used in syntheses of arginyl peptides. Recently, the *p*-methoxybenzenesulfonyl group (Nishimura and Fujino, 1976), and the mesitylen-2-sulfonyl group (Yajima *et al.*, 1978a,b) were employed. Because of the easier cleavage of these protecting groups by acids, they will probably replace the tosyl group. The benzenesulfonyl group, which was proposed by Milne and Peng (1957), has not yet been applied in peptide chemistry. It can be removed by sodium in liquid ammonia or by raney nickel at room temperature.

The tosyl and the 4-methoxybenzenesulfonyl groups are stable to catalytic hydrogenation and several acids such as trifluoroacetic acid and HBr in acetic acid. The tosyl group is usually split off by sodium in liquid ammonia (Schwyzer and Li, 1958) or by HF–anisole at 0°C (Mazur and Plume, 1968). Besides the formation of ornithine, which occurs to a large extent if the amino group of Arg(Tos) is unprotected (Guttmann, 1963), sodium in liquid ammonia causes a number of side reactions with peptides that are outlined in Table XVII. The discovery by Hofmann and Yajima (1961) that the Lys-Pro bond in protected deacetyl-α-MSH is reductively cleaved during the treatment with sodium in liquid ammonia especially concerns Lys-Pro and Arg-Pro containing peptides. Thus, model peptides (Guttmann, 1963) and naturally occurring peptides such as insulin B-chain (Wilchek *et al.*, 1965; Benisek and Cole, 1965; Benisek *et al.*, 1967), ribonuclease (Sarid and Patchornik, 1963), apoferredoxin (Benisek *et al.*, 1967), evolidine (Ressler and Kashelikar, 1966), gonadoliberin (König and Geiger, 1975), gramicidin S, and ovalbumin (Wilchek *et al.*, 1965) could be split in this way. In the case of acetylglycyl-proline, it was noticed that the carboxyl group of

Table XVII Side Reactions of Sodium in Liquid Ammonia Treatment of Peptides

Side-reaction	References
Desulfuration	Zahn *et al.* (1967)
Racemization	Hope (1968); Hanby *et al.* (1950); Brenner and Pfister (1951); Arnstein and Morris (1960)
Met → homocysteine	Hope (1968)
Decomposition of Trp	Bajusz and Medzihradszky (1963)
Ac-Pro → Ac-δ-aminovaleric acid	Ramachandran (1965)
Cleavage of the X-Pro-bond	Hofmann and Yajima (1961)

Table XVIII Comparison of Deprotection Conditions for N^G-Tosyl-arginine and N^G-*p*-Methyloxybenzenesulfonylarginine

Protecting group	% deprotection: Reaction time (35–40 min at room temperature)		Reaction time (60 min at 0°C)
	CH_3SO_3H–anisole	CH_3SO_3H–CF_3COOH (1:1) anisole	10 equiv boron tristrifluoroacetate in CF_3COOH
Arg(Tos)	24.1	8.2	11.5
Arg(Mbs)	99.1	92.4	100.0

glycine was reduced to the corresponding alcohol and aldehyde mixture (Benisek *et al.*, 1967).

Because of these reactions caused by sodium in liquid ammonia, liquid HF in the presence of anisole has been preferred for deprotection of Arg(Tos)-peptides, although it too can cause severe side reactions (see Section VIII,A). Trifluoromethanesulfonic acid and anisole at 40°C (Yajima *et al.*, 1974) and pyridinium polyhydrogen fluoride and anisole at room temperature (Matsuura *et al.*, 1976) remove the tosyl group within 1 hr. Boron tribromide (1 *M*) was ineffective. Good results were obtained with fluorosulfonic acid and anisole which deblocks Arg(Tos) quantitatively within 15 min at 20°C (Yajima *et al.*, 1975a). The trifluoromethanesulfonic acid and anisole procedure has been applied to a synthesis of neurotensin (Yajima *et al.*, 1975b). The *p*-methoxybenzenesulfonyl (Mbs) group, which can be removed under somewhat milder conditions, should be of considerable interest. Table XVIII compares the conditions of deprotection of Arg(Tos) and Arg(Mbs) (Nishimura and Fujino, 1976). The properties of the recently introduced mesitylene-2-sulfonyl (Mts) group are comparable to those of the *p*-methoxybenzenesulfonyl group. But the Mts group is more labile against hydrogen bromide in acetic acid (Yajima *et al.*, 1978a,b).

C. Acyl Groups

There are two feasible ways of guanidine protection by acyl groups: 1). monoacylation at the ω-position, and 2). diacylation of the δ,ω- and ω,ω-positions, respectively.

1. Monoacylation

The only advantage of monoacylation over nitro or tosyl protection lies in easier deprotection. Lactam formation of the activated derivatives remains possible. Zervas *et al.* (1957) synthesized Z-Arg(Z)-OH, which can be transformed to H-Arg(Z)-OH and H-Arg(Z)-OBzl, the latter via the corresponding *N*-carboxyanhydride as intermediate. The N^{ω}-benzyloxycarbonyl group can be removed by catalytic hydrogenation or by HBr in acetic acid. But it should be mentioned that the benzyloxycarbonyl group linked to the guanidine function is more resistant to acidolysis than when linked to an α-amine function (Guttmann and Pless, 1965). This explains the complete stability against trifluoroacetic acid in methylene chloride at room temperature (Losse and Rüger, 1973).

The corresponding 4-nitrobenzyloxycarbonyl (4Nz) derivative proposed by Guttmann and Pless (1965) is more stable against acidolysis than the N^{ω}-benzyloxycarbonyl residue. The 4Nz group at the guanidine group is totally resistant to HBr in acetic acid or to trifluoroacetic acid but is easily removed by hydrogenation.

Paulay and Bajusz (1965) prepared Z-Arg(Boc)-OH, a logical continuation of side-chain protection based on the *tert*-butyl group. It has been frequently used since the ACTH synthesis by Schwyzer and Sieber (1963).

Guttmann and Pless (1965) left the route of the urethane-type protecting groups and found in 2-isopropoxycarbonyl-3,4,5,6-tetrachlorobenzoyl (**54**) a suitable protecting group for the guanidine function of arginine. It is stable to anhydrous acidolysis (e.g., HBr in acetic acid or trifluoroacetic acid) and catalytic hydrogenolysis and is removed by acidic hydrolysis (0.1 *N* HCl or

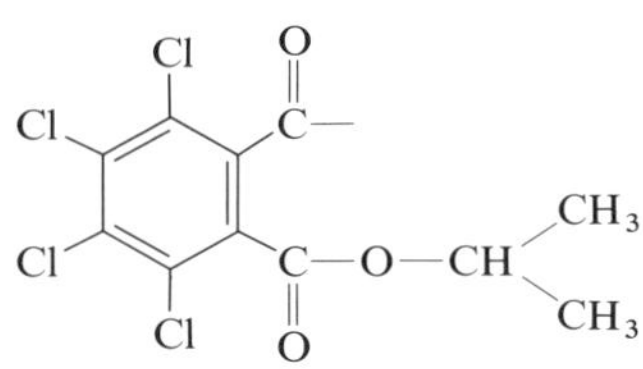

(**54**)

aqueous acetic acid for 1 hr at 100°). Treatment with alkali should be avoided because of the saponification of the isopropyl ester group. The ensuing 2-carboxy-3,4,5,6-tetrachlorobenzoyl group is in fact deprotected as easily as the isopropyl ester, but further peptide coupling may cause side reactions. Hydrazinolysis should also be omitted, because the protecting group is slowly cleaved by hydrazine.

Another well-known protecting group, the trifluoroacetyl, has been investigated by Guttmann and Pless (1965). It is easily removable by ammonia, alkali, hydrazine, or 0.1 *N* HCl at 100°C. Although it is stable in anhydrous acids, this group is overall too unstable for permanent guanidine protection.

2. N^{δ},N^{ω}-Diacylation

In acylating the guanidine group, two diacylated isomeres should theoretically be observed. Indeed, Zervas *et al.* (1957) found a mixture of two isomeric compounds when they prepared Z-Arg(Z_2)-OH. The N^{ω},$N^{\omega'}$-dibenzyloxycarbonyl derivative is not stable and decomposes to Z-Arg(Z)-OH. The δ-acyl group is susceptible to alkali treatment and thus can be removed in nearly quantitative yield (Zervas *et al.*, 1959).

The preparation of Z-Arg(Z_2)-OH has been improved by Wünsch and Wendlberger (1967). Peptide coupling with this advantageous arginine derivative may be accomplished via activated esters or by means of other common peptide coupling procedures. Unfortunately, the N^{α}-benzyloxycarbonyl group cannot be removed selectively. Usually all three benzyloxycarbonyl groups are split off in one step by catalytic hydrogenation, and before coupling of the next amino acid the diacyl protection is replaced by protonation. Furthermore, fluorosulfonic acid and methanesulfonic acid in the presence of anisole have been used to deprotect Z-Arg(Z_2)-OH (Yajima *et al.*, 1975a). Z-Arg(Z_2)-OH can be transformed (via the corresponding *N*-carboxy anhydride into H-Arg(Z_2)-OH (Zervas *et al.*, 1957), which was used by Gros *et al.* (1961) in synthesizing Boc-Arg(Z_2)-OH. Weygand and Nintz (1965) started from Moz-Arg-OH to prepare Moz-Arg(Z_2)-OH. In the same way Boc-Arg(Z_2)-OH has been synthesized by Losse and Rüger (1973). The N^{α}-*tert*-butyloxycarbonyl as well as the N^{α}-4-methyloxybenzyloxycarbonyl (Moz) group may selectively be removed by trifluoroacetic acid. Thus, permanent guanidine N^{δ},N^{ω}-dibenzyloxycarbonyl protection is possible.

Jäger and Geiger (1970, 1973a,c) synthesized Z-Arg(Adc)$_2$-OH and Z-Arg(Ibc)$_2$-OH in much better yield than Z-Arg(Z_2)-OH by means of of corresponding chloroformates. The isobornyloxycarbonyl (Ibc) as well as the admantyloxycarbonyl (Adc) group is stable to catalytic hydrogenation and can be removed with trifluoroacetic acid similar to the tert-butyloxycarbonyl group, thus adapting δ,ω-diacylated arginine to the "Schwyzer (1963) strategy."

D. Trityl Group (Trt)

Gazis *et al.* (1963) investigated the tritylation of the guanidine group. Reaction of H-Arg-OMe with increasing amounts of trityl chloride and alkali yielded Trt-Arg-OMe → Trt-Arg(Trt)-OMe → Trt-Arg(Trt)$_2$-OMe → Trt-Arg(Trt)$_3$-OMe. Because of the steric hindrance of the trityl group the methyl ester could not be saponified. Starting with Boc-Arg-OH, Losse and Rüger (1973) synthesized Boc-Arg(Trt)-OH. As yet this guanidine protection has not been applied to peptide synthesis.

E. Reaction Products with Bifunctional Aldehydes and Ketones

Bifunctional aldehydes and ketones are used for more or less specific modification of arginine residues in proteins. Malonaldehyde and its derivatives react irreversibly with the guanidine group of arginine to form N^5-(2-pyrimidyl)ornithine derivatives (King, 1966; Sigor *et al.*, 1971). Reversible reagents are 1,2-cyclohexanedione (Toi *et al.* 1965, 1967; Liu *et al.*, 1968; Patthy and Smith, 1975), oligomers of 2,3-butanedione (Yankeelov *et al.*, 1966, 1968; Yankeelov, 1970; Grossberg and Pressman, 1968), and phenylglyoxal (Takahashi, 1968). Further vicinal dioxo compounds that are used for arginine modification of proteins are benzil (Itano and Gottlieb, 1963; Toi *et al.*, 1965), glyoxal (Nakaya *et al.*, 1967, 1969), and 2,4-dinitrophenylglyoxal (Wilchek and Miron, 1979).

Conditions for being a useful N^G-protective group are fulfilled by 1,2-cyclohexanedione in the presence of borate buffer (Patthy and Smith, 1975). In neutral and slightly alkaline borate buffer, 1,2-cyclohexanedione reacts with free arginine to form a single amino acid derivative, N^7,N^8-(1,2-dihydroxycyclohex-1,2-enyl)arginine (=DHCH-arginine) which is stable in acetic solutions and in borate buffers at pH 8 to 9. This blocking group can be removed by hydroxylamine at pH 7.0 with quantitative recovery of arginine. Consistent with the presence of vicinal *cis*-hydroxyl groups in the molecule, it forms a complex with borate. Evidence for an intact cyclohexane ring comes from the fact that DHCH-arginine is slowly hydrolyzed to yield cyclohexanedione and arginine on removal of borate. This procedure was successfully applied for the reversible blocking of arginine in the B-chain of insulin during the semisynthesis of human insulin by trypsin-catalyzed condensation of threonine with des-B 30-alanine porcine insulin (Morihara *et al.*, 1979).

F. Conclusions

The kind of guanidine protection depends *a priori* on the specific strategy for the synthesis of the desired peptide. If hydrogenation has to be the last deprotection step, Arg(NO_2), Arg(Z), Arg(4Nz), and Arg(Z_2) are employed. Arg(NO_2) is only recommended for syntheses of short peptides since the nitro group is a source of several side reactions (see Table XV). The prerequisite for hydrogenation is solubility of the protected peptide in a suitable solvent such as aqueous acetic acid, methanol, dimethylformamide, *N*-methylpyrrolidone, and mixtures thereof. In some cases aqueous 2,2,2-trifluoroethanol exhibits an amazing capacity to dissolve large, protected peptides. Liquid ammonia is a promising novel solvent for catalytic hydrogenation (Meienhofer and Kuromizu, 1974). Anyway, hydrogenation as a deprotecting procedure for large peptides is always risky since there is no guarantee of the completeness of the reaction.

Sodium in liquid ammonia, which has frequently been used to remove the tosyl- and benzyl-type protecting groups, should only be employed in peptides lacking Arg-Pro and Lys-Pro, unless the cleavage of these bonds can be suppressed by the addition of sodium amide (Katsoyannis *et al.*, 1971). Other possible damage of the peptides has to be kept in mind (see Table XVII).

Using the HF–anisole deprotection technique, comparative investigations of Yamashiro *et al.* (1972) during the synthesis of human growth hormone (17–23)-peptide revealed that the nitro group was inferior to the tosyl residue for guanidine group protection. This was confirmed by Arnold *et al.* (1973) and Flouret *et al.* (1973) who experienced similar results in synthesizing gonadoliberin. The recently introduced *p*-methoxybenzenesulfonyl group, which is more easily removable by strong acids than the tosyl group, may possibly replace the tosyl group. In cases where trifluoroacetic acid is employed for deblocking of protected peptides, Arg$(Adoc)_2$ proved to be the best group, but unfortunately it is quite expensive. A good compromise is the use of Z-Arg(Z_2)-OH. With this compound arginine can always be introduced into a peptide with good results. After catalytic hydrogenolysis, protonation takes over as a protection of the free guanido group. For example, with this technique glucagon (Wünsch and Wendlberger, 1968), secretin (Wünsch, 1972; Jäger *et al.*, 1974), and gonadoliberin (Geiger *et al.*, 1971) have been synthesized.

VIII. IMIDAZOLE PROTECTING GROUPS

Synthesis of histidine-containing peptides with an unprotected imidazole residue raises no great difficulties as long as histidine is not the COOH-terminal amino acid. Activation of the carboxyl group of N^{α}-acylated histidine may lead to cyclic products (imidazolides) with the imidazole ring as demonstrated by Sheehan *et al.* (1959), [Eq. (38)] (see also Veber, 1975). This

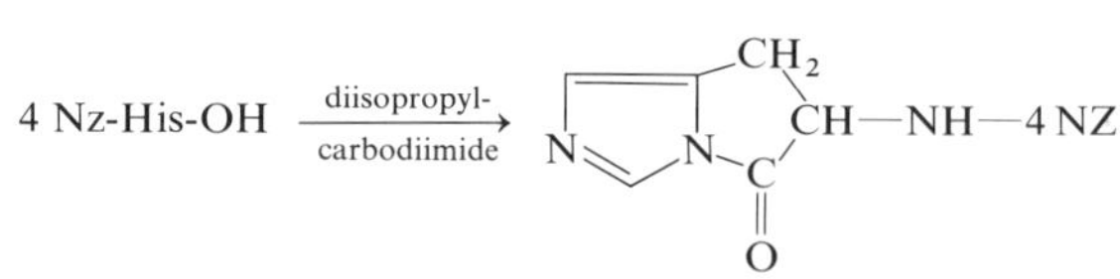

(38)

cyclization may be the reason for the low coupling yields of Boc-His-OH and dicyclohexylcarbodiimide, compared with N^{im}-protected *tert*-butyloxycarbonyl histidine. It could not be improved even by additional application of *N*-hydroxysuccinimide (Schaich *et al.*, 1973). Nevertheless, this procedure has occasionally been employed, especially in solid-phase peptide synthesis (Gutte and Merrifield, 1969; Loffet, 1967). N^{im}-Unprotected histidine, is very susceptible to racemization due to the basic imidazole ring. Lin *et al.* (1972) and Terada *et al.* (1978) observed racemization of Boc-His-OH during coupling with dicyclohexylcarbodiimide. Danger increases when COOH-terminal histidine peptides are activated. Even with the use of DCC–HOBt (König and Geiger, 1970a), a coupling method with a decreased tendency of racemization, Z-Gln(Ddm)-His-Pro-NH_2 was contaminated with a small amount of D-histidine, which could be removed in this case by crystallization (König and Geiger, 1972). The application of the azide method, recommended by Holley and Sondheimer (1954) as well as Honzl and Rudinger (1961) for activating histidine derivatives is especially in the case of peptides with carboxyl-terminal histidine prone to racemization. It appears to be accompanied by the cyclization described above; on the other hand the cyclic product may induce racemization (Veber, 1975; Merrifield and Woolley, 1956; Schneider, 1960a,b).

While treating histidine peptides with dicyclohexylcarbodiimide, Rink and Riniker (1974) observed that the reagent could react with the imidazole ring by formation of an N^{im}-amidinohistidine derivative. This reaction was catalyzed by 1-hydroxybenzotriazole. This side product of unprotected histidine derivatives can be easily cleaved by methanol, preferably containing acetic acid or pyridine hydrochloride.

Imidazole protecting groups bearing electron-withdrawing substituents are expected to prevent racemization during peptide coupling. But except for the recently investigated N^{π}-phenacyl residue (Jones and Ramage, 1978), all other established N^{im}-acylations probably lead to τ-substitution, which may not prevent cyclization with accompanying racemization. Presently, there is a choice of eight types of N^{im}-blocking groups: 1). benzyl, 2). diphenylmethyl and triphenylmethyl, 3). nitrophenyl, 4). acetal, 5). urethane forming, 6). piperidinocarbonyl, 7). tosyl, and 8). phenacyl groups.

A. Benzyl Groups

H-His(Bzl)-OH has been prepared by du Vigneaud and Behrens (1937) by means of benzyl chloride and stoichiometric amounts of sodium in liquid ammonia. Sodium in liquid ammonia is also the most suitable reagent for removal of this residue (for side reactions, see Table XVII). Another possibility for cleaving the benzyl group is catalytic hydrogenation (Theodoropoulos, 1956); but it requires long reaction times and simultaneous hydrogenation of aromatic amino acids has been observed (Jorgensen and Windridge, 1970;

Windridge and Jorgensen, 1971b). The recently introduced 1,4-cyclohexadiene as an effective hydrogen donor requires the efficient palladium black catalyst for successful cleavage of the N^{im}-benzyl group (Felix, 1978).

Since the benzyl residue is not capable of reducing the basicity of the imidazole ring, His(Bzl) is prone to racemization. Even N^{α}-urethane-protected His(Bzl) racemizes especially using solid-phase synthesis in considerable amounts (Windridge and Jorgensen, 1971a). Angiotensin II analogues prepared by solid-phase peptide synthesis were found to contain about 10% D-histidine (Jorgensen and Windridge, 1970). Application of DCC–HOBt (König and Geiger, 1970a) to solid-phase synthesis (Windridge and Jorgensen, 1971b) resulted in quantitative coupling and a decrease of the D-histidine content from 11% to <0.3%. Extensive racemization occurred with Boc-His(Bzl)-OPcp and with Boc-His(Bzl)-OH, employing EDAC as condensing reagent (Khosla *et al.*, 1972). The racemization rate of Z-His(Bzl)-OPcp was found to be almost 7000 times faster than Z-Ala-OPcp (Kovacs *et al.*, 1979). *Thirty-seven* percent racemization was observed synthesizing Z-His(Bzl)-Ser(Bzl)-OBzl in dimethylformamide by the DCC method (Terada *et al.*, 1978).

The 2-nitrobenzyl residue, a photolabile protecting group, has been investigated by Kalbag and Roeske (1975). It is inert to acid (e.g., trifluoroacetic acid or HCl in dioxane) and alkali, but is slowly cleaved by catalytic hydrogenation. To remove the N^{im}-2-nitrobenzyl group, a dioxane solution was irradiated with an Hg lamp for 1 hr.

B. Diphenyl- and Triphenylmethyl Groups

So far three different protecting groups of this type have been described: trityl (**55**), diphenylmethyl (**56**), and diphenyl-4-pyridylmethyl (**57**), in order of their acid stability. Trt-His(Trt)-OH was first synthesized by Amiard *et al.*

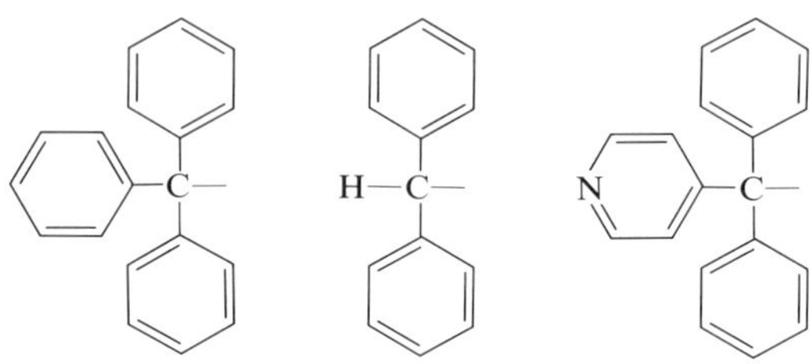

(**55**) (**56**) (**57**)

(1955a) via the corresponding methyl ester and subsequent saponification. Stelakatos *et al.* (1959) succeeded in preparing this compound by direct tritylation of histidine. Losse and Krychowski (1970) obtained the N^{α}-*tert*-butyloxycarbonyl derivative by tritylation of Boc-His-OH. For the coupling of Trt-His(Trt)-OH with amino acid esters or peptide esters, the dicyclohexylcarbodiimide method (Amiard *et al.*, 1955b; Boissonnas *et al.*, 1958) or the DCC–HOBt method (Bosshard, 1971) have been recommended. The N^{im}-trityl group is more stable to acids than the N^{α}-trityl group. However, selective removal of the N^{α}-trityl protection was not possible with the usual hot acetic acid. One equivalent of HCl in hot acetic acid cleaves both trityl groups (Boissonnas *et al.*, 1958). Similarly, Bosshard (1971) was unable to cleave the N^{α}-trityl group selectively by use of 80% aqueous acetic acid. Best results were obtained with 4 *N* HCl in ethyl acetate, but the reaction product was contaminated with 5–10% N^{im}-deprotected material. On the other hand, when a solution of Trt-His(Trt)-OMe hydrochloride in absolute alcohol was briefly heated on a steam bath, detritylation occurred only at the α-amino position (Stelakatos *et al.*, 1959). Using Boc-His(Trt)-OH, Losse and Krychowski (1970) showed that the N^{im}-trityl residue is stable in 1 *N* HCl–AcOH for at least 20 hr at room temperature whereas the N^{α}-*tert*-butyloxycarbonyl group is removed in 30 min.

Further investigations of Losse and Krychowski (1971) showed the N^{im}-trityl group to be very stable to nucleophiles such as hydrazine, mercaptoethanol, or alkali, but cleavable by catalytic hydrogenation or acidolysis at room temperature [e.g., by HBr–AcOH (2 hr), trifluoroacetic acid (30 min), or formic acid (2 min)]. The diphenymethyl (Dpm) group is more stable for masking the N^{im}-function than is the trityl group. Losse and Krychowski (1970, 1971) synthesized Boc-His(Dpm)-OH in a manner analogous to the tritylation procedure. The N^{im}-Dpm protection is stable to alkali, hydrazine, or mercaptoethanol, but is removed by 6 *N* HBr–AcOH (3 hr), trifluoroacetic acid (1 hr), or formic acid (10 min).

The insertion of a basic residue such as pyridyl into an acid-labile group greatly increases the stability to acid (Coyle *et al.*, 1975). Thus, Coyle and Young (1976a,b) synthesized N^{α}-benzyloxycarbonyl, N^{im}-diphenyl-4-pyridylmethylhistidine, and the corresponding N^{α}-*tert*-butyloxycarbonyl derivative. As expected, the N^{im}-diphenyl-4-pyridylmethyl protection is very stable to acidolysis and cannot be removed by either trifluoroacetic acid or 45% HBr in acetic acid. Cleavage is accomplished by reductive methods such as hydrogenation, zinc in acetic acid, or electrolytic reduction.

C. Dinitro- and Trinitrophenyl Groups

The well-known reagent 2,4-dinitrofluorobenzene (Sanger, 1945) reacts not only with NH_2-terminal amino acids but also with ω-amino groups, sulfydryl groups, hydroxyl groups, and the imino group of histidine. Siepmann and Zahn (1964) were the first to describe the synthesis of Z-His(Dnp)-OH and H-His(Dnp)-OH, but they did not apply these derivatives to peptide synthesis. The easy removal of the 2,4-dinitrophenyl (Dnp) group by thiolysis at pH 7–8 with mercaptanes such as 2-mercaptoethanol or mercaptoacetic acid under nonaqueous and aqueous conditions was discovered by Shaltiel (1967). The thiolytic cleavage is accompanied by a considerable increase in uv-absorption at 335–360 nm. This observation can be utilized for monitoring the rate and extent of the reaction which depends strongly on the sequence, on the solvent, and of course on the applied mercaptan (Shaltiel and Fridkin, 1970). The N^{im}-2,4-dinitrophenyl group is very stable in neutral or strongly acidic media (Shaltiel and Fridkin, 1970) such as trifluoroacetic acid, formic acid, HBr–HOAc (Losse and Krychowski, 1971), and even under the conditions of total hydrolysis of peptides in 6 *N* HCl (Siepmann and Zahn, 1964). At pH 12.2, Boc-His(Dnp)-OH decomposes within 20 hr, whereas the extent of decomposition at pH 10.8 does not exceed 5% (Shaltiel and Fridkin, 1970). The N^{im}-Dnp group may also be translocated onto hydrazine (Losse and Krychowski, 1971, Zheltukhina *et al.*, 1976) and even onto amino acids [if, for example, His(Dnp) and Gly are dissolved in $NaHCO_3$ solution and heated to 60°C for 6 hr (Siepmann and Zahn, 1964)]. This effect may be responsible for side reactions during the synthesis of His(Dnp)-containing peptides. Nevertheless, Boc-His(Dnp)-OH (Chillemi and Merrifield, 1969) was successfully adopted in solid-phase peptide synthesis. Furthermore starting with Z-His(Dnp)-OH the corresponding *N*-carboxyanhydride was synthesized and polymerized by Fridkin and Shaltiel (1971).

An important property of the N^{im}-2,4-dinitrophenyl group is its electron-withdrawing capacity which reduces the basicity of the imidazole ring. Racemization was decreased compared to Boc-His(Bzl)-OH in solid-phase synthesis (Lin *et al.*, 1972). The location of the 2,4-dinitrophenyl group has been proved to be the τ-nitrogen of the imidazole moiety (Bell and Jones, 1974), and therefore racemization via cyclization may be possible. In contrast to Lin *et al.* (1972) and confirming the above mentioned thesis, we detected no remarkable difference in racemization if peptides with COOH-terminal histidine or N^{im}-Dnp-histidine were activated by DCC–HOBt. For instance, about 20% [D-His^2]gonadoliberin was formed when pGlu-His-OH or pGlu-His(Dnp)-OH were coupled with the corresponding octapeptide using DCC–HOBt. In the case of pGlu-His(Dnp)-OH, the racemization could be reduced to below 2% if DCC–3-hydroxy-4-oxo-3,4-dihydro-1,2,3-benzotriazine (König and Geiger, 1970c) was used. Applying this method, with pGlu-His-OH, 12% [D-His^2] gonadoliberin was formed (W. König, R. Geiger, and V. Teetz, unpublished results). Also Beyerman *et al.* (1974) using the *r*epetitive *e*xcess *m*ixed *a*nhydride (REMA) method observed about the same extent of racemization (6% D His) during coupling of Boc-His(Dnp)-OH with H-Leu-Gly-Leu-Val-NH_2 as was found with Boc-His(Bzl)-OH OH (7% D His). In addition, colored by-products were formed in this step of synthesis.

The 2,4-dinitrophenyl group has been also used as a handle in solid-phase peptide synthesis. Glass *et al.* (1972) synthesized thyroliberin (TRH) with histidine bound to a solid support via an N^{im}-dinitrophenyl bridge. Losse and Krychoski (1971) synthesized N^{α}-*tert*-butyloxycarbonyl-N^{im}-2,4,6-trinitrophenyl (Tnp) histidine. Similar to the 2,4-dinitrophenyl group, the N^{im}-Tnp group is stable to acidolysis but slightly more susceptible to nucleophiles such as mercaptans or hydrazine.

D. Acetal Groups

Weygand *et al.* (1966a, 1967) found, that the 2,2,2-trifluoro-1-acylamino-ethyl group (58) is a suitable N^{im}-protecting group. Because of the electron-withdrawing effect of the CF_3 group, the basicity of the imidazole moiety is strongly diminished and its solubility in organic solvents is improved. Removal and stability of this kind of protecting group depend on the adjacent acyl residue. After the acyl group has been cleaved off, the resulting 2,2,2-trifluoro-1-aminoethyl derivative is easily hydrolyzed to 2,2,2-trifluoro-acetaldehyde hydrate, ammonia, and the N^{im}-unprotected histidine derivative [Eq. (39)].

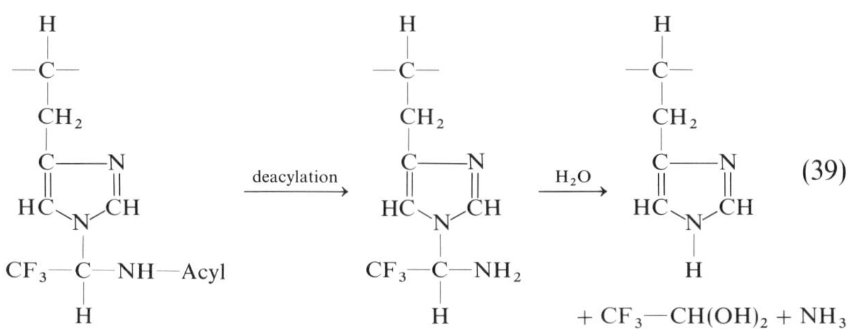

(39)

(58)

Two types of these protecting groups have been synthesized: 1). the 2,2,2-trifluoro-1-benzyloxycarbonylaminoethyl group (Ztf), which was introduced by 2,2,2-trifluoro-1-chloro-*N*-benzyloxycarbonylethylamine (**59**); and 2). the 2,2,2-trifluoro-1-*tert*-butyloxycarbonylaminoethyl group (Btf), which can be introduced by 2,2,2-trifluoro-1-hydroxy-*N*-*tert*-butyloxycarbonylethylamine (**60**) with trifluoroacetic acid anhydride in pyridine. Both protecting groups

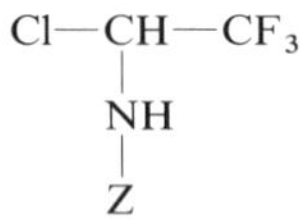

(59)

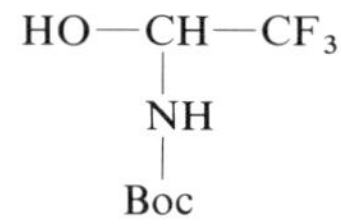

(60)

are stable to dilute alkali at 20°C as normally used for saponification of alkyl esters. The benzyloxycarbonyl-containing blocking group can be removed with HBr–AcOH or catalytic hydrogenation and the *tert*-butyloxycarbonyl-containing derivative is cleaved by trifluoroacetic acid at room temperature. Disadvantages of these protecting residues are the newly introduced asymmetric center and the pronounced lability of the starting materials to hydrolysis. In spite of the diminished basicity of the imidazole moiety racemization may occur. Beyerman *et al.* (1974) found, as with the 2,4-dinitrophenyl group, 4% racemization if Boc-His(Ztf)-OH was coupled with H-Leu-Gln-Gly-Leu-Val-NH_2 by the mixed anhydride method. Along with the expected hexapeptide, a by-product (about 60%) was formed which was determined to be Ztf-Leu-Gln-Gly-Leu-Val-NH_2.

By analogy with the Ztf group, the 1-benzyloxycarbonyl-2,2,2-trichloroethyl group has also been applied for N^{im}-protection of histidine; but this residue is sensitive to alkali (Weygand *et al.*, 1966d).

To overcome the difficulties with the new asymmetric center Seltzman and Chapman (1974) developed another histidine protecting group, based on hexafluoroacetone rather than trifluoroacetaldehyde. Reaction of the silver salt of Z-His-OMe with hexafluoroacetone, followed by reaction with

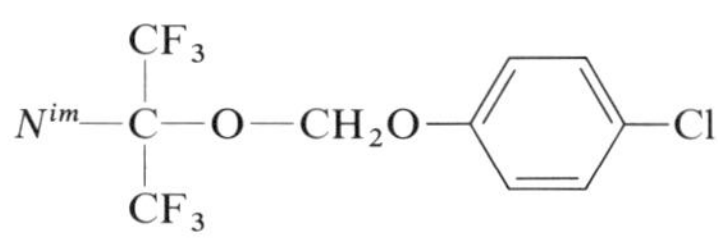

(61)

α-*p*-dichloroanisole, yielded the corresponding crystalline N^{im}-protected histidine derivative (**61**). This protecting group is stable to alkali and is removed with 2.4 *N* HCl in aqueous acetic acid at room temperature.

E. Urethane Groups

Investigations by Gerngross (1919) showed that diacylated histidine esters easily lose the N^{im}-acyl group by hydrolysis. More easily than water, nucleophiles such as amino acids cleave the N^{im}-acyl group to form acylamino acids (Bergmann and Zervas, 1928).

Urethane groups are more stable to hydrolysis and nucleophilic attack. Presently, the benzyloxycarbonyl (Z), 4-methoxybenzylocarbonyl(Moz), *tert*-butyloxycarbonyl (Boc), adamantyloxycarbonyl (Adc), and isobutyloxycarbonyl(Ioc) groups are used for N^{im}-protection. The basicity of the imidazole ring is reduced and the solubility of such histidine derivatives in organic solvents is improved. Racemization with N^{α},N^{im}-urethane protected histidine was not detected using the mixed anhydride method, but partial deprotection of N^{im}-urethane was observed during hydrolysis of the excess mixed anhydride when using the REMA procedure (Beyerman *et al.*, 1974). Also coupling reagents as DCC, DCC–HOSu, DCC–HOBt and Woodward's reagent can be used without noticeable racemization (Terada *et al.*, 1978).

1. Benzyloxycarbonyl Protection (Z)

Crude Z-His(Z)-OH is reported to be unstable at room temperature. More stable compounds were obtained by crystallization from methanol or ethanol containing one mole of alcohol (Patchornik *et al.*, 1957). N^{im}-Benzyloxycarbonyl is readily and selectively removed by nucleophilic reagents such as sodium methoxide or amines. With ammonia and aniline (Patchornik *et al.*, 1957) as well as with amino acid esters (Sakiyama *et al.*, 1958) the corresponding benzylurethanes are formed. Therefore, transcarbobenzoxylation during peptide synthesis cannot be excluded. The corresponding esters and peptides seem to be more resistant to hydrolysis than Z-His(Z)-OH (Sakiyama, 1962). Whereas Z-His(Z)-ONp could not be crystallized (Sakiyama, 1962), Boc-His(Z)-ONp is described as a stable crystalline substance (Schaich *et al.*, 1973).

Although the N^{im}-benzyloxycarbonyl group can be removed with HBr–AcOH or with concentrated HCl at 37°C (Akabori *et al.*, 1958), the N^{im}-protection seems more resistant to acidolysis with strong acids than does N^{α}-protection. On the other hand, the N^{im}-benzyloxycarbonyl group has been removed from Z-His(Z)-ONb by boiling acetic acid within a few minutes (Elliott and Morris, 1960). Possibly, protonation of the imidazole ring is responsible for the higher stability of the N^{im}-benzyloxycarbonyl group in strong acid. Sakiyama (1962) compared the mixed anhydride, the *p*-nitrophenyl ester, and the dicyclohexylcarbodiimide methods in synthesizing Z-His(Z) peptides. The most satisfactory results were obtained with dicyclohexylcarbodiimide.

2. 4-Methyloxybenzyloxycarbonyl Group (Moz)

Schaich *et al.* (1973) synthesized Moz-His(Moz)-OH and the corresponding crystalline nitrophenyl ester, which was coupled in good yield with Gly-Gly-Ala-resin.

3. tert-Butyloxycarbonyl Group (Boc)

Schnabel *et al.* (1968, 1971) synthesized Boc-His(Boc)-OH by means of the highly reactive *tert*-butyloxycarbonyl fluoride at pH 7–8. Boc-His(Boc)-OH was also prepared via *tert*-butyloxycarbonyl azide with MgO (Fridkin and Goren, 1971), under pH-stat conditions at pH 9–9.5 (Yamashiro *et al.*, 1972) or using di-*tert*-butyl pyrocarbonate and crystallized from benzene and tetrachloromethane (Pozdmev, 1978). The N^{im}-*tert*-butyloxycarbonyl group was found to be much more stable toward alkali than the N^{im}-benzyloxycarbonyl group.

Z-His(Boc)-OH has been synthesized by Fridkin and Goren (1971). Eckstein (1976) observed slow decomposition of this compound at room temperature to give Z-His-OH. Crystalline Boc-His(Boc)-ONp, which has been synthesized by Schnabel *et al.* (1968), Fridkin and Goren (1971), and Schaich *et al.* (1973), was reportedly more stable than Boc-His(Boc)-OH.

The N^{im}-*tert*-butyloxycarbonyl group can be removed selectively with methanolic hydrazine hydrate or methanolic ammonia (Schnabel *et al.*, 1971). As usual, it is also cleaved with trifluoroacetic acid, mineral acids in organic solvents (Schnabel *et al.*, 1968), or anhydrous HF, but not with HCl in dioxane (Fridkin and Goren, 1971). Yamashiro *et al.* (1972) employed Boc-His(Boc)-OH in solid-phase peptide synthesis where both blocking groups were cleaved simultaneously by 50% trifluoroacetic acid in methylene chloride. Using Bpoc-His(Boc)-OH and 0.05–0.1 *N* HCl in dioxane as deprotecting reagent for the 2-(4-biphenylyl)propyl(2)oxycarbonyl (Bpoc) group, partial deblocking of the histidine side chain was observed.

4. Adamantyloxycarbonyl Group (Adc)

Adc-His(Adc)-OH was prepared by Haas *et al.* (1966). The N^{im}-Adc group is cleaved with trifluoroacetic acid as described for the N^{α}-Adc residue. It appears to be more stable than the Boc group since no decomposition of Z-His(Adc)-OH has been seen as in the case of Z-His(Boc)-OH (Tilak *et al.*, 1971; Eckstein, 1976). Adc-His(Adc)-OH was successfully applied in the synthesis of glucagon and secretin (Wünsch *et al.*, 1968; Wünsch, 1972).

5. Isobutyloxycarbonyl Group (Ioc)

Grønvald *et al.* (1979) investigated the acylation of the imidazolyl moiety of histidine by isobutylchloroformate, which occurs to various degrees during mixed anhydride coupling. It is easily introduced into N^{α}-protected histidine derivatives. The N^{im}-Ioc group is stable against trifluoroacetic acid, 2 *N* HBr in acetic acid, and during catalytic hydrogenolysis. The protecting group may be removed by pyridine, HCl in methanol, and alkali.

F. Piperidinocarbonyl Group

Jäger *et al.* (1968) investigated N^{im}-carbamoyl residues and recommended the piperidinocarbonyl group, supposedly more resistant to nucleophiles than urethane N^{im}-protecting groups. This moiety remains unaffected by acidolysis and catalytic hydrogenation, but hydrazine and aqueous or alcoholic alkali easily remove the group from the imidazole ring. The introduction of this group is easily performed by reaction of N^{α}-acylhistidine esters or N^{α}-acylhistidine peptide esters with piperidinocarbonyl chloride in absolute pyridine at 65°C [Eq. (40)].

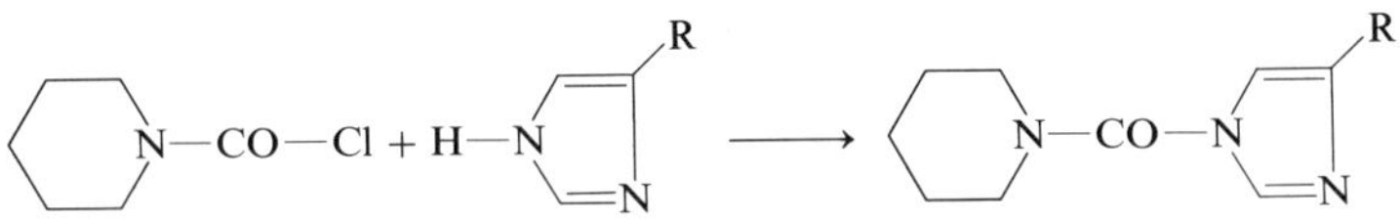

(40)

G. Tosyl Group (Tos)

Di-2-naphthylsulfonylhistidine, the first tosyl-like N^{im}-protected compound, has been prepared by Pauly (1904). However, that masking was never utilized in peptide chemistry. Sixty-five years later, Sakakibara and Fujii (1969) introduced the tosyl group into N^{α}-protected histidine derivatives by means of tosyl chloride under the conditions of the Schotten–Baumann reaction. The N^{im}-tosyl group was found to be easily removable with HF or alkali. Also, pyridinium polyhydrogen fluoride (Matsuura *et al.*, 1976), trifluoromethanesulfonic acid (Yajima *et al.*, 1974), and fluorosulfonic acid (Yajima *et al.*, 1975a) enable N^{im}-tosyl deprotection. It has been stated that N^{im}-tosyl is stable against trifluoroacetic acid, but Schwartz and Katsoyannis (1973) observed partial detosylation of the histidine residue. Partial cleavage was also reported for the use of 2 *N* HBr–AcOH and for 4 *N* HCl–dioxane. As Fujii and Sakakibara (1974) found, the N^{im}-tosyl group is stable against catalytic hydrogenation but labile to the nucleophilic attack of amine groups; therefore, a small amount of the tosyl group was cleaved during the removal of N^{α}-benzyloxycarbonyl groups. Also 1-hydroxybenzotriazole in tetrahydrofuran removed the N^{im}-tosyl group completely within 1 hr at room temperature, forming 1-(tosyloxy)benzotriazole. Beyerman *et al.* (1974) who investigated the racemization of Boc-His(Tos)-OH via the mixed anhydride methode (no racemization was found), cleaved the N^{im}-tosyl group with pyridine hydrochloride in dimethylformamide. As shown above, the N^{im}-tosyl residue is very susceptible to weak acids, and therefore the slow decomposition of N^{α}-protected N^{im}-tosylhistidines during storage at room temperature may be due to the acidity of the molecule's own carboxyl group. Storage of these compounds in the form of their DCHA salts was proposed to be a practical solution (Fujii and Sakakibara, 1974). Synthesizing peptides with N^{im}-tosyl protection mixed anhydrides, dicyclohexylcarbodiimide, or active esters may be employed (Fujii *et al.*, 1976). Using dicyclohexylcarbodiimide, Z-His(Tos)-OH was condensed with H-Ser(Bzl)-OBzl without racemization (Terada *et al.*, 1978).

The "dansyl" group (1-dimethylaminonaphthalene-5-sulfonyl) was also used for N^{im}-protection, but precise conditions for deprotection have not been given (Schaich *et al.*, 1973).

H. Phenacyl Group

Jones and Ramage (1978) synthesized N^{α}-Z-N^{τ}-phenacylhistidine and N^{α}-Z-N^{π}-phenacylhistidine via an intermediately formed N^{α}-Z-N^{τ}-Trt-histidine. As suggested by Veber (1975), only the N^{τ}-isomer was prone to racemization during peptide coupling. The N^{π}-phenacyl group is uneffected by hydrogen bromide in acetic acid, trifluoroacetic acid, and trifluoromethanesulfonic acid, and can be split off by photolysis, or better with zinc dust in aqueous acetic acid. The ketonic group in *N*-phenacylimidazoles is unreactive, but reaction occurs with hydrazine in methanol (Fletcher *et al.* 1979).

I. Conclusions

The kind of protection to be employed for the imidazole moiety of histidine during peptide synthesis depends on the strategy planned. N^{im}-Benzyl-protected peptides can be synthesized without problems, but deprotection by sodium in liquid ammonia or hydrogenation frequently gives rise to by-products. If hydrazinolysis is allowed for final deprotection, the piperidinooxycarbonyl group, which is stable to acidolysis and hydrogenolysis, may be recommended. The behavior of the thiolysis-sensitive dinitrophenyl moiety is quite similar. However, it appears to be more susceptible to nucleophilic attack of amines and to racemization. Vice versa, N^{im}-trityl- and diphenylmethyl-type protecting groups are stable to nucleophiles, but selective deprotection of acid-labile N^{α}-protecting groups seems to be only feasible by applying the diphenyl-4-pyridylmethyl moiety for N^{im}-protection. The most stable urethane-type group for the imidazole ring is the adamantyloxycarbonyl group (Adc) which is resistant to catalytic hydrogenation.

But the use of the less expensive N^{im}-*tert*-butyloxycarbonyl group may also be possible. Especially if histidine is in the amino terminal position, Boc-His(Boc)-OH or Adc-His(Adc)-OH may be preferably utilized. The N^{im}-tosyl group provides no advantages over Adc or Boc since stability to nucleophiles is markedly decreased and specific acidolysis of N^{α}-protection is obviously not yet possible.

IX. N^{pyr}-PROTECTION OF β-(3-PYRAZOLYL)ALANINE

In order to investigate structure–activity relationships, histidine was frequently substituted by β-(1-pyrazolyl)alanine (**62**) and β-(3-pyrazolyl)-alanine (**63**).

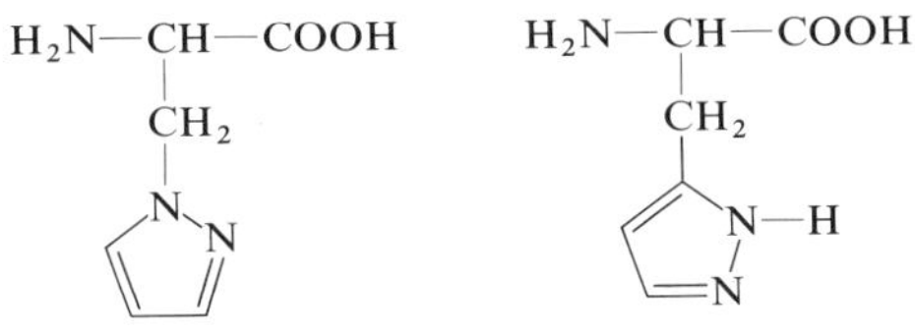

(**62**) (**63**)

The naturally occurring β-(1-pyrazolyl)alanine (**62**) (Shinano and Kaya, 1957) has been N^{α}-acylated with benzyloxycarbonyl chloride or *tert*-butyloxycarbonyl azide to yield the corresponding crystalline derivatives (Hofmann *et al.*, 1968a). Further acylation, i.e., of the pyrazolyl moiety, is not possible, and therefore β-(1-pyrazolyl)alanine-containing peptides were synthesized without N^{pyr}-protection (Hofmann and Bohn, 1966; Coy *et al.*, 1975).

On the other hand, β-(3-pyrazolyl)alanine (**63**) behaves like histidine since the unnatural amino acid contains a hydrogen at the pyrazole nitrogen that can easily be substituted. Therefore, N^{α}-protected derivatives give rise to cyclization, as was observed by Hofmann *et al.* (1968a) [Eq. (41)]. In contrast

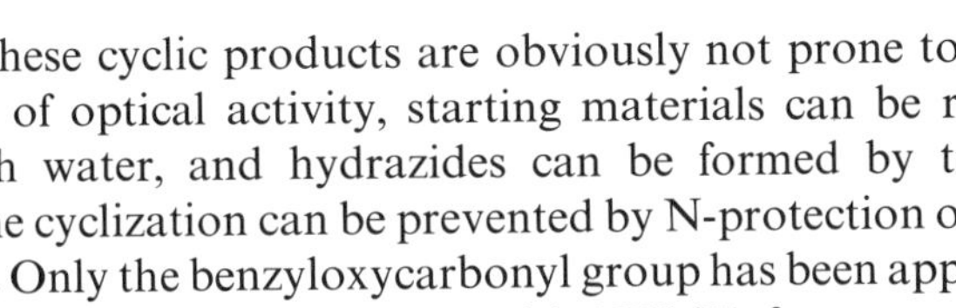

(41)

X = Z, Boc

to histidine, these cyclic products are obviously not prone to racemization. Without loss of optical activity, starting materials can be regenerated on refluxing with water, and hydrazides can be formed by treatment with hydrazine. The cyclization can be prevented by N-protection of the pyrazolyl moiety (N^{pyr}). Only the benzyloxycarbonyl group has been applied, since *tert*-butyloxycarbonyl azide did not react with N^{pyr} (Hofmann *et al.*, 1968a).

N^{α},N^{pyr}-Dibenzyloxycarbonyl-β-(3-pyrazolyl)alanine appears to possess much greater stability than Z-His(Z)-OH. The corresponding crystalline *p*-nitrophenyl ester is said to be the most useful compound for the synthesis of peptides containing β-(3-pyrazolyl)alanine (Hofmann *et al.*, 1967, 1968a; Andreatta and Hofmann, 1968). But, in most cases N^{pyr}-unprotected derivatives have been used to synthesize correspondingly substituted peptide analogues (Hofmann and Bohn, 1966; Finn and Hofmann, 1967; Hofmann *et al.*, 1968b; Hofmann and Bowers, 1970; Gillessen *et al.*, 1971; Coy *et al.*, 1974).

X. INDOLE PROTECTING GROUPS

Peptide coupling with tryptophan in the NH_2- or COOH-terminal position causes no difficulties. However, during acid treatment for removal of protecting groups decomposition of tryptophan frequently occurs. At acidic pH, the indole moiety is susceptible to oxidation and to attack by cations. A mixture of dimethyl sulfoxide and concentrated HCl, for example, quantitatively oxidizes tryptophan derivatives to oxindolylalanine-containing compounds (Savige and Fontana, 1976) [Eq. (42)].

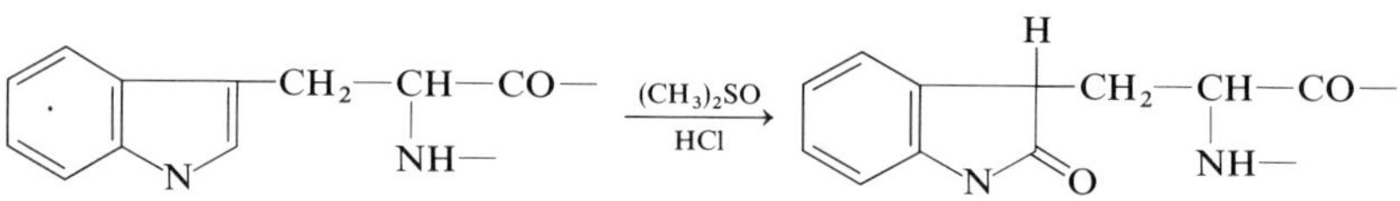

(42)

Undesired *tert*-butylation of the indole moiety has been confirmed by Alakhov *et al.* (1970) and Löw *et al.* (1978b), if *tert*-butyloxycarbonyl groups or *tert*-butyl esters are cleaved by trifluoroacetic acid or HF in tryptophan-containing peptides. This side reaction can be suppressed by using 95% trifluoroacetic acid, formic acid or 2 *N* HCl in either ethyl acetate, methanol or acetic acid. The *tert*-butylation of free tryptophan under simulated conditions of acidolytic cleavage of *tert*-butyl groups yielded as a main product 2,5,7-tri-*tert*-butyltryptophan (Löw *et al.*, 1978a). In the case of Z-Trp-OBzl, thirteen by-products could be observed (Jaeger *et al.* 1978b). Synthetic N^{in}-*tert*-butyltryptophan was found to be identical to a modified tryptophan, which was isolated from a synthetic gastrin analogue (Jaeger *et al.* 1978a). To minimize oxidation and alkylation, mercaptoethanol (Blake and Li, 1968), dithiothreitol (Li and Yamashiro, 1970) or 1,2-ethanedithiol (Sharp *et al.* 1973) have been added as scavengers; however the formation of by-products cannot be entirely prevented (Ohno *et al.*, 1972a,b). In a large comparative investigation of scavengers, 1,2-ethandithiol was recommended to be the best additive to prevent *tert*-butylation of tryptophan (Lundt *et al.* 1978). Also, cleavage of the 2-nitrophenylsulfenyl group by HCl in dioxane is accompanied by nearly quantitative formation of *S*-(2-nitrophenyl)thio-tryptophan derivatives. This side reaction may be avoided by addition of suitable indole derivatives such as indole, *N*-methylindole, or 2- or 3-methylindole in excess (Wünsch and Drees, 1967; Wünsch *et al.*, 1967).

Despite these side reactions, few investigations have been concerned with N^{in}-protecting groups.

A. Formyl Group (For)

Previero *et al.* (1967) reported that tryptophan and tryptophan-containing peptides react in HCl–formic acid to yield, specifically, the corresponding N^{i}-formyltryptophan derivatives, which can subsequently be deformylated in aqueous solution at pH 9–10.

Ohno *et al.* (1972a,b) applied this observation to solid-phase peptide synthesis, eliminating the *tert*-butyloxycarbonyl groups by treatment with HCl in formic acid and removing the N^{in}-formyl group, which was formed during this procedure, by 0.1 *N* aqueous piperidine. This reaction sequence gave better results than direct deprotection with HCl–AcOH–mercapto-ethanol as the deblocking reagent.

Boc-Trp(For)-OH was independently synthesized by Ohno *et al.* (1973) and by Yamashiro and Li (1973) and utilized in solid-phase peptide synthesis. Trp(For) has been shown to be stable under both acidic and basic conditions during solid-phase synthesis, including treatment with HF. Depending on the deprotection conditions, the formation of N^{α}-formyl peptides has been detected after removal of the N^{in}-formyl group (see Table XIX). Possibly, hydrazine in dimethylformamide, which also removes the N^{i}-formyl group (Ohno *et al.*, 1972a, 1973), would minimize this adverse reaction. Löw and Kisfaludy (1979) applied the N^{in}-formyl group in conventional peptide synthesis and found that this protecting group prevents the *tert*-butylation of tryptophan residues during acidolytic removal of *tert*-butyloxycarbonyl groups and allows the preparation of the tryptophan *tert*-butyl ester. However, the formation of 2,3-dihydroindol derivatives during catalytic hydrogenation of N^{in}-formyltryptophan derivatives and the susceptibility to saponification and hydrazinolysis limits the application of the N^{in}-formyl group in conventional peptide synthesis.

Table XIX Deformylation of H-Gly-Ala-Arg-Gly-Ala-Trp(For)-Gly-OH

Deprotection conditions	Formation of the side product For-Gly-Ala-Arg-Gly-Ala-Trp-Gly-OH (approx %)
0.01 *M* NH_4HCO_3 buffer, pH 9	50
1 *M* NH_4HCO_3 buffer, pH 9	8.5
Liquid NH_3	19
Liquid NH_3 + hydroxylamine	5

B. Benzyloxycarbonyl Group (Z)

The widely utilized benzyloxycarbonyl group has been introduced into the side chain of a N^{α}-protected tryptophan derivative by Chorev and Klausner (1976) using benzyl *p*-nitrophenyl carbonate in the presence of unsolvated fluoride anions generated by crown ethers. The compound is stable to 4 *N* HCl in ethyl acetate, 0.12 *N* HCl in formic acid, and short exposure to trifluoroacetic acid. It can be removed by catalytic hydrogenation, hydrazine, or liquid HF. Recently the 2,4-dichlorobenzyloxycarbonyl group, which is more resistant to acids, has also been employed for protection of the indole ring (Klausner and Chorev, 1977).

C. *tert*-Butyloxycarbonyl Group (Boc)

So far, only 3-formylindole, a component of the "four component condensation" method, has been protected with the *tert*-butyloxycarbonyl group, which proved to be more advantageous than the corresponding N^{in}-formyl compound (Wackerle and Ugi, 1975). In conclusion, the lability of tryptophan against both oxidation and cations is reduced by N^{in}-protection. But only a few experimental data are available concerning blocking the indole moiety of tryptophan. Thus, it is still too early to pass judgment on the value of this technique.

XI. CONCLUDING REMARKS

Most of the N-protecting groups that have been described in the preceeding sections originate from only a few basic species. A main guideline for principles concerning the selection of protecting group is the cleavage condition to which a given peptides proved to be sufficiently stable. Special properties of the amino acids involved have to be considered as well as maximal selectivity of deblocking conditions for groups serving as easily reversible N^{α}-protection and as permanent protection of functional amino acid side chains. The goal of devising ideal protection, meeting all requirements of peptide synthesis, has not been realized so far. However, the remarkable number of different types and sophisticated modifications yield a respectable pool of available methods. Tables XX and XXI give a general survey of feasible combinations, considering a series of residues and their cleavage conditions, as commonly used. Of course, such tabulations cannot express the many differentiated properties of protecting groups that are encountered in practice.

Because of the stability of a great number of peptides and proteins against acidolysis in aqueous or anhydrous medium, much attention has been devoted to acid-labile protecting groups of the urethane type. Thus, highly differentiated acid lability becomes the most frequently employed principle of cleavage, complemented by hydrogenation and, more recently, by β-elimination.

Special, distinct methods are only rarely applied. For reasons of economy, the rule is to manage peptide synthesis by means of as few protecting residues as possible. Some important strategies have already been outlined in Table X, Section IV. For the classical fragment condensation technique, the "Schwyzer (1963) Strategy" is predominantly used, employing *tert*-butyl-type and comparable groups for side-chain and benzyl-type groups for N^{α}-protection. Hydrogenolysis of benzyloxycarbonyl groups is performed mainly with shorter peptides. The presence of sulfur-containing amino acids is no longer an absolute restriction, since hydrogenation of even methionine-containing peptides is possible in the presence of bases like cyclohexylamine and increased catalyst (Medzihradszky and Medzihradszky-Schweiger, 1965). However, no more than one or two such steps should be scheduled during the synthesis of a peptide. Hydrogenation can be carried out even with peptide chains including *S*-benzylcysteine if liquid ammonia is used as a solvent (Meienhofer and Kuromizu, 1974).

In order to synthesize longer peptides, it may become necessary to switch from benzyl-type to highly acid-labile groups (e.g., Bpoc, Ddz, Trt), which, of course, can also be introduced in the beginning, if hydrogenation is less advisable.

A reverse strategy is preferred in solid-phase peptide synthesis: here the benzyloxycarbonyl or 2-chlorobenzyloxycarbonyl groups, showing enhanced stability towards acids (Erickson and Merrifield, 1974), are typical side-chain protecting groups, whereas N^{α}-protection is provided by the *tert*-butyloxycarbonyl group. Yet, there is a growing tendency to use more acid-labile pairs of groups such as Boc combined with Bpoc, since hydrogen fluoride, fluorosulfonic acid, etc. were found to cause severe damage in long peptide chains during cleavage of benzyl-type and related residues. Thus, for instance, a 50% loss of activity was observed after treatment of reduced bovine trypsin inhibitor with HF (Rocchi, 1976).

Furthermore, the inferior reactivity of *tert*-butyl cations compared to benzyl cations has to be taken into account. On the whole, the formation of carbonium ions during deprotection of peptides remains a serious obstacle, since trapping by anisole or mercaptoethanol, for example, cannot totally inhibit all cation-induced side reactions. Attempts to substitute cation-forming groups by safer residues are still at the very beginning. Problems encountered with side-chain nitrogen functions as well as susceptibilities of

Table XX Cleavage Conditions for Amine-Protecting Groups

Protecting group(s)	Abbrev	HCl in trifluoroethanol, pH 4	2–5% TFA in CH_2Cl_2 or dilute AcOH	Cold TFA or dilute CH_3SO_3H	HBr–AcOH	BTFA
Triphenylmethyl (Trityl)	Trt	+	+	+	+	+
2-(4-Biphenylyl)propyl(2)oxycarbonyl	Bpoc	−	+	+	+	+
2-(3,5-Dimethyloxyphenyl)propyl(2)oxycarbonyl	Ddz	−	+	+	+	+
tert-Butyloxycarbonyl, Adamantyloxycarbonyl, *tert*-Amyloxycarbonyl, Isobornyloxycarbonyl, Diphenylphosphinyl	Boc, Adc, Aoc, Ibc, Dpp	−	−[b]	+	+	+
4-Methyloxybenzyloxycarbonyl	Moz	−	−[b]	+	+	+
Diphenylmethyloxycarbonyl	Doc	−	−	+	+	+
Di-2-pyridylmethyloxycarbonyl	Dpc	−	−	−	−	
Benzyloxycarbonyl	Z	−	−	−[b]	+	+
2-, 4-Chlorobenzyloxycarbonyl	2 Cz, 4 Cz	−	−	−	+	+
4-Nitro-, 4-Phenylazo- benzyloxycarbonyl	4 Nz, Paz	−	−	−	(−)	
Dihydroxyborylbenzyloxycarbonyl	Dobz	−	−	−[b]	+	+
γ-Picolyl (Isonicotinyl) oxycarbonyl	Inc	−	−	−	−	−
2-Iodo-, 2,2,2-Trichloro- ethyl oxycarbonyl	Iec, Tcc	−	−	−	−	
Methylsulfonylethyloxycarbonyl	Msc	−	−	−	−	−
Formyl	For	−	−	−	−	−
Trifluoroacetyl	Tfa	−	−	−	−	−
Phthaloyl	Pht	−	−	−	−	−
4-Toluenesulfonyl (Tosyl)	Tos	−	−	−	−[b]	−
2-Nitrophenylsulfenyl	Nps		(+)	−	+	+

[a] Symbols: + = cleavage; (+) = cleavable, but not recommended; ± = uncertain; (−) = insufficiently stable; − = stable/sufficiently stable; sr = side reaction(s); BFTA = boron tris(trifluoroacetate).
[b] Unstable under stronger cleavage conditions.
[c] Cleavable by H_2O_2.
[d] Nucleophiles in weak acid and indole.

Table XXI Cleavage Conditions for Amide, Guanido, and Imidazole Protecting Groups[a]

Protecting group	Abbrev	HCl in trifluoroethanol, pH 4	2–5% TFA in CH_2Cl_2 or dilute AcOH	Cold TFA or dilute CH_3SO_3H	HBr–AcOH	BTFA	CH_3SO_3H
Amide:							
4,4′-Dimethyloxybenzhydryl	Ddm	−	(−)	+	(+)		
2,4-Dimethyloxybenzyl	Dmob	±		+			
Guanido:							
Nitro	NO_2	−	−	−	−	(+)	(+)
4-Toluenesulfonyl (Tosyl)	Tos	−	−	−	−	+	(+)
4-Methyloxybenzenesulfonyl	Mbs	−	−	−	−	+	+
Bisbenzyloxycarbonyl	Z_2	−	−	−	(+)	+	+
Bisadamantyloxycarbonyl	Adc_2	−	−[b]	+	+	+	+
Imidazole:							
Benzyl	Bzl	−	−	−	−	−	−
Triphenylmethyl (Trityl)	Trt	−	−	(+)	+		
2,4-Dinitrophenyl	Dnp	−	−	−	−		
Benzyloxycarbonyl	Z	−	−	−	+	+	+
tert-Butyloxycarbonyl	Boc	−	−	+	+	+	+
4-Toluenesulfonyl (Tosyl)	Tos	−	−	(−)	(−)		(+)

[a] Symbols explained in Table XX.
[b] Unstable under stronger cleavage conditions.
[c] Cleavage by formic acid.
[d] Cleavage by thiolysis.
[e] Cleavage by boiling AcOH.

CH_3SO_3H	CF_3SO_3H	FSO_3H	HF	H_2–Pd	H_2–Pd in liq NH_3	Zn in dilute AcOH	OH^-	N_2H_4	Na in liq NH_3	*hv*	e^-	Special conditions
+	+	+	+	+	+	−	−	−	+	−		
+	+	+	+	+	+	−	−	−	+	−		
+	+	+	+	−	+?	−	−	−	+	+		
+	+	+	+	−	−	−	−	−	−	−	−	
+	+	+	+	+	+	−	−	−	+			
+	+	+	+	+	+	−	−	−	+	−		
				+	+	+	−	−	+	−	+	
+	+	+	+	+	+	−	−	−	+	−	−	
+	+	+	+	+	+	−	−	−	+	−		
		+	+	+	+	sr			+		sr	
+	+	+	+	+	+	−	−	−	+	−		[c]
−				+	+	+	−	−	+	−	+	
−				(−)	(−)	+	sr	(−)	(−)		+	
−				−		−	+	(+)		−	−	
−	−	−		±			−	+		−		[c]
−	−	−	−	−		−	+	+	(+)	−	−	
−	−	−	−	−	sr?	−	sr	+	sr?	−	−	
−	−	−	−	−	−	−	−	−	+	−	+	
+	+	+	+			(−)	(−)	(+)				[d]

CF_3SO_3H	FSO_3H	HF	H_2–Pd	H_2–Pd in liq NH_3	Zn in dilute AcOH	OH^-	N_2H_4	Na in liq NH_3	*hv*	e^-	Special conditions
+	+	+	−		−	−	−	(−)			
+	+	+	−		−	−	−	(−)			
(+)	+	+	+	+	+	−	sr	(+)		+	
+	+	+	−	−	−	−	−	+	−	+?	
+	+	+	−	−	−	−	−	+		+?	
+	+	+	+	+	−	sr	sr	+	−		
+	+	+	−	−	−	sr	sr	(−)	−	−	
−	−	−	+	+	−	−	−	+	−	−	
			+	+		−	−	+	−	−	[c]
				sr	sr	(+)	+				[d]
+	+	+	+	+		+	+	+			[e]
+	+	+	−	−	−	+	+		−	−	
+	+	+	−			+	+		−		

sensitive amino acid residues under deblocking conditions have been discussed in the corresponding sections (see Table XVI). The danger of racemization within the peptide chain at alkaline pH has also been mentioned. However, hitherto existing results with groups such as the methylsulfonylethyloxycarbonyl (Msc) or 9-fluorenylmethyloxycarbonyl (Fmoc) group which are cleaved in alkaline medium by β-elimination, gave no evidence of racemization.

The favorable properties of the 9-fluorenylmethyloxycarbonyl group led to a novel strategy to use this residue for N^{α}-protection and *tert*-butyl-type groups for side-chain protection in solid-phase synthesis (Chang and Meienhofer, 1978; Atherton *et al.*, 1978).

The low solubility of many protected peptides is an additional problem for syntheses in solution. Sometimes it can be overcome by employing such exotic solvents as hexafluoroisopropanol. Solubilization of protecting groups, however, is obviously a superior tool. Investigations aimed at devising such residues deserve great attention. The conditions for the major reaction, as well as side reactions, during the introduction and elimination of protecting groups cover a wide range and can frequently not be generalized or predicted.

Readers seeking further knowledge and insight are well advised to study the papers presented at both the American and European peptide symposia which appear in print as the "Peptides" series.

Acknowledgment

We wish to thank the colleagues who provided us with reprints, manuscripts, and valuable comments. We are also greatly indebted to Dr. R. Obermeier for many discussions and much advice during the preparation of this chapter as well as to Dr. V. Teetz for reading the manuscript.

REFERENCES

Abdipranoto, A., Hope, A. P., and Halpern, B. (1977). *Aust. J. Chem.* **30**, 2711–2715.

Akabori, S., Okawa, K., and Sakiyama, F. (1958). *Nature (London)* **181**, 772–773.

Akabori, S., Sakakibara, S., and Shimonishi, Y. (1961). *Bull. Chem. Soc. Jpn.* **35**, 1966–1970.

Alakhov, Yu. B., Kiryushkin, A. A., and Lipkin, V. M. (1970). *J. Chem. Soc. Chem. Commun.* pp. 406–407.

Amiard, G., Heymès, R., and Velluz, L. (1955a). *Bull. Soc. Chim. Fr.* pp. 191–193.

Amiard, G., Heymès, R., and Velluz, L. (1955b). *Bull. Soc. Chim. Fr.* pp. 1464–1465.

Anderson, G. W. (1953). *J. Am. Chem. Soc.* **75**, 6081.

Anderson, G. W., and McGregor, A. C. (1957). *J. Am. Chem. Soc.* **79**, 6180–6183.

Andreatta, R., and Hofmann, K. (1968). *J. Am. Chem. Soc.* **90**, 7334–7338.

Arnold, W. H., White, W., and Flouret, G. (1973). *J. Med. Chem.* **16**, 1054–1055.

Arnstein, H. R. V., and Morris, D. (1960). *Biochem. J.* **76**, 318–323.

Arold, H., and Reissmann, S. (1968). *Z. Chem.* **8**, 107.

Arshady, R., Atherton, E., and Sheppard, R. C. (1979). *Tetrahedron Lett.*, pp. 1521–1524.
Atherton, E., Fox, H., Harkiss, D., Logan, C. J., and Sheppard, R. C. (1978). *J. Chem. Soc. Chem. Commun.*, pp. 537–540.
Atherton, E., Bury, C., Sheppard, R. C., and Williams, B. J. (1979). *Tetrahedron Lett.*, pp. 3041–3042.
Baba, T., Sugiyama, H., and Seto, S. (1973). *Chem. Pharm. Bull.* **21**, 207–209.
Bailey, W. J., and Griffith, D. R. (1964). 147th Ntl. Meeting, *Am. Chem. Soc.*, p. 15Q.
Bajusz, S., and Fauszt, I. (1972). *Acta Chim. Acad. Sci. Hung.* **75**, 419–422.
Bajusz, S., and Medzihradszky, K. (1963). *In* "Peptides 1962" (G. T. Young, ed.), pp. 49–52. Pergamon, Oxford.
Bajusz, S., Turán, A., Fauszt, I., and Juhász, A. (1973). *In* "Peptides 1972" (H. Hanson and H. D. Jakubke, eds.), pp. 93–96. North-Holland Publ., Amsterdam.
Barany, G., and Merrifield, R. B. (1977). *J. Am. Chem. Soc.* **99**, 7363–7365.
Battersby, A. R., and Robinson, J. C. (1955). *J. Chem. Soc.*, pp. 259–269.
Bell, J. R., and Jones, J. H. (1974), *J. Chem. Soc. Perkin I*, pp. 2336–2339.
Benisek, W. F., and Cole, R. D. (1965). *Biochem. Biophys. Res. Commun.* **20**, 655–660.
Benisek, W. F., Raftery, M. A., and Cole, R. D. (1967). *Biochemistry* **6**, 3780–3790.
Ben Ishai, D., and Berger, A. (1952). *J. Org. Chem.* **17**, 1564–1570.
Bentley, P. M., and Morley, J. S. (1966). *J. Chem. Soc. C*, pp. 60–64.
Bergmann, M., and Zervas, L. (1928). *Hoppe Seyler's Z. Physiol. Chem.* **175**, 145–153.
Bergmann, M., and Zervas, L. (1932). *Ber. Dtsch. Chem. Ges.* **65**, 1192–1201.
Bergmann, M., Ennslin, H., and Zervas, L. (1925). *Ber. Dtsch. Chem. Ges.* **58**, 1034–1043.
Bergmann, M., Zervas, L., and Rinke, H. (1934). *Hoppe Seyler's Z. Physiol. Chem.* **224**, 40–44.
Berse, C., Piché, L., and Uchiyama, A. (1960). *Can. J. Chem.* **38**, 1946–1950.
Beyerman, H. C., Hirt, J., Kranenburg, P., Syrier, J. L. M., and van Zon, A. (1974). *Recl. Trav. Chim. Pays-Bas* **93**, 256–257.
Bezas, B., and Zervas, L. (1961). *J. Am. Chem. Soc.* **83**, 719–722.
Birr, C., Lochinger, W., Stahnke, C., and Lang, P. (1972). *Justus Liebig's Ann. Chem.* **763**, 162–172.
Bláha, K., and Rudinger, J. (1965). *Collect. Czech. Chem. Commun.* **30**, 599–604.
Blake, J., and Li, C. H. (1968). *J. Am. Chem. Soc.* **90**, 5882–5884.
Bodanszky, M. (1960). *Ann. N. Y. Acad. Sci.* **88**, 655–664.
Bodanszky, M., and du Vigneaud, V. (1959). *J. Am. Chem. Soc.* **81**, 5688–5691.
Bodanszky, M., and Sheehan, J. T. (1960). *Chem. Ind.* (*London*), pp. 1268–1269.
Bodanszky, M., Klausner, Y. S., and Bodanszky, A. (1975). *J. Org. Chem.* **40**, 1507–1508.
Bodanszky, M., Deshmane, S. S., and Martinez, J. (1979). *J. Org. Chem.* **44**, 1622–1625.
Bodanszky, M. Ondetti, M. A., Birkhimer, C. A., and Thomas, P. L. (1964). *J. Am. Chem. Soc.* **86**, 4452–4459.
Boissonnas, R. A., and Preitner, G. (1953). *Helv. Chim. Acta* **36**, 875–886.
Boissonnas, R. A., Guttmann, S., Huguenin, R. L., Jaquenoud, P.-A., and Sandrin, E. (1958). *Helv. Chim. Acta* **41**, 1867–1882.
Bosshard, H. R. (1971). *Helv. Chim. Acta* **54**, 951–958.
Bovarnick, M., and Clarke, H. T. (1938). *J. Am. Chem. Soc.* **60**, 2426–2430.
Brady, S. F., Hirschmann, R., and Veber F. (1977). *J. Org. Chem.* **42**, 143–146.
Brenner, M., and Pfister, R. W. (1951). *Helv. Chim. Acta* **34**, 2085–2096.
Broadbent, W., Morley, J. S., and Stone, B. E. (1967). *Chem. Soc. C*, pp. 2632–2636.
Brugger, M. (1965). Dissertation, Zürich.
Büchi, H. (1965). Dissertation, Zürich.
Busse, W. D., Hansen, S. R., and Carpenter, F. H. (1974). *J. Am. Chem. Soc.* **96**, 5949–5950.
Callahan, F. M., Anderson, G. W., Paul, R., and Zimmermann, J. E. (1963). *J. Am. Chem. Soc.* **85**, 201–207.

Camble, R., Dupuis, G., Kawasaki, K., Romovacek, H., Yanaihara, N., and Hofmann, K. (1972). *J. Am. Chem. Soc.* **94**, 2091–2100.
Carpenter, F. H., and Gish, D. T. (1952). *J. Am. Chem. Soc.* **74**, 3818–3821.
Carpino, L. A. (1966). *Chem. Commun.*, pp. 858–859.
Carpino, L. A., and Han, G. Y. (1970). *J. Am. Chem. Soc.* **92**, 5748–5749.
Carpino, L. A,, Williams, J. R., and Lopusinski, A. (1978). *J. Chem. Soc., Chem. Commun.*, pp. 450–451.
Chamberlin, J. W. (1966). *J. Org. Chem.* **31**, 1658–1660.
Chang, C.-D., and Meienhofer, J. (1978). *Int. J. Pept. Protein Res.* **11**, 246–249.
Channing, D. M., Turner, P. B., and Young, G. T. (1951). *Nature (London)* **167**, 487–488.
Chillemi, F., and Merrifield, R. B. (1969). *Biochemistry* **8**, 4344–4346.
Chillemi, F., Bernardi, L., and Bosisio, G. (1964). *Gazz. Chim. Ital.* **94**, 891–901.
Chino, N., Masui, Y., and Sakakibara, S. (1978). *In* "Peptide Chemistry, 1977" (T. Shiba ed.), pp. 27–32, Protein Research Foundation, Osaka (Japan).
Chorev, M., and Klausner, Y. S. (1976). *J. Chem. Soc., Chem. Commun.* p. 596.
Clubb, M. E., Scopes, P. M., and Young, G. T. (1960). *Chimia* **14**, 373.
Coy, D. H., Coy, E. J., Hirotsu, Y., and Schally, A. V. (1974). *J. Med. Chem.* **17**, 140–142.
Coy, D. H., Yoshihiro, H., Redding, T. W., Coy, E. J., and Schally, A. V. (1975). *J. Med. Chem.* **18**, 948–949.
Coyle, S., and Young, G. T. (1976a). *In* "Peptides 1976" (A. Loffet, ed.), pp. 205–208. Editions de l'Université de Bruxelles, Belgium.
Coyle, S., and Young, G. T. (1976b). *J. Chem. Soc., Chem. Commun.*, pp. 980–981.
Coyle, S., Keller, O., and Young, G. T. (1975). *J. Chem. Soc., Chem. Commun.*, pp. 939–940.
Cuiban, F. (1971). *Tetrahedron Lett.*, pp. 2471–2472.
Dane, E., and Dockner, T. (1964). *Angew. Chem.* **76**, 342.
Dane, E., and Dockner, T. (1965). *Chem. Ber.* **98**, 789–796.
Dane, E., Drees, F., and Dockner, T. (1962). *Angew. Chem., Int. Ed. Engl.* **1**, 658.
D'Angeli, F., Filiva, F., and Scoffone, E. (1965). *Tetrahedron Lett.*, pp. 605–608.
Di Bello, C., Filiva, F., and D'Angeli, F. (1971). *J. Org. Chem.* **36**, 1818–1820.
Dorman, L. C., Nelson, D. A., and Chow, R. C. L. (1972). *Prog. Pept. Res.* **2**, 65–68.
du Vigneand, V., and Behrens, O. K. (1937). *J. Biol. Chem.* **117**, 27–36.
Eberle, A., Fauchère, J. L., Tesser, G. I., and Schwyzer, R. (1975). *Helv. Chim. Acta* **58**, 2106–2129.
Eckert, H., and Ugi, I. (1976). *Angew. Chem., Int. Ed. Engl.* **15**, 681.
Eckstein, H. (1976). *Justus Liebig's Ann. Chem.* pp. 1289–1294.
Eckstein, H., Sievers, R. E., and Bayer, E. (1973). *Justus Liebig's Ann. Chem.* pp. 1467–1470.
Elliott D. F., and Morris, D. (1960). *Chimia* **14**, 373–374.
Entwistle, I. D. (1979). *Tetrahedron Lett.* pp. 555–558.
Erickson, B. W., and Merrifield, R. B. (1972). *J. Am. Chem. Soc.* **95**, 3750–3763.
Erickson, B. W., and Merrifield, R. B. (1974). *Isr. J. Chem.* **12**, 79–85.
Feinberg, R. S., and Merrifield, R. B. (1975). *J. Am. Chem. Soc.* **97**, 3485–3496.
Felix, A. M. (1978). *J. Org. Chem.* **39**, 1427–1429.
Felix, A. M., Heimer, E. P., Lambros, T. J., Tzougraki, C., and Meienhofer, J. (1978). *J. Org. Chem.* **43**, 4194–4196.
Finn, F. M., and Hofmann, K. (1967). *J. Am. Chem. Soc.* **89**, 5298–5300.
Fischer, E. (1915). *Ber. Dtsch. Chem. Ges.* **48**, 93–102.
Fischer, E., and Otto, E. (1903). *Ber. Dtsch. Chem. Ges.* **36**, 2106–2116.
Fischer, E., and Warburg, O. (1905). *Ber. Dtsch. Chem. Ges.* **38**, 3997–4005.
Fletcher, A. R., Jones, J. H., Ramage, W. I., and Stachulski, A. V. (1979). *In* "Peptides 1978" (I. Z. Siemion and G. Kupryszewski, eds.) pp. 168–171, Wroclaw University Press, Wroclaw, Poland.

Flouret, G. R., Arnold, W. H., Cole, J. W., Morgan, R. L., White, W. F., Hedlund, M. T., and Rippel, R. H. (1973). *J. Med. Chem.* **16**, 369–373.
Fontana, A., Marchiori, F., Moroder, L., and Scoffone, E. (1966). *Tetrahedron Lett.*, pp. 2985–2987.
Fosse, M. R., and Haller, M. A. (1907). *C. R. Hebd. Seances Acad. Sci.* **145**, 813–815.
Frankel, M., Ladkany, D., Gilon, C., and Wolman, Y. (1966). *Tetrahedron Lett.*, pp. 4765–4768.
Fridkin, M., and Goren, H. J. (1971). *Can. J. Chem.* **49**, 1578–1581.
Fridkin, M., and Shaltiel, S. (1971). *Arch. Biochem. Biophys.* **147**, 767–771.
Fruton, J. S. (1949). *Adv. Protein Chem.* **5**, 1–82.
Fujii, T., and Sakakibara, S. (1974). *Bull. Chem. Soc. Jpn.* **47**, 3146–3151.
Fujii, T., Kimura, T., and Sakakibara, S. (1976). *Bull. Chem. Soc. Jpn.* **49**, 1595–1601.
Fujino, M., and Hatanaka, C. (1967). *Chem. Pharm. Bull.* **15**, 2015–2016.
Fujino, M., and Shinagawa, S. (1972). *Chem. Pharm. Bull.* **20**, 1021–2025.
Fujino, M., Shinagawa, S., Nishimura, O., and Fukuda, T. (1972). *Chem. Pharm. Bull.* **20**, 1017–1025.
Fujino, M., Fukuda, T., Kobayashi, S., and Obayashi, M. (1973). *Chem. Pharm. Bull.* **21**, 87–91.
Fukuda, T., and Fujino, M. (1978). *In* "Peptide Chemistry 1977" (T. Shiba ed.), pp. 19–22, Protein Research Foundation, Osaka (Japan).
Fukuda, T., Kitada, C. and Fujino, M. (1978). *J. Chem. Soc. Chem. Commun.*, pp. 220–221.
Gante, J. (1966). *Fortschr. Chem. Forsch.* **6**, 358–381.
Gazis, E., Bezas, B., Stelakatos, G. C., and Zervas, L. (1963). *In* "Peptides 1962" (G. T. Young, ed.), pp. 17–21. Pergamon, Oxford.
Geiger, R., and Langner, D. (1975). *In* "Peptides 1974" (Y. Wolman, ed.), pp. 159–163, Wiley, New York.
Geiger, R., and Obermeier, R. (1974/77) U.S. Pat. 4,014,861.
Geiger, R., and Siedel, W. (1968). *Chem. Ber.* **101**, 3386–3391.
Geiger, R., Obermeier, R., and Tesser, G. I. (1975). *Chem. Ber.* **108**, 2758–2763.
Geiger, R., Sturm, K., and Siedel, W. (1964). *Chem. Ber.* **97**, 1207–1213.
Geiger, R., König, W., Wissmann, H., Geisen, K., and Enzmann, F. (1971). *Biochem. Biophys. Res. Commun.* **45**, 767–773.
Gerngross, O. (1919). *Hoppe Seyler's Z. Physiol. Chem.* **108**, 50–63.
Gillessen, D., Piva, F., Steiner, H., and Studer, R. O. (1971). *Helv. Chim. Acta* **54**, 1335–1342.
Gish, D. T., and Carpenter, F. H. (1953). *J. Am. Chem. Soc.* **75**, 5872–5877.
Gish, D. T., Katsoyannis, P. G., Hess, G. P., and Stedman, R. J. (1956). *J. Am. Chem. Soc.* **78**, 5954.
Glass, J. D., Schwartz, J. L., and Walter, R. (1972). *J. Am. Chem. Soc.* **94**, 6209–6211.
Goerdeler, J., and Holst, A. (1959). *Angew. Chem.* **71**, 775.
Grimshaw, J. (1965). *J. Chem. Soc.*, pp. 7136–7141.
Grønvald, F. C., Johansen, N. L., and Lundt, B. F. (1979). *In* "Peptides, Structure and Biological Function" (E. Gross and J. Meienhofer, eds.) pp. 309–312. Pierce Chemical Co. Rockford, Illinois.
Gros, C., de Garilhe, P., Costopanagiotis, A., and Schwyzer, R. (1961). *Helv. Chim. Acta* **44**, 2024–2048.
Gross, E., and Witkop, B. (1962). *J. Biol. Chem.* **237**, 1856–1860.
Grossberg, A. L., and Pressman, D. (1968). *Biochemistry* **7**, 272–279.
Gutte, B., and Merrifield, R. B. (1969). *J. Am. Chem. Soc.* **91**, 501–502.
Guttmann, S. (1963). *In* "Peptides 1962" (G. T. Young, ed.), pp. 41–47 Pergamon, Oxford.
Guttmann, S., and Pless, J. (1965). *Acta Chim. Acad. Sci. Hung.* **44**, 23–30.
Haas, W. L., Krumkalns, E. V., and Gerzon, K. (1966). *J. Am. Chem. Soc.* **88**, 1988–1992.
Habeeb, A. F. S. A., and Bennet, J. C. (1971). *Biochim. Biophys. Acta* **251**, 181–184.
Halpern, B. (1965). *Aust. J. Chem.* **18**, 417–421.

Halpern, B., and James, L. B. (1964a). *Nature* (*London*) **202**, 592–593.

Halpern, B., and James, L. B. (1964b). *Aust. J. Chem.* **17**, 1282–1287.

Hanby, W. E., Waley, S. G., and Watson, J. (1950). *J. Chem. Soc.*, pp. 3239–3249.

Hardy, P. M., Rydon, H. N., and Thompson, R. C. (1972). *J. Chem. Soc.*, *Perkin Trans. I* pp. 5–12.

Hayakawa, T., Fujiwara, Y., and Noguchi, J. (1967). *Bull. Chem. Soc. Jpn.* **40**, 1205–1208.

Helferich, B., Moog, L., and Jünger, A. (1925). *Ber. Dtsch. Chem. Ges.* **58**, 872–886.

Hillmann-Elies, A., Hillmann, G., and Jatzkewitz, J. (1953). *Z. Naturforsch.*, *Teil B* **8**, 445–446.

Hirschmann, R. (1971). *Intra-Sci. Chem. Rep.* **5**, 203–228.

Hirschmann, R., Nutt, R. F., Veber, D. F., Vitali, R. A., Varga, S. L., Jacob, T. A., Holly, F. W., and Denkewalter, R. G. (1969). *J. Am. Chem. Soc.* **91**, 507–508.

Hofmann, K., and Bohn, H. (1966). *J. Am. Chem. Soc.* **88**, 5914–5919.

Hofmann, K., and Bowers, C. Y. (1970). *J. Med. Chem.* **13**, 1099–1101.

Hofmann, K., and Yajima, H. (1961). *J. Am. Chem. Soc.* **83**, 2289–2293.

Hofmann, K., Lindenmann, A., Magee, M. Z., and Haokhan, N. (1952). *J. Am. Chem. Soc.* **74**, 470–476.

Hofmann, K., Stutz, E., Spühler, G., Yajima, H., and Schwartz, E. T. (1960). *J. Am. Chem. Soc.* **82**, 3727–3732.

Hofmann, K., Yajima, H., Liu, T. J., Yanaihara, N., and Lande, S. (1961a). *J. Am. Chem. Soc.* **83**, 487–489.

Hofmann, K., Liu, T. Y., Yajima, H., Yanaihara, N., and Lande, S. (1961b). *J. Am. Chem. Soc.* **83**, 2294–2299.

Hofmann, K., Haas, W., Smithers, M. J., Wells, R. D., Wolman, Y., Yanaihara, N., and Zanetti, G. (1965). *J. Am. Chem. Soc.* **87**, 620–631.

Hofmann, K., Bohn, H., and Andreatta, R. (1967). *J. Am. Chem. Soc.* **89**, 7126–7127.

Hofmann, K., Andreatta, R., Buckley, J. P., Hageman, W. E., and Shapiro, A. P. (1968a). *J. Am. Chem. Soc.* **90**, 1654–1655.

Hofmann, K., Andreatta, R., and Bohn, H. (1968b). *J. Am. Chem. Soc.* **90**, 6207–6212.

Holley, R. W., and Holley, A. D. (1952). *J. Am. Chem. Soc.* **74**, 3069–3074.

Holley, R. W., and Sondheimer, E. (1954). *J. Am. Chem. Soc.* **76**, 1326–1328.

Honzl, J., and Rudinger, J. (1961). *Collect. Czech. Chem. Commun.* **26**, 2333–2344.

Hope, A. P. and Halpern, B. (1976). *Australian J. Chem.* **29**, 1591–1603.

Hope, D. B. (1968). *In* "Peptides 1968" (E. Bricas, ed.), pp. 111–117 North-Holland Publ., Amsterdam.

Horner, L., and Neumann, H. (1965). *Chem. Ber.* **98**, 3462–3469.

Hruby, V. J., Muscio, F. A., Brown, W., and Gitu, P. M. (1972). *In* Chemistry and Biology of Peptides" (J. Meienhofer, ed.), pp. 331–333. Ann. Arbor Sci. Publ., Ann Arbor, Mich.

Hubbuch, A., Danho, W., and Zahn, H. (1977). *In* Peptides, (M. Goodman and J. Meienhofer eds.), pp. 540–542, John Wiley and Sons, New York.

Ikeda, S., Tonegawa, F., Shikano, E., and Shinozaki, K. (1979). *Bull. Chem. Soc. Jap.* **52**, 1431–1436.

Iselin, B. M. (1966). *In* "Peptides 1963" (L. Zervas, ed.), pp. 27–33. Pergamon, Oxford.

Itano, H. A., and Gottlieb, A. J. (1963). *Biochem. Biophys. Res. Commun.* **12**, 405–408.

Ito, H., Ogawa, K., Iida, T., and Ichikizaki, I. (1978). *Chem. Pharm. Bull.* **26**, 296–299.

Itoh, M., Hagiwara, D. and Kamiya, T. (1975). *Tetrahedron Lett.* pp. 4393–4394.

Itoh, M., Hagiwara, D., and Kamiya, T. (1977). *Bull Chem. Soc. Jap.* **50**, 718–721.

Jackson, A. G., Kenner, G. W., Moore, G. A., Ramage, R., and Thorpe, W. D. (1976). *Tetrahedron Lett.* pp. 3627–3630.

Jaeger, E., Thamm, P., Knof, S., Wünsch, E. Löw, M., and Kisfaludy, L. (1978a). *Hoppe-Seyler's Z. Physiol. Chem.* **359**, 1617–1628.

Jaeger, E., Thamm, P., Knof. S., and Wünsch, E. (1978b). *Hoppe-Seyler's Z. Physiol. Chem.* **359**, 1629–1636.

Jäger, G. (1973). *Chem. Ber.* **106**, 206–210.
Jäger, G., and Geiger, R. (1970). *Chem. Ber.* **103**, 1727–1747.
Jäger, G., and Geiger, R. (1973a). *In* "Peptides 1971" (H. Nesvadba, ed.), pp. 78–81. North-Holland Publ., Amsterdam.
Jäger, G., and Geiger, R. (1973b) *Justus Liebig's Ann. Chem.* pp. 1535–1544.
Jäger, G., and Geiger, R. (1973c) *Justus Liebig's Ann. Chem.* pp. 1928–1933.
Jäger, G., Geiger, R., and Siedel, W. (1968). *Chem. Ber.* **101**, 3537–3544.
Jäger, G., König, W., Wissmann, H., and Geiger, R. (1974). *Chem. Ber.* **107**, 215–231.
Jones, J. H., and Ramage, W. I. (1978). *J. Chem. Soc. Chem. Comm.*, pp. 472–473.
Jorgensen, E. C., and Windridge, G. C. (1970). *J. Med. Chem.* **13**, 352–356.
Kader, A. T., and Stirling, C. J. M. (1964). *J. Chem. Soc.* pp. 258–266.
Kalbacher, H., and Voelter, W. (1978). *Angew. Chem. Int. Ed.* **17**, 944–945.
Kalbag, S., and Roeske, R. W. (1975). *J. Am. Chem. Soc.* **97**, 440–441.
Kappeler, H., and Schwyzer, R. (1961). *Helv. Chim. Acta* **44**, 1136–1141.
Kasafirek, E. (1972). *Tetrahedron Lett.* pp. 2021–2024.
Kashelikar, D. V., and Ressler, C. (1964). *J. Am. Chem. Soc.* **86**, 2467–2473.
Katsoyannis, P. G. (1966). *Am. J. Med.* **40**, 652–661.
Katsoyannis, P. G., Zalut, C., Tometsko, A., Tilak, M., Johnson, S., and Trakatellis, A. C. (1971). *J. Am. Chem. Soc.* **93**, 5871–5877.
Kawasaki, K., Camble, R., Dupuis, G., Romovacek, H., Storey, H. T., Yanaihara, C., and Hofmann, K. (1973). *J. Am. Chem. Soc.* **95**, 6815–6824.
Keller, O., and Rudinger, J. (1975). *Helv. Chim. Acta* **58**, 531–541.
Kemp, D. S., and Hoyng, C. F. (1975). *Tetrahedron Lett.* pp. 4625–4628.
Kemp, D. S., and Roberts, D. C. (1975). *Tetrahedron Lett.* pp. 4629–4632.
Kenner, G. W., Moore, G. A., and Ramage, R. (1976). *Tetrahedron Lett.* pp. 3623–3626.
Kessler, W., and Iselin, B. (1966). *Helv. Chim. Acta* **49**, 1330–1344.
Khosla, M. C., Smeby, R. R., and Bumpus, F. M. (1972). *In* "Chemistry and Biology of Peptides" (J. Meienhofer, ed.), pp. 227–230. Ann Arbor Sci. Publ. Ann Arbor, Michigan.
King, T. P. (1966). *Biochemistry* **5**, 3454–3459.
Kisfaludy, L., and Dualszky, S. (1960). *Acta Chim. Acad. Sci. Hung.* **24**, 301–309.
König, W., and Geiger, R. (1970a). *Chem. Ber.* **103**, 788–798.
König, W., and Geiger, R. (1970b). *Chem. Ber.* **103**, 2041–2051.
König, W., and Geiger, R. (1972). *Chem. Ber.* **105**, 2872–2882.
König, W., and Geiger, R. (1973). *Chem. Ber.* **106**, 3626–3635.
König, W., and Geiger, R. (1975). *In* Hypothalamic Hormones—Structure, Synthesis and Biological Activity" (D. Gupta and W. Voelter, eds.), pp. 93–106. Verlag Chemie, Weinheim.
König, W., and Volk, A. (1977). *Chem. Ber.* **110**, 1–11.
König, W., and Kernebeck, K. (1979). *Liebigs Ann. Chem.* pp. 227–247.
König, W., Volk, A., and Teetz, V. (1979). *In* "Peptides 1978". (I. Z. Siemion and G. Kupryszewski, eds.), pp. 619–623, Wroclaw University Press, Wroclaw, Poland.
Klausner, Y. S., and Chorev, M. (1977). *J. Chem. Soc. Perkin I*, pp. 627–631.
Koul, A. K., Bachhawat, I. M., Prashad, B., Ramegowda, N. S., Mathur, A. K., and Mathur, N. K. (1973). *Tetrahedron* **29**, 625–628.
Kovacs, J., Hsieh, Y., Holleran, E. M., and Ting, Y. F. (1979). *In* "Peptides 1978", (I. Z. Siemion and G. Kupryszewski, eds.), pp. 159–164, Wroclaw University Press, Wroclaw, Poland.
Kunz, H. (1976a). *Chem. Ber.* **109**, 2670–2683.
Kunz, H. (1976b). *Chem. Ber.* **109**, 3693–3706.
Kunz, H. (1978). *Angew. Chem. Int. Ed.* **17**, 67–68.
Künzi, H., and Studer, R. O. (1975). *Helv. Chim. Acta* **58**, 139–146.
Künzi, H., Manneberg, M., and Studer, R. O. (1974). *Helv. Chim. Acta* **57**, 566–572.
Kurtz, A. C. (1941). *J. Biol. Chem.* **140**, 705–710.

Kuwata, S., and Watanabe, H. (1965). *Bull. Chem. Soc. Jpn.* **38**, 676–677.

Ledger, R., and Stewart, F. H. C. (1965). *Aust. J. Chem.* **18**, 933–935.

Le Corre, G., Guibe-Jampel, E., and Wakselman, M. (1978). *Tetrahedron* **34**, 3105–3112.

Lenard, J., and Hess, G. P. (1964). *J. Biol. Chem.* **239**, 3275–3281.

Lenard, J., Schally, A. V., and Hess, G. P. (1964). *Biochem. Biophys. Res. Commun.* **14**, 498–502.

Li, C. H., and Yamashiro, D. (1970). *J. Am. Chem. Soc.* **92**, 7608–7609.

Li, C. H., Meienhofer, J., Schnabel, E., Chung, D., Lo, T.-B., and Ramachandran, J. (1961). *J. Am. Chem. Soc.* **83**, 4449–4457.

Liberek, B. (1962). *Bull. Acad. Pol. Sci., Ser. Sci. Chim.* **10**, 227–231.

Lin, M. C., Gutte, B., Caldi, D. G., Moore, S., and Merrifield, R. B. (1972). *J. Biol. Chem.* **247**, 4768–4774.

Liu, W.-H., Feinstein, G., Osuga, D. T., Haynes, R., and Feeney, R. E. (1968). *Biochemistry* **7**, 2886–2892.

Löw, M., and Kisfaludy, L. (1979). *Hoppe-Seyler's Z. Physiol. Chem.* **360**, 13–18.

Löw, M., Kisfaludy, L., Jaeger, E., Thamm, P., Knof, S., and Wünsch, E. (1978a). *Hoppe-Seyler's Z. Physiol. Chem.* **359**, 1637–1642.

Löw, M. Kisfaludy, L. and Sohàr, P. (1978b). *Hoppe-Seyler's Z. Physiol. Chem.* **359**, 1643–1651.

Loffet, A. (1967). *Experientia* **23**, 406–407.

Losse, G., and Krychowski, J. (1970). *J. Prakt. Chem.* **312**, 1097–1104.

Losse, G., and Krychowski, J. (1971). *Tetrahedron Lett.* pp. 4121–4124.

Losse, G., and Nadolski, D. (1964). *J. Prakt. Chem.* **24**, 118–124.

Losse, G., and Rüger, C. (1973). *Z. Chem.* **13**, 344–345.

Losse, G., and Zönnchen, W. (1960). *Justus Liebig's Ann. Chem.* **636**, 140–143.

Losse, G., Jeschkeit, H., and Willenberg, E. (1964). *Angew. Chem. Int. Ed. Engl.* **3**, 307.

Lundt, B. F., Johansen, N. L., Vølund, A., and Markussen, J. (1978). *Int. J. Pept. Protein Res.* **12**, 258–268.

McDermott, J. R., and Benoiton, N. L. (1973). *Can. J. Chem.* **51**, 2555–2556.

McIntyre, F. C. (1947). *J. Am. Chem. Soc.* **69**, 1377–1381.

McKay, A. F. (1952). *Chem. Rev.* **51**, 301–346.

McKay, F. C., and Albertson, N. F. (1957). *J. Am. Chem. Soc.* **79**, 4686–4690.

Masaki, H., Kitahara, T., Kurita, H., and Ohta, M. (1968). *J. Am. Chem. Soc.* **90**, 4508–4509.

Matsuura, S., Niu, C.-H., and Cohen, J. S. (1976). *J. Chem. Soc., Chem. Commun.* pp. 451–452.

Mazur, R. H., and Plume, G. (1968). *Experientia* **24**, 661.

Medzihradszky, K., and Medzihradszky-Schweiger, H. (1965). *Acta Chim. Acad. Sci. Hung.* **44**, 15–18.

Meienhofer, J., and Kuromizu, K. (1974). *Tetrahedron Lett.* pp. 3259–3262.

Merrifield, R. B. (1963). *J. Am. Chem. Soc.* **85**, 2149–2154.

Merrifield, R. B., and Woolley, D. W. (1956). *J. Am. Chem. Soc.* **78**, 4646–4649.

Merrifield, R. B. and Bach, A. E. (1978). *J. Org. Chem.* **43**, 4808–4816.

Meyers, C., and Glass, J. D. (1975). *Proc. Natl. Acad. Sci.* **72**, 2193–2196.

Meyers, C., Havran, R. T., Schwartz, J. L., and Walter, R. (1969). *Chem. Ind. (London)* pp. 136–137.

Milne, H. B., and Peng, C.-H. (1957). *J. Am. Chem. Soc.* **79**, 639–644.

Miyamoto, M., Kawamatsu, Y., Shinohara, M., and Meno, Y. (1961). *J. Pharm. Soc. Jpn.* **81**, 439–442.

Morihara, K., Oka, T., Tsuzuki, H. Inouye, K., and Sakakibara, S. (1979). *In* "Peptides, Structure and Biological Function" (E. Gross and J. Meienhofer, eds.) pp. 617–620. Pierce Chemical Co., Rockford, Illinois.

Moroder, L., Borin, G., Marchiori, F., and Scoffone, E. (1974). *Justus Liebig's Ann. Chem.* pp. 213–224.

Moroder, L., Wackerle, L., and Wünsch, E. (1976). *Hoppe-Seyler's Z. Physiol. Chem.* **357**, 1647–1650.

Nakaya, K., Horinishi, H., and Shibata, K. (1967). *J. Biochem.* (*Tokyo*) **61**, 345–351.
Nakaya, K., Suzuki, T., Takenaka, O., and Shibata, K. (1969). *Biochim. Biophys. Acta* **194**, 301–309.
Nefkens, G. H., Tesser, G. I., and Nivard, R. J. (1960). *Rec. Trav. Chim. Pays-Bas* **79**, 688–698.
Niedrich, H., and Grupe, R. (1965). *J. Prakt. Chem.* **27**, 108–112.
Nishimura, O., and Fujino, M. (1976). *Chem. Pharm. Bull.* **24**, 1568–1575.
Ohno, M., Tsukamoto, S. and Izumiya, N. (1972a). *J. Chem. Soc., Chem. Commun.* pp. 663–664.
Ohno, M., Tsukamoto, S., Makisumi, S., and Izumiya, N. (1972b). *Bull. Chem. Soc. Jpn.* **45**, 2852–2855.
Ohno, M., Tsukamoto, S., Sato, S., and Izumiya, N. (1973). *Bull. Chem. Soc. Jpn.* **46**, 3280–3285.
Okawa, K., and Hase, S. (1963). *Bull. Chem. Soc. Jpn.* **36**, 754.
Okumura, K., Iwasaki, T., Matsuoka, M., and Matsumoto, K. (1971). *Chem. Ind.* (*London*) p. 929.
Olofson, R. A., Yamamoto, Y. S., and Wancowicz, D. J. (1977) *Tetrahedron Lett.* pp. 1563–1566.
Orlowski, R. C., and Walter, R. (1976). *J. Org. Chem.* **41**, 3701–3705.
Panetta, C. A. (1969). *J. Org. Chem.* **34**, 2773–2775.
Panetta, C. A., and Rahman, A. U. (1971). *J. Org. Chem.* **36**, 2250–2252.
Patchornik, A., Berger, A., and Katchalski, E. (1957). *J. Am. Chem. Soc.* **79**, 6416–6420.
Patchornik, A., Amit, B., and Woodward, R. B. (1970). *J. Am. Chem. Soc.* **92**, 6333–6335.
Patthy, L., and Smith, E. L. (1975). *J. Biol. Chem.* **250**, 557–564.
Paul, R., and Kende, A. S. (1964). *J. Am. Chem. Soc.* **86**, 4162–4166.
Paul, R., Anderson, G. W., and Callahan, F. M. (1961). *J. Org. Chem.* **26**, 3347–3350.
Paulay, Z., and Bajusz, S. (1965). *Acta Chim. Acad. Sci. Hung.* **43**, 147–148.
Pauly, H. (1904). *Hoppe-Seyler's Z. Physiol. Chem.* **42**, 508–518.
Photaki, I., and Yiotakis, A. (1976). *J. Chem. Soc, Perkin Trans 1* pp. 259–264.
Pietta, P. G., and Marshall, G. (1970). *Chem. Commun.* pp. 650–651.
Pietta, P. G., Cavallo, P. F., and Marshall, G. R. (1971). *J. Org. Chem.* **36**, 3966–3970.
Pietta, P. G., Cavallo, P. F., Marshall, G. R., and Pace, M. (1973). *Gazz. Chim. Ital.* **103**, 483–489.
Pietta, P. G., Cavallo, P. F., Takahashi, K., and Marshall, G. R. (1974). *J. Org. Chem.* **39**, 44–48.
Pietta, P. G., Biondi, P. A., and Brenna, O. (1976). *J. Org. Chem.* **41**, 703–704.
Pless, J., and Bauer, W. (1973). *Angew. Chem.* **85**, 142.
Pless, J., and Guttmann, S. (1967). *In* "Peptides 1966" (H. C. Beyerman, A. van de Linde, and W. Maassen van den Brink, ed.), pp. 50–54, North-Holland Publ., Amsterdam.
Pozdmev, V. F. (1978). *Zh. Obshch. Khim.* **48**, 476–477.
Preston, J., and Weinstein, B. (1967). *Experientia* **23**, 265–266.
Previero, A., Coletti-Previero, M. A., and Cavadore, J-C., (1967). *Biochim. Biophys. Acta* **147**, 453–461.
Ramachandran, J. (1965). *Nature*/(*London*) **206**, 927–928.
Ressler, C., and Kashelikar, D. V. (1966). *J. Am. Chem. Soc.* **88**, 2025–2035.
Ressler, C., and Ratzkin, H. (1961). *J. Org. Chem.* **26**, 3356–3360.
Rich, D. H., and Gurwara, S. K. (1975). *Tetrahedron Lett.* pp. 301–304.
Riniker, B., Kamber, B., and Sieber, P. (1975). *Helv. Chim. Acta* **58**, 1086–1094.
Rink, H., and Riniker, B. (1974). *Helv. Chim. Acta* **57**, 831–836.
Rivaille, P., Robinson, A., Kemen, M., and Milhaud, G. (1971). *Helv. Chim. Acta* **54**, 2772–2775.
Rocchi, R. (1976). *In* "Peptides 1976" (A. Loffet, ed.), p. 231. Editions de l'Université de Bruxelles, Belgium.
Roeske, R. (1963). *J. Org. Chem.* **28**, 1251–1253.

Rudinger, J. (1973). *In* "The Chemistry of Polypeptides" (P. G. Katsoyannis, ed.), pp. 87–123. Plenum, New York.
Rudinger, J., and Pravda, Z. (1958). *Collect. Czech. Chem. Commun.* **23**, 1947–1957.
Sakakibara, S., and Fujii, T. (1969). *Bull. Chem. Soc. Jpn.* **42**, 1466.
Sakakibara, S., Shin, K. H., and Hess, G. P. (1962). *J. Am. Chem. Soc.* **84**, 4921–4928.
Sakakibara, S., Shin, M., Fujino, M., Shimonishi, Y., Inone, S., and Inukai, N. (1965). *Bull. Chem. Soc. Jpn.* **38**, 1522–1525.
Sakakibara, S., Shimonishi, Y., Kishida, Y., Okada, M., and Sugihara, H. (1967). *Bull. Chem. Soc. Jpn.* **40**, 2164–2167.
Sakakibara, S., Fukuda, T., Kishida, Y. and Honda, I. (1970). *Bull. Chem. Soc. Jpn.* **43**, 3322.
Sakiyama, F. (1962). *Bull. Chem. Soc. Jpn.* **35**, 1943–1950.
Sakiyama, F., Okawa, K., and Yamakawa, T. (1958). *Bull. Chem. Soc. Jpn.* **31**, 926–932.
Sanger, F. (1945). *Biochem. J.* **39**, 507–515.
Sano, S., and Kawanishi, S. (1975). *J. Am. Chem. Soc.* **97**, 3480–3484.
Sarid, S., and Patchornik, A. (1963). *Isr. J. Chem.* **1**, 63–64.
Saunders, J., and Offord, R. E. (1977). *Biochem. J.* **165**, 479–486.
Savige, W. E., and Fontana, A. (1976). *J. Chem. Soc. Chem. Commun.* pp. 599–600.
Schafer, D. J., and Young, G. T. (1971). *J. Chem. Soc. C* pp. 46–49.
Schaich, E., Fretzdorff, A.-M., and Schneider, F. (1973). *Hoppe-Seyler's Z. Physiol. Chem.* **354**, 897–902.
Schallenberg, E. E., and Calvin, M. (1955). *J. Am. Chem. Soc.* **77**, 2779–2783.
Schnabel, E. (1967). *Justus Liebig's Ann. Chem.* **702**, 188–196.
Schnabel, E., and Schüssler, H. (1965). *Justus Liebig's Ann. Chem.* **686**, 229–238.
Schnabel, E., Herzog, H., Hoffmann, P., Klauke, E., and Ugi, J. (1968). *Justus Liebig's Ann. Chem.* **716**, 175–185.
Schnabel, E., Stoltefuss, J., Offe, H. A., and Klauke, E. (1971) *Justus Liebig's Ann. Chem.* **743**, 57–68.
Schneider, F. (1960a). *Hoppe-Seyler's Z. Physiol. Chem.* **320**, 82–91.
Schneider, F. (1960b). *Hoppe-Seyler's Z. Physiol. Chem.* **321**, 38.
Schröder, E., and Lübke, K. (1965). "The Peptides" Vol. 1. *Academic Press*, New York.
Schröder, E., and Lübke. K. (1966). "The Peptides," Vol. 2. *Academic Press*, New York.
Schwartz, G. P., and Katsoyannis, G. (1973). *J. Chem. Soc., Perkin Trans. 1* pp. 2894–2901.
Schwyzer, R. (1959). *Angew. Chem.* **71**, 742.
Schwyzer, R., and Li, C. H. (1958). *Nature* (*London*) **182**, 1669–1670.
Schwyzer, R., and Rittel, W. (1961). *Helv. Chim. Acta* **44**, 159–169.
Schwyzer, R., and Sieber, P. (1960). *Helv. Chim. Acta* **43**, 1910–1915.
Schwyzer, R., and Sieber, P. (1963). *Nature* (*London*) **199**, 172–174.
Schwyzer, R., Sieber, P. and Kappeler, H. (1959). *Helv. Chim. Acta* **42**, 2622–2624.
Schwyzer, R., Sieber, P., and Zatskó, K. (1958). *Helv. Chim. Acta* **41**, 491–498.
Schwyzer, R., Rittel, W., Kappeler, H., and Iselin, B. (1960). *Angew. Chem.* **72**, 915–917.
Schwyzer, R., Costopanagiotis, A. and Sieber, P. (1963). *Helv. Chim. Acta* **46**, 870–889.
Scopes, M. P., Walshaw, K. B., Welford, M., and Young, G. T. (1965). *J. Chem. Soc.* pp. 782–786.
Seltzman, H. H., and Chapman, T. M. (1974). *Tetrahedron Lett.* pp. 2637–2640.
Shaltiel, S. (1967). *Biochem. Biophys. Res. Commun.* **29**, 178–183.
Shaltiel, S., and Fridkin, M. (1970). *Biochemistry* **9**, 5122–5127.
Sharp, J. J., Robinson, A. B., and Kamen, M. D. (1973). *J. Am. Chem. Soc.* **95**, 6097–6108.
Sheehan, J. C., and Yang, D.-D. H. (1958). *J. Am. Chem. Soc.* **80**, 1154–1158, and 1158–1164.
Sheehan, J. C., Hasspacher, K., and Yeh, Y. L. (1959). *J. Am. Chem. Soc.* **81**, 6086.
Shimonishi, Y., Sakakibara, S., and Akabori, S. (1962). *Bull. Chem. Soc. Jpn.* **35**, 1966–1970.
Shinano, S., and Kaya, T. (1957). *J. Agric. Chem. Soc. Jpn.* **31**, 759–762.

Sieber, P., and Iselin, B. (1968). *Helv. Chim. Acta* **51**, 614–622.
Sieber, P., Kamber, B., Hartmann, A., Jöhl, A., Riniker, B., and Rittel, W. (1974). *Helv. Chim. Acta* **57**, 2617–2621.
Siepmann, E., and Zahn, H. (1964). *Biochim. Biophys. Acta* **82**, 412–415.
Sifferd, R. H., and du Vigneaud, V. (1935). *J. Biol. Chem.* **108**, 753–761.
Sigor, A., Bonora, G. M., Biondi, L., Nisato, D., Marzotto, A., and Scoffone, E. (1971). *Biochemistry* **10**, 2748–2752.
Smeets, P., Granger, M., van Nispen, J. B., Bloemendal, H., and Tesser, G. I. (1977). *Int. J. Pept. Protein. Res* **9**, 52–56.
Sondheimer, E., and Holley, R. W. (1954). *J. Am. Chem. Soc.* **76**, 2467–2470.
Staab, H. A., and Benz, W. (1961). *Angew. Chem* **73**, 66.
Stammer, C. H. (1961). *J. Org. Chem.* **26**, 2556–2560.
Steglich, W., and Batz, H.-G. (1971). *Angew. Chem., Int. Ed. Engl.* **10**, 75–76.
Steglich, W., Mayer, D., Barocio de la Lama, X., Tanner, H., and Weygand, F. (1967). *In* "Peptides 1966" (H. C. Beyerman, A. van de Linde, and W. Maassen van den Brink, eds.), pp. 67–72, North-Holland Publ., Amsterdam.
Stelakatos, G. C., Theodoropoulos, D. M., and Zervas, L. (1959). *J. Am. Chem. Soc* **81**, 2884–2887.
Stephens, C. R., Bianco, E. J., and Pilgrim, F. J. (1955). *J. Am. Chem. Soc.* **77**, 1701–1702.
Stevenson, D., and Young, G. T. (1967). *Chem. Commun.* pp. 900–901.
Stevenson, D., and Young, G. T. (1969). *J. Chem. Soc. C* pp. 2389–2398.
Stoll, A., and Petrzilka, T. (1952). *Helv. Chim. Acta* **35**, 589–607.
Storey, H. T., and Hofmann, K. (1973). *In* "Peptides 1971", (H. Nesvada, ed.), pp. 227–236. North-Holland Publ., Amsterdam.
Storey, H. T., Beacham, J., Cernosek, S. F., Finn, F. M., Yanaihara, C., and Hofmann, K. (1972). *J. Am. Chem. Soc.* **94**, 6170–6178.
Sturm, K., Geiger, R., and Siedel, W. (1963). *Chem. Ber.* **96**, 609–618.
Suzuki, K., Endo, N., Nitta, K., and Sosaki, Y. (1978). *Chem. Pharm. Bull.* **26**, 2198–2204.
Takahashi, K. (1968). *J. Biol. Chem.* **243**, 6171–6179.
Tarbell, D. S., Yamamoto, Y., and Pope, B. M. (1972). *Proc. Natl. Acad. Sci. U.S.A.* **69**, 730–732.
Terada, S., Kawabata, A., Mitsuyasu, N. Aoyagi, H., and Izumiya, N. (1978). *Bull. Chem. Soc. Jap.* **51**, 3409–3410.
Tesser, G. I., and Balvert-Geers, I. C. (1975). *Int. J. Pept. Protin Res.* **7**, 295–305.
Theodoropoulos, D. (1956). *J. Org. Chem.* **21**, 1550–1551.
Tilak, M. A., Russell, R., and Hendricks, M. L. (1971). *Org. Prep. Proced. Int.* **3**, 17–20.
Toi, K., Bynum, E., Norris, E., and Itano, H. A. (1965). *J. Biol. Chem.* **240**, PC 3455–3457.
Toi, K., Bynum, E., Norris, E., and Itano, H. A. (1967). *J. Biol. Chem.* **242**, 1036–1043.
Tomida, I., and Kayahara, H. (1976). *Agr. Biol. Chem.* **40**, 875–881.
Tomida, I., Senda, S., Kuwabara, T., and Katayama, K. (1976). *Agric. Biol. Chem. Jpn.* **40**, 2033–2036.
Tritsch, G. L., and Woolley, D. W. (1960). *J. Am. Chem. Soc.* **82**, 2787–2793.
Tun-Kyi, A. (1978). *Helv. Chim. Acta* **61**, 1086–1090.
Tun-Kyi, A., and Schwyzer, R. (1976). *Helv. Chim. Acta* **59**, 1642–1646.
Turán, A., Patthy, A., and Bajusz, S. (1975). *Acta Chim. Acad. Sci. Hung.* **85**, 327–332.
van Orden, H. O., and Smith, E. L. (1954). *J. Biol. Chem.* **208**, 751–764.
Veber, D. F. (1975). *In* "Peptides—Chemistry, Structure, Biology" (R. Walter and J. Meienhofer, eds.), pp. 301–316. Ann Arbor Sci. Publ., Ann Arbor, Michigan.
Veber, D. F., Brady, S. F., and Hirschmann, R. (1972). *In* Chemistry and Biology of Peptides" (J. Meienhofer, ed.), pp. 315–319. Ann Arbor Sci. Publ., Ann Arbor, Michigan.
Veber, D. F., Paleveda, Jr., W. J., Lee, Y. C., and Hirschmann, R. (1977). *J. Org. Chem.* **42**, 3286–3288.

Vogler, K., and Lanz, P. (1960). *Helv. Chim. Acta* **43**, 270–279.
Wackerle, L., and Ugi, I. (1975). *Synthesis* pp. 598–599.
Wakselman, M., and Guibé-Jampel, E. (1973). *Chem. Commun.* pp. 593–594.
Wang, S. S., and Merrifield, R. B. (1969). *J. Am. Chem. Soc.* **91**, 6488–6491.
Weygand, F., and Czendes, E. (1952). *Angew. Chem.* **64**, 136.
Weygand, F., and Frauendorfer, E. (1970). *Chem. Ber.* **103**, 2437–2449.
Weygand, F., and Geiger, R. (1956). *Chem. Ber.* **89**, 647–652.
Weygand, F., and Geiger, R. (1959). *Chem. Ber.* **92**, 2099–2106.
Weygand, F., and Hunger, K. (1962). *Chem. Ber.* **95**, 1–6.
Weygand, F., and Nintz, E. (1965). *Z. Naturforsch., Teil B* **20**, 429–434.
Weygand, F., and Röpsch, A. (1959). *Chem. Ber.* **92**, 2095–2099.
Weygand, F., and Steglich, W. (1959a). *Z. Naturforsch., Teil B* **14**, 472–473.
Weygand, F., and Steglich, W. (1959b). *Chem. Ber.* **92**, 313–319.
Weygand, F., Prox, A., Fessel, H. H., and Sun, K. K. (1965). *Z. Naturforsch., Teil B* **20**, 1169–1182.
Weygand, F., Steglich, W., Bjarnason, J., Akhtar, R., and Khan, N. M. (1966a). *Tetrahedron Lett.* pp. 3483–3487.
Weygand, F., Prox, A., and König, W. (1966b). *Chem. Ber.* **99**, 1451–1460.
Weygand, F., Hoffmann, D., and Wünsch, E. (1966c). *Z. Naturforsch., Teil B* **21**, 426–428.
Weygand, F., Steglich, W., and Pietta, P. (1966d). *Tetrahedron Lett.* pp. 3751–3754.
Weygand, F., Steglich, W., and Pietta, P. (1967). *Chem. Ber.* **100**, 3841–3849.
Weygand, F., Steglich, W., Bjarnason, J., Akhtar, R., and Chytil, N. (1968a). *Chem. Ber.* **101**, 3623–3641.
Weygand, F., Steglich, W., and Bjarnason, J. (1968b). *Chem. Ber.* **101**, 3642–3648.
Wieland, T., and Schäfer, W. (1952). *Justus Liebig's Ann. Chem.* **576**, 104–109.
Wilchek, M., and Miron, T. (1979). *In* "Peptides, Structure and Biological Function" (E. Gross and J. Meienhofer, eds.) pp. 49–57, Pierce Chemical Co., Rockford, Illinois.
Wilchek, M., Sarid, S., and Patchornik, A. (1965). *Biochim. Biophys. Acta* **104**, 616–618.
Windholz, T. B., and Johnston, D. B. R. (1967). *Tetrahedron Lett.* pp. 2555–2557.
Windridge, G. C., and Jorgensen, E. C. (1971a). *Intra-Sci. Chem. Rep.* **5**, 375–380.
Windridge, G. C., and Jorgensen, E. C. (1971b). *J. Am. Chem. Soc.* **93**, 6318–6319.
Wolters, E. Th. M., Tesser, G. I., and Nivard, R. J. F. (1974). *J. Org. Chem.* **39**, 3388–3392.
Woodward, R. B., Heusler, K., Gosteli, J., Naegeli, P., Oppolzer, W., Ramage, R., Ranganthan, S., and Vorbrüggen, H. (1966). *J. Am. Chem. Soc.* **88**, 852–853.
Wünsch, E. (1967). *Z. Naturforsch., Teil B* **22**, 1269–1276.
Wünsch, E. (1972). *Naturwissenschaften* **59**, 239–246.
Wünsch, E. (1973). *In* "The Chemistry of Polypeptides" (P. G. Katsoyannis, ed.), pp. 279–295. Plenum, New York.
Wünsch, E., and Drees, F. (1967). *Chem. Ber.* **100**, 816–819.
Wünsch, E., and Spangenberg, R. (1971). *Chem. Ber.* **104**, 2427–2429.
Wünsch, E., and Spangenberg, R. (1972). *Chem. Ber.* **105**, 740–742.
Wünsch, E., and Wendlberger, G. (1964). *Chem. Ber.* **97**, 2504–2514.
Wünsch, E., and Wendlberger, G. (1967). *Chem. Ber.* **100**, 160–172.
Wünsch, E., and Wendlberger, G. (1968). *Chem. Ber.* **101**, 3659–3663.
Wünsch, E., and Zwick, A. (1964). *Chem. Ber.* **97**, 3305–3311.
Wünsch, E., Fontana, A., and Drees, F. (1967). *Z. Naturforsch., Teil B* **22**, 607–609.
Wünsch, E., Zwick, A., and Jäger, E. (1968). *Chem. Ber.* **101**, 336–340.
Wünsch, E., Deffner, M., Deimer, K.-H., Jaeger, E., Stelzel, P., Thamm, P., and Wendlberger, G. (1974). *In* "Methoden der organischen Chemie (Houben-Weyl) (E. Mueller et al., eds.), Vol. XV, part 2. Thieme, Stuttgart.

Yajima, H., Kawasaki, K., Kinomura, Y., Oshima, T., Kimoto, S., and Okamoto, M. (1968). *Chem. Pharm. Bull.* **16**, 1342–1350.

Yajima, H., Watanabe, H., and Okamoto, M. (1971). *Chem. Pharm. Bull.* **19**, 2185–2189.

Yajima, H., Fujii, N., Ogawa, H., and Kawatani, H. (1974). *J. Chem. Soc., Chem. Commun.* pp. 107–108.

Yajima, H., Kiso, Y., Ogawa, H. Fujii, N., and Irie, H. (1975a). *Chem. Pharm. Bull.* **23**, 1164–1166.

Yajima, H., Kitagawa, H., Segawa, T., Nakano, M., and Kataoka, K. (1975b). *Chem. Pharm. Bull.* **23**, 3299–3301.

Yajima, H., Ogawa, H., Fujii, N. and Funakoshi, S. (1977). *Chem. Pharm. Bull.* **25**, 740–747.

Yajima, H., Takeyama, M., Kanaki, J., and Mitani, K. (1978a). *J. Chem. Soc. Chem. Comm.* pp. 482–483.

Yajima, H., Takeyama, M., Kanaki, J., Nishimura, O., and Fujino, M. (1978b). *Chem. Pharm. Bull.* **26**, 3752–3757.

Yankeelov, J. A., Jr., Kochert, M., Page, J., and Westphal, A. (1966). *Fed. Proc.* **25**, 590.

Yankeelov, J. A., Jr., Mitchell, C. D., and Crawford, T. H. (1968). *J. Am. Chem. Soc.* **90**, 1664–1666.

Yankeelov, J. A., Jr. (1970). *Biochemistry* **9**, 2433–2439.

Yamashiro, D., and Li, C. H. (1972). *Int. J. Pept. Protein Chem.* **4**, 181–185.

Yamashiro, D., and Li, C. H. (1973). *J. Org. Chem.* **38**, 2594–2597.

Yamashiro, D., Blake, J., and Li, C. H. (1972). *J. Am. Chem. Soc.* **94**, 2855–2859.

Zahn, H., and Falkenburg, H. R. (1960). *Justus, Liebig's Ann. Chem.* **636**, 117–131.

Zahn, H., and Fölsche, E. T. J. (1969). *Chem. Ber.* **102**, 2158–2159.

Zahn, H., and Umlauf, E. (1954). *Hoppe-Seyler's Z. Physiol. Chem.* **297**, 127–145.

Zahn, H., Okuda, T., and Shimonishi, Y. (1967). *In* "Peptides 1966" (H. C. Beyerman, A. van de Linde, and W. Maassen van den Brink, eds.), pp. 108–112. North-Holland Publ., Amsterdam.

Zaoral, M. (1965). *Collect. Czech. Chem. Commun.* **30**, 1853–1868.

Zervas, L., and Bergmann, M. (1928). *Ber. Dtsch. Chem. Ges.* **61**, 1195–1203.

Zervas, L., and Dilaris, L. (1955). *J. Am. Chem. Soc.* **77**, 5354–5357.

Zervas, L., and Dilaris, L. (1956). *Chem. Ber.* **89**, 925–933.

Zervas, L., and Theodoropoulos, D. M. (1956). *J. Am. Chem. Soc.* **78**, 1359–1363.

Zervas, L., Winitz, M., and Greenstein, J. P. (1957). *J. Org. Chem.* **22**, 1515–1521.

Zervas, L., Otani, T. T., Winitz, M., and Greenstein, J. P. (1959). *J. Am. Chem. Soc.* **81**, 2878–2884.

Zervas, L., Borovas, D., and Gazis, E. (1963). *J. Am. Chem. Soc.* **85**, 3660–3666.

Zheltukhina, G. A., Filippovich, E. I., Lagutkina, V. I., Sidorova, M. V., and Evstigneeva, R. P. (1976). *Zh. Obshch. Khim.* **46**, 2633–2634.

Chapter 2

Carboxyl Protecting Groups

ROGER W. ROESKE

THE PEPTIDES, VOLUME 3

ISBN 0-12-304203-8

I. INTRODUCTION

The purposes of carboxyl protection during peptide synthesis are to promote the solubility of reactants, to avoid reaction of the carboxylate ion with the acylating agent and to facilitate the purification of products. The most important considerations in choosing a carboxyl protecting group are its stability during synthesis and the conditions required for its removal after it has served its purpose. Therefore this chapter is arranged according to the most significant chemical properties of the protecting groups, rather than strictly according to their structures. There is considerable overlap in this classification scheme—many of the protecting groups can be removed by several methods—but they are listed here according to the primary, or most useful method.

II. SYNTHESIS WITHOUT PROTECTION OF CARBOXYL GROUPS

It is often possible to carry out a synthesis without protecting the carboxyl group(s) of the amine component at all, by deprotonating the amino group with a tertiary amine or with tetramethylguanidine and carrying out the coupling reaction in an organic solvent such as dioxane, dimethylformamide, dimethoxyethane, or hexamethylphosphoramide, or in a mixture of water and one of these solvents. Although the acid chloride and the mixed anhydride methods have frequently been used in this way, the active ester coupling method seems to give the best yields in such cases and has the added advantage that any unreacted active ester is usually easy to separate from the product. But the side reactions, e.g., partial hydrolysis of the active ester and reaction of the carboxylate ion with the active ester to give a mixed anhydride, produce minor impurities which are liable to contaminate the product. Thus, the success of this approach often depends on the tendency of the product or its salt with dicyclohexylamine (DCHA) to crystallize. Equation (1) (Rosamund and Ferger, 1976) is an example of the many successful peptide syntheses that have been carried out without protection of carboxyl groups.

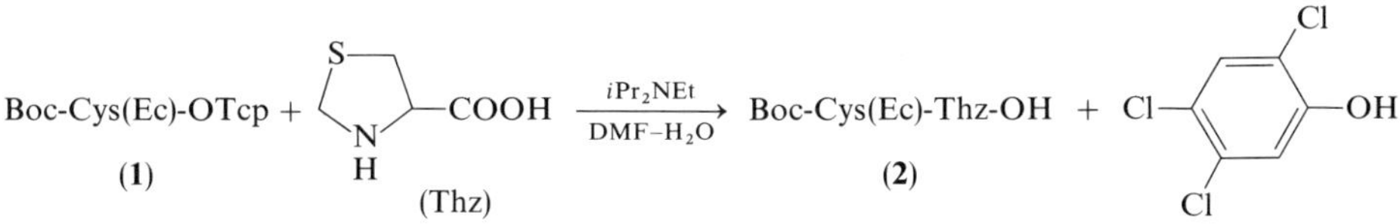

(73%, purified via DCHA salt)

Boc = *tert*-butyloxycarbonyl
Ec = ethylcarbamoyl
Thz = thiazolidine-2-carboxylic acid
OTcp = 2,4,5-trichlorophenyl ester
iPr$_2$NEt = diisopropylethylamine

That the carboxylate ion does compete with the amine group for the acylating agent has been shown in several cases. Bodanszky (1968) found that acetate ion reacts with nitrophenyl esters, and Kopple *et al.* (1969) showed that the reaction of benzyloxycarbonyl chloride (Z—Cl) with several tripeptides in aqueous base or in dimethylformamide (DMF) and triethylamine gave almost exclusively benzyloxycarbonyl tripeptide oligomers, presumably via reaction of the COOH-terminal carboxylate with the Z—Cl to give a mixed anhydride. This interesting reaction is sequence-dependent, for no oligomers were detected when Z-Gly-Gly-Gly-OH was prepared in this way.

During the preparation of the COOH-terminal pentapeptide (**5**) of cholecystokinin, the reaction of an unprotected β-carboxylate of aspartic acid with Boc-Gly-ONp to form a mixed anhydride was followed by both intramolecular aspartimide formation (**6**) and by intermolecular acylation of the tetrapeptide amide (**4**) to produce **7** (Natarajan and Bodanszky, 1976). The side reactions could be minimized by avoiding an excess of both the active ester and the tertiary amine (Scheme 1).

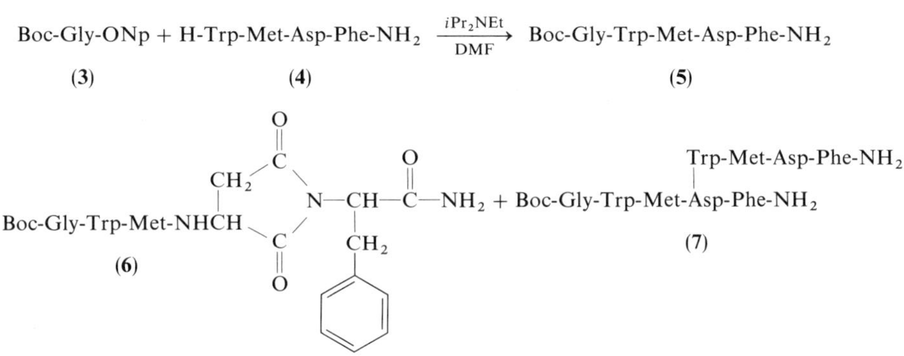

Scheme 1

III. TRIMETHYLSILYL ESTERS

Trimethylsilyl (Tms) esters have several advantages for temporary protection of carboxyl groups; they are easily formed and easily cleaved (Birkhofer and Ritter, 1965). Because they are immediately hydrolyzed on contact with aqueous media, whether basic, acidic, or neutral, they carry out only two of the three functions of a carboxyl protecting group, i.e., they solubilize the amine component and they prevent reaction of the carboxylate ion with the acylating agent. These esters are formed *in situ* from amino acids or peptides or their salts by treatment with various silylating agents, e.g., trimethylchlorosilane in methylene chloride or chloroform with an equivalent of triethylamine or with trimethylsilyl-*N*-methylacetamide or bis-*N*,*O*-trimethylsilylacetamide in dioxane or dimethylformamide (Kricheldorf, 1972). Sufficient reagent is used to silylate both the carboxyl and the amino groups, to give the intermediate (**9**) which is not isolated, but coupled with an activated carboxyl component by a variety of methods. The active ester method as illustrated in Scheme 2 is preferred for ease of work-up.

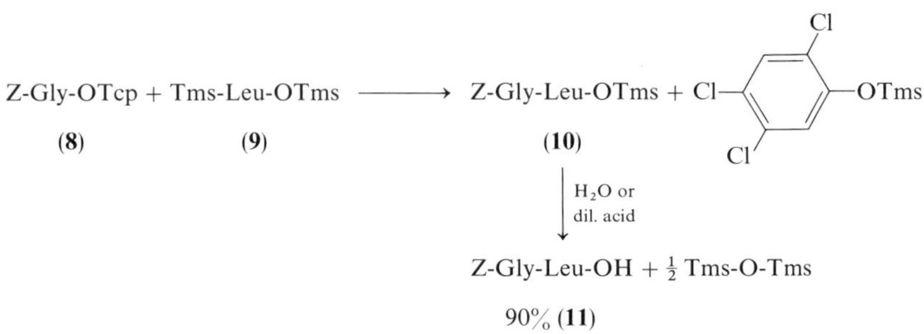

Scheme 2

The nucleophilic properties of trimethylsilylamino acid trimethylsilyl esters have not been studied in detail. Kricheldorf (1972) reported that, although they react readily with *p*-nitrophenyl, 2,4,5-trichlorophenyl, and *N*-hydroxysuccinimide esters, they do not react at all with the less active cyanomethyl esters. On the other hand, Brenner (1968) found that trimethylsilylamino acid trimethylsilyl esters react extremely rapidly with acylating agents. Regardless of the mechanism by which they react, these intermediates appear to be very attractive in peptide synthesis, and it seems surprising that they have not been used more frequently in the way described by Schnabel and Oberdorf (1968) in the synthesis of pGlu-Ala-Asp(O*t*Bu)-Pro-OH, that is, by the stepwise addition of amino acids to a peptide that is silylated and desilylated at each step. A possible complication in such a scheme is the silylation of the peptide bonds by excess silylating agent to form *N*,*O*-trimethylsilyl derivatives (**13**) which can be acylated (**14**), as described by Davies *et al.* (1973) and which are shown in Scheme 3.

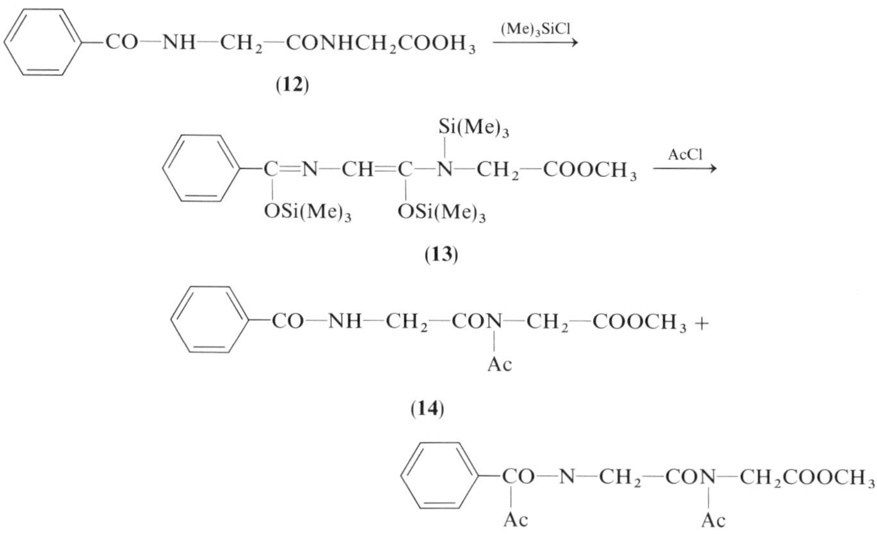

Scheme 3

The trimethylsilyl group has been used for the temporary protection of carboxyl groups and for the solubilization of arginine during the preparation of tribenzyloxycarbonyl-L-arginine (Smithwick and Shuman, 1974) and in the preparation of N^{α}-trifluoroacetylamino acids and N^{α}-*o*-nitrophenylsulfenylamino acids (Kricheldorf and Fehrle, 1974).

Trimethylsilyl esters are intermediates in a new method of cleaving methyl esters by reaction with trimethylsilyl iodide (see the section on methyl esters).

IV. PROTECTING GROUPS CLEAVED BY MILD ACID TREATMENT

A. General Remarks

Acid-labile protecting groups are among the most useful in peptide synthesis, since the treatment of a protected peptide with mild acid causes fewer side reactions than most other methods of deprotection. In fact, until a few years ago, almost all selective removal of amine and carboxyl protecting groups was based on graded acidolysis which, however, suffers from the weakness that its selectivity is not total, but relative (Kuromizu and Meienhofer, 1974). This has led recently to the development of other carboxyl protecting groups that can be cleaved under neutral or mildly basic conditions. But the search for reagents and groups that will allow a more precise manipulation of acid-labile protecting groups continues. An outstanding example of the use of selective acidolysis of protecting groups is the synthesis of human insulin (Sieber *et al.*, 1974, 1976; Kamber *et al.*, 1976), and an important aspect of this work is the finding that the solvent used to effect acidolysis makes a great deal of difference in getting selective cleavage (Riniker *et al.*, 1975).

The following esters are listed in order of their increasing resistance to acidolysis: triphenylmethyl (OTrt) $<$ diphenylmethyl (benzhydryl, ODpm) $\approx$ *tert*-butyl (O*t*Bu) $\approx$ *p*-methoxybenzyl (OMob) $\approx$ pentamethylbenzyl (OPmb) $<$ trimethylbenzyl (OTmb) $\approx$ phthalimidomethyl (OPtm) $\approx$ 9-anthrylmethyl (OAm) $<$ benzyl (OBzl). Acidolysis is not necessarily the most useful method for cleaving all of the above esters in all synthetic schemes: thus, the benzyl ester is generally removed by catalytic hydrogenolysis except in syntheses by the solid-phase method; the 9-anthrylmethyl ester is cleaved by base, hydrogenolysis, or thiolate anion as well as by acid; and phthalimidomethyl esters can be cleaved readily by thiolate anion (Sheehan and Daves, 1964) and by hydrazine (Nefkens, 1962). In this section only the *tert*-butyl, diphenylmethyl, pentamethylbenzyl, and trimethylbenzyl esters are discussed, along with substituted benzyl esters cleaved by acidolysis in the solid-phase method. The *p*-methoxybenzyl (Stewart, 1968; Stelekatos and Argyropoulos, 1970) and triphenylmethyl esters have been used only rarely.

B. *tert*-Butyl Esters

tert-Butyl esters are cleaved by mild acid conditions, e.g., hydrogen chloride in organic solvents, trifluoroacetic acid at room temperature, or hydrogen bromide in acetic acid. These esters are especially useful in combination with an amine protecting group that is removed by hydrogenolysis (e.g., benzyloxycarbonyl), since they are quite stable under these conditions. The addition of one equivalent of HCl to the hydrogenation mixture, to protonate the newly formed amine, does not cleave a *tert*-butyl ester (Sieber *et al.*, 1970) unless the hydrogenolysis is very slow. *tert*-Butyl esters remain intact under the very mildly acidic conditions that cleave the 2-(*p*-biphenylyl)-propyl(2) oxycarbonyl (Bpoc), *o*-nitrophenylsulfenyl (Nps), and Trityl (Trt) amine protecting groups. Extensive use has been made of these combinations in the synthesis of long peptides, especially when sulfur-containing amino acids are present, since hydrogenolysis of the benzyloxycarbonyl group is then impractical. For example, in the synthesis of human insulin, Sieber *et al.*

(1976) and Kamber *et al.* (1976) selectively removed the Bpoc and trityl groups while keeping the *tert*-butyl ester groups intact.

The usefulness of the *tert*-butyl ester as a carboxyl protecting group would be enhanced if it could be used in combination with the *tert*-butyloxycarbonyl (Boc) group, i.e., if the Boc group could be removed completely without cleaving the ester. There are reports of this reaction having been carried out successfully but it cannot yet be regarded as a reliable procedure in all peptides. Hiskey *et al.* (1972a) were able to remove the *tert*-butyloxycarbonyl group selectively from the model peptide Boc-Cys(Trt)-Cys(Dpm)-Gly-Phe-Gly-O*t*Bu by treatment with boron trifluoride–etherate in acetic acid at 0°C, by stopping the reaction before all of the starting material had disappeared, but this procedure was not practical on a large scale. Similarly, attempts to remove the *tert*-butyloxycarbonyl group selectively from Boc-Trp-Met-Asp(O*t*Bu)-Phe-NH_2, e.g., with 50% aqueous trifluoroacetic acid for 70 min. at room temperature, were only moderately successful and an alternative synthetic scheme was chosen. The *tert*-butyloxycarbonyl group was removed selectively from Boc-Phe-Ala-Glu(O*t*Bu)-O*t*Bu and several related peptides in yields of 47–87% by allowing them to remain on a column of Zeo Carb 225 (H^+) for six hours and then eluting with a 1:1 mixture of methanol and 10% ammonium hydroxide (Gray and Khoujah, 1969). Apparently this technique has not been investigated further; it may be limited by the sparing solubility of larger protected peptides in aqueous solvents and by the basic conditions required to elute the product. The perchlorate of H-Gly-Ile-Val-Glu(O*t*Bu)-OBzl was obtained in 100% yield when Boc-Gly-Ile-Val-Glu(O*t*Bu)-OBzl was treated with one equivalent of trimethylsilyl perchlorate in benzene–acetonitrile at 24°C for five minutes (Vorbrüggen and Krolikiewicz, 1975). It is hoped that the scope and limitations of this interesting reagent will be explored.

Perhaps the most promising development in this area has been the finding (Goodacre *et al.*, 1975) that one equivalent of *p*-toluenesulfonic acid in ether–ethanol cleaved the *tert*-butyloxycarbonyl group but not the *tert*-butyl ester from Boc-DAla-DAla-O*t*Bu in three hours at room temperature. The *p*-toluenesulfonate of the product was obtained in 93% yield. Several related peptides were treated in the same way, with reaction times varying from three to twenty-four hours and with yields from 81 to 95%.

tert-Butyl esters are of course resistant to attack by nucleophiles and remain intact when a peptide is subjected to mild alkali, hydrazine, ammonia, and similar reagents. Amino acid *tert*-butyl esters can be distilled and stored unchanged for months and dipeptide *tert*-butyl esters do not cyclize readily to form diketopiperazines.

An exception to their lack of reactivity is the intramolecular reaction of ω-*tert*-butyl esters of aspartic and glutamic acid derivatives in which the alpha carbonyl is in an amide, hydrazide, or peptide link. Although mixed esters such as **16** can be saponified selectively, the treatment of a peptide

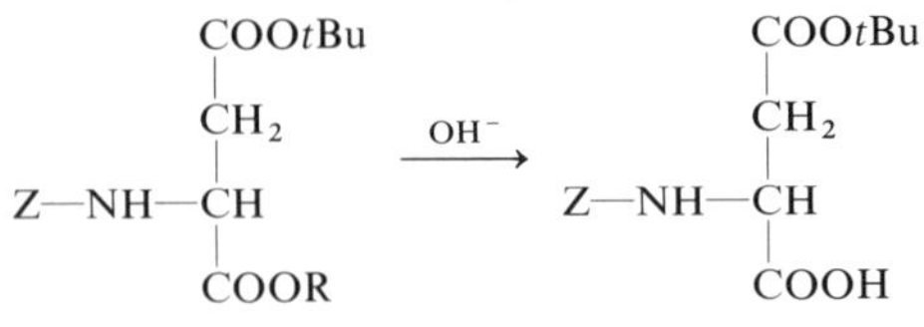

(**16**)

(2)

(**17**) with one equivalent of dilute base leads quickly to the aspartimide (**18**) which, on longer standing in base, reacts further to give mixtures of α- and β-aspartyl peptides.

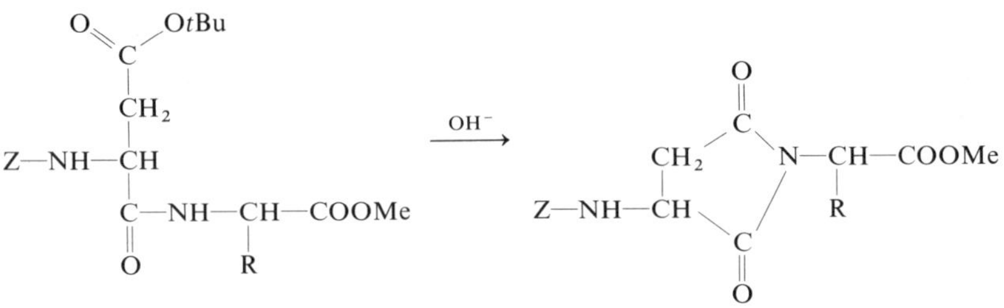

(**17**) (**18**)

(3)

This behavior of aspartic and glutamic acid derivatives has been known for a long time; Sondheimer and Holley showed in 1954 that the attempted saponification of benzyloxycarbonyl-L-asparagine methyl ester gave benzyloxycarbonyl-L-aspartimide in five minutes and that on prolonged treatment, followed by hydrogenation, one obtained a mixture of asparagine and isoasparagine. Similar reactions occurred with benzyloxycarbonyl-L-glutamine methyl ester. Related observations on aspartic and glutamic peptides were made by Battersby and Robinson (1955) and by Clayton *et al.* (1956). It was hoped that this intramolecular reaction could be prevented by using a *tert*-butyl ester, but surprisingly this was not the case. Treatment of benzyloxycarbonyl-L-asparagine *tert*-butyl ester with one equivalent of sodium hydroxide in methanol for 3 min at room temperature gave benzyloxycarbonyl-L-aspartimide in 40% yield (Roeske, 1963). Similarly, when Z-Asp(O*t*Bu)-OMe (**19**) was treated with hydrazine, the main product had the analysis corresponding to the dihydrazide (**21**) (Wünsch and Zwick, 1963). Presumably, the monohydrazide cyclizes to the aminoimide (**20**) which reacts further to give the dihydrazide.

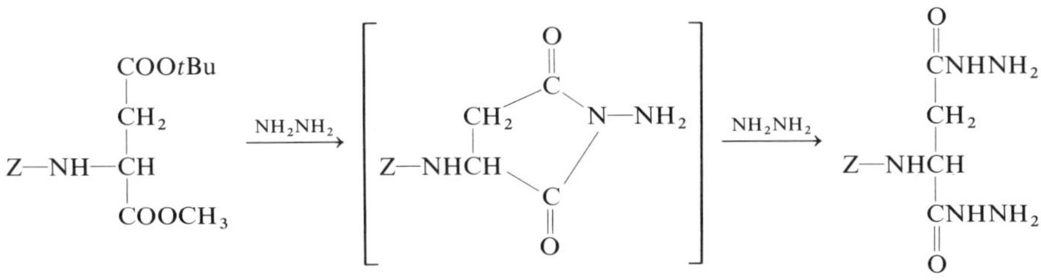

(19) **(20)** **(21)**

(4)

Hence, the treatment of any peptide that contains an ester of aspartic acid with alkali, hydrazine, ammonia, or related nucleophiles is likely to give a mixture of products because of the intramolecular cyclization. An example from the synthesis of melanotropin (MSH): long treatment of Boc-Asp(O*t*Bu)-Ser-Gly-Pro-Tyr-Lys(Boc)-Met-OMe with excess hydrazine hydrate in methanol gave several by-products. By running the reaction under milder conditions at room temperature for four hours with a fivefold excess of hydrazine hydrate, the by-products were reduced to 10%, some starting material was recovered, and the desired COOH-terminal hydrazide was isolated in a yield of 60% (Schwyer *et al.*, 1963). The severity of the problem may depend on the sequence: Z-Thr(O*t*Bu)-Thr(O*t*Bu)-Thr(O*t*Bu)-Gln-Asp(O*t*Bu)-Phe-OMe was converted to the crystalline COOH-terminal hydrazide in 80% yield (Sieber *et al.*, 1970).

The corresponding side reactions with C^{γ}-*tert*-butyl glutamic acid residues in similar reactions are much less severe, and there are many examples of the selective saponification of a COOH-terminal methyl ester of a peptide containing a glutamic γ-*tert*-butyl ester residue. Side reactions can occur, however, as in this unusual sequence **(22)** from human gastrin.

Boc-Glu(O*t*Bu)-Glu(O*t*Bu)-Glu(O*t*Bu)-Glu(O*t*Bu)-Glu(O*t*Bu)-Glu(O*t*Bu)-Ala-Tyr-Gly-OMe

(22)

Attempted alkaline hydrolysis of the COOH-terminal methyl ester under a variety of conditions was always accompanied by some by-products (Beacham *et al.*, 1967).

tert-Butyl esters of amino acids cannot be prepared in good yield by the general methods of ester formation discussed in Section IX, but they can be made by reaction with isobutene in an organic solvent, either from the free amino acids in the presence of concentrated sulfuric acid or *p*-toluenesulfonic acid (Roeske, 1959, 1963) or by reaction of the benzyloxycarbonyl derivative with isobutene and an acid catalyst, followed by removal of the amine protecting group (Anderson and Callahan, 1960).

tert-Butyl acetate has been used also as a source of *tert*-butylcarbonium ions for the esterification of both free amino acids and *N*-protected amino acids (Taschner *et al.*, 1961b; Chimiak *et al.*, 1972). *ter*-Butyl esters have been prepared by reaction of *N*-benzyloxycarbonyl and *N*-phthaloylamino acids with $POCl_3$ and *tert*-butanol in pyridine (Taschner *et al.*, 1961a).

ω-*tert*-Butyl esters of aspartic and glutamic acid can be obtained in good yield by reaction of the α-benzyl esters with isobutene, followed by hydrogenolysis of the α-protecting group. These esters survive reaction of the α-amine and α-carboxyl group with phosgene to form the ω-*tert*-butyl-*N*-carboxyl anhydrides of aspartic and glutamic acid (Roeske, 1963) [Eq. (5)].

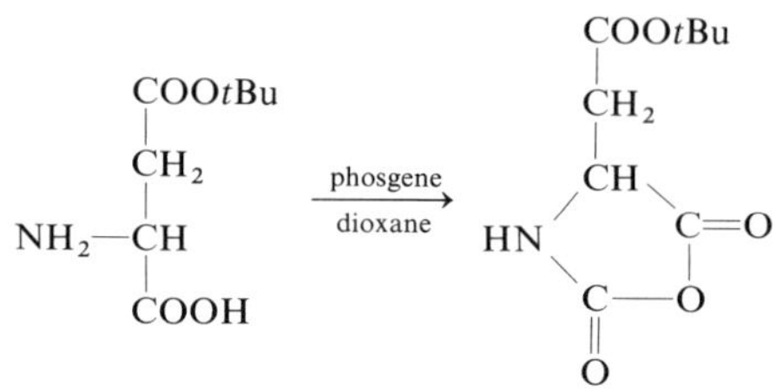

(5)

C. Diphenylmethyl Esters

Diphenylmethyl esters (ODpm) are an alternative to *tert*-butyl esters for carboxyl protection, but they have been used only sparingly. They are about as labile in acid as the latter (Hiskey and Adams, 1965)—thus, they are cleaved completely by acidic conditions that do not affect the *N*-benzyloxycarbonyl group, but they are stable under the conditions required for removal of the *o*-nitrophenylsulfenyl group. Stelakatos *et al.*, (1966) used trifluoroacetic acid (TFA) at 20°C for 30 min to prepare benzyloxycarbonyl dipeptides from the corresponding diphenylmethyl esters, and Hiskey *et al.* (1967) cleaved the diphenylmethyl ester group from Pht-Cys(Dpm)-Gly-ODpm with a fivefold excess of BF_3–etherate in AcOH–$CHCl_3$ at 20°C for 1 hr.

They are easily cleaved by hydrogenolysis and were used in this way as carboxyl protecting groups in steroid and terpene work by Hardegger *et al.* (1948) who prepared them by reaction of carboxylic acids with diphenyldiazomethane. Diphenylmethyl esters of amino acids can be prepared directly by reaction of diphenyldiazomethane with the *p*-toluenesulfonate or β-naphthalenesulfonate salts of amino acids in dimethyformamide at 50°C for 10 min (Aboderin *et al.*, 1965).

In spite of the obvious steric hindrance, diphenylmethyl esters react fairly readily with nucleophiles, e.g., Bz-Gly-ODpm was converted quantitatively to Bz-Gly-$NHNH_2$ by reaction with two equivalents of 85% hydrazine hydrate in refluxing methanol for 1 hr and was saponified quantitatively by two equivalents of NaOH in dioxane at 25°C for 1 hr (Hiskey and Adams, 1965).

The need to prepare diphenyldiazomethane has probably been a deterrent to the use of diphenylmethyl esters but a new method of preparing this reagent quickly in excellent yield from benzophenone hydrazone is now available (Adamson *et al.*, 1975). Alternatively, the isolation of diphenyldiazomethane can be avoided; diphenymethyl esters have been prepared directly from N-protected amino acids and peptides by oxidation of benzophenone hydrazone with peracetic acid in the presence of the protected carboxyl component and a trace of iodine [Eq. (6)]. Bywood *et al.*, 1975). Sulfides are oxidized to sulfoxides under these conditions.

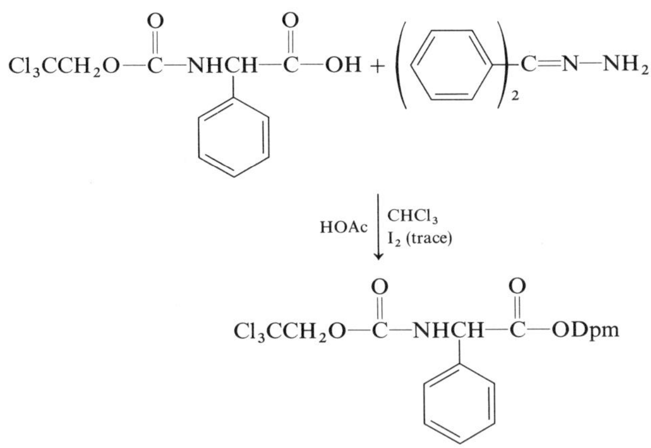

(6)

(91%)

D. Trimethylbenzyl Esters

The 2,4,6-trimethylbenzyl ester (OTmb) protecting group is slightly more stable to acid than the *tert*-butyl ester and so provides a greater margin of safety when it is used in combination with temporary amine protecting groups that are more acid-labile. It has been used especially in combination with the *o*-nitrophenylsulfenyl group (F. H. C. Stewart, 1967a) and even with the *tert*-butyloxycarbonyl group, but Ledger and Stewart (1968), while trying to remove the *tert*-butyloxycarbonyl group selectively from a series Boc-AA-OTmb always detected traces of the free amino acid in the product, indicating some cleavage of the trimethylbenzyl ester. An example of its use is in the synthesis of a protected A(14–21) fragment of ovine insulin (**23**), where it remained intact as the COOH-terminal protecting group while the *tert*-butyloxycarbonyl, *o*-nitrophenysulfenyl and 2-(*p*-biphenylyl)propyl(2)oxy-carbonyl groups were used for temporary protection of the amine groups in various steps (Hiskey *et al.*, 1972b).

Bpoc-Tyr(*t*Bu)-Gln-Leu-Glu(O*t*Bu)-Asn-Tyr(*t*Bu)-Cys(Trt)-Asn-OTmb

(23)

Trimethylbenzyl esters are cleaved by treatment with 2 *N* hydrogen bromide in glacial acetic acid for 10 min at room temperature or by anhydrous trifluoroacetic acid at room temperature for one hour. Although they are cleaved more rapidly in trifluoroacetic acid than are benzyloxycarbonyl groups, the difference in lability is probably not sufficient to allow their selective removal in most cases (F. H. C. Stewart, 1966).

Trimethylbenzyl esters cannot be prepared in good yield from free amino acids by direct esterification with the corresponding alcohol; generally, they have been prepared by treating *o*-nitrophenylsulfenyl amino acids with chloromethylmesitylene and triethylamine, followed by removal of the *o*-nitrophenylsulfenyl group by reaction with slightly more than 2 equivalents of HCl in MeOH (F. H. C. Stewart, 1966). More recently, they have been prepared by temporarily protecting the amino group with ethyl acetoacetate. The resulting enamine (**24**) is alkylated with chloromethylmesitylene and

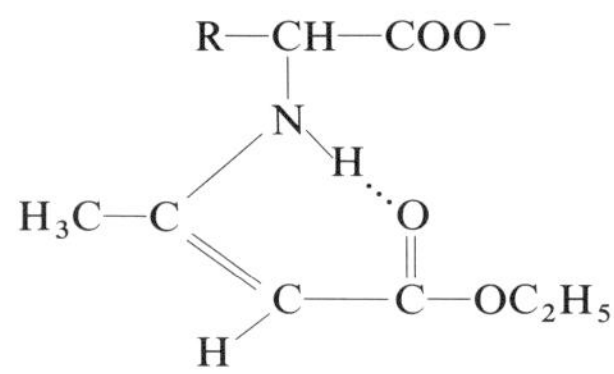

(**24**)

the N-protecting group is removed with HCl in MeOH. The three reactions are carried out without isolation of intermediates and proceed in good yield (Maclaren, 1972).

E. Phthalimidomethyl Esters

Phthalimidomethyl esters (OPtm) undergo acidolysis by treatment with 2 *N* HBr in AcOH for 10 to 15 min (Nefkens *et al.*, 1963), that is, under the

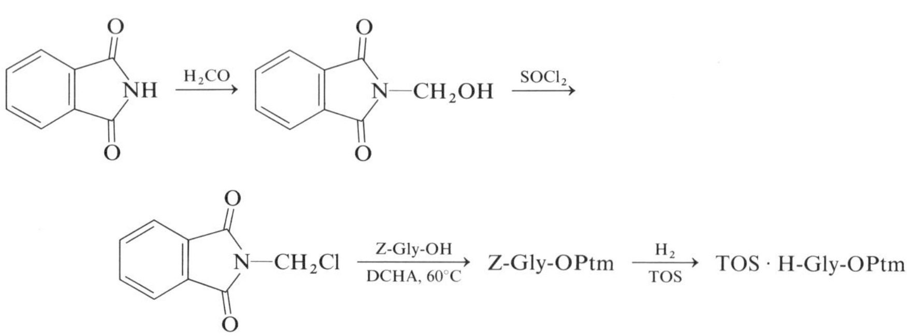

Scheme 4

same conditions as trimethylbenzyl esters. They are not cleaved under the usual conditions of catalytic hydrogenation, but readily undergo alkyl–oxygen cleavage by one equivalent of sodium thiophenoxide in dimethylformamide at room temperature for 30 min (Sheehan and Daves, 1964). Reaction of the esters with hydrazine hydrate for three hours at room temperature produces the free carboxylic acid and phthalhydrazide.

In spite of this extraordinary versatility, they have not been used much in peptide synthesis, perhaps because their preparation (Scheme 4) involves several steps (Nefkens *et al.*, 1963).

F. Acid-Labile Carboxyl Protecting Groups in Solid-Phase Peptide Synthesis (SPS)

The conditions of solid-phase synthesis, as compared to synthesis in solution, require a greater difference in acid lability between the carboxyl protecting groups, designed to remain intact during the synthesis, and the α-amine protecting group that is removed in each cycle. The usual COOH-terminal protecting group in solid-phase synthesis is a *p*-alkylbenzyl ester (**26**) which, if it were used in solution synthesis, would be entirely compatible

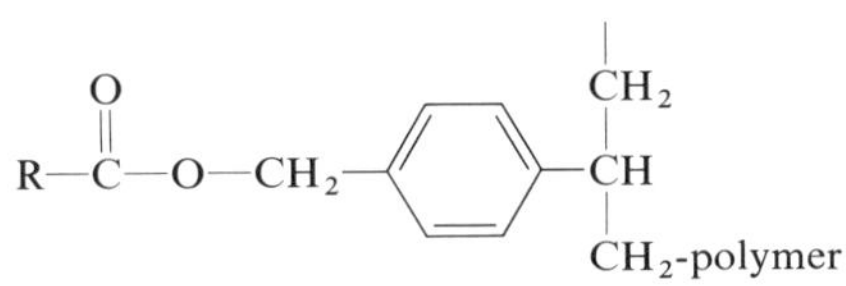

(**26**)

with an N^{α}-*tert*-butyloxycarbonyl group, but which is cleaved to a considerable degree during repeated exposure to 50% TFA in $MeCl_2$, a commonly used reagent for amine deprotection. To minimize the loss of peptide from the resin during synthesis, Mitchell *et al.* (1976) have used a *p*-acetamidobenzyl ester (Pam)* (**27**) which is 100 times more stable to 50% TFA in CH_2Cl_2 than **26**, but as **26**, is cleaved by treatment with HF.

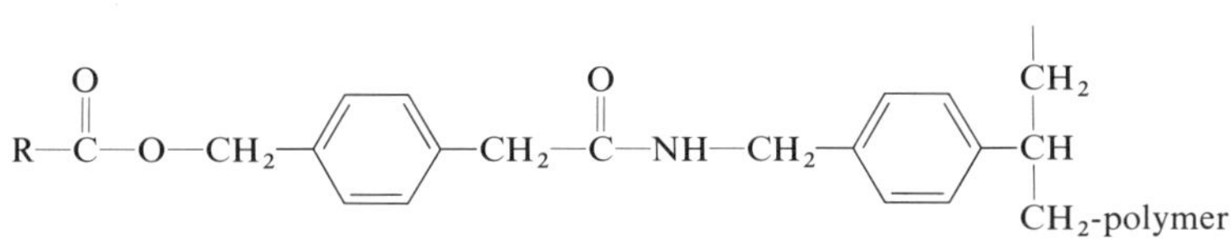

(**27**)

The *p*-acetamidobenzyl ester has been used also by Sparrow (1976) to link the first amino acid to a long spacer chain attached to the polystrene support (**28**).

* Pam is also termed phenylacetamidomethyl.

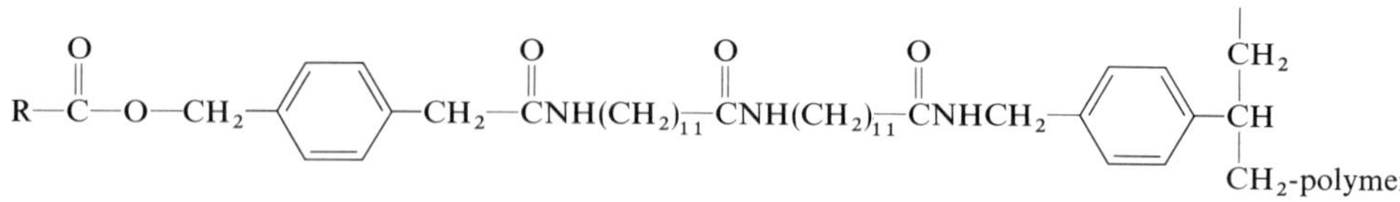

(28)

The benzyl ester, used routinely in solid-phase synthesis for the protection of the β- and γ-carboxyls of aspartic and glutamic acids, respectively, is lost to the extent of about 0.01% per cycle (Erickson and Merrifield, 1973). For the synthesis of long peptides containing many aspartic and glutamic acid residues, e.g., acyl carrier protein from *E. coli*, the *p*-chlorobenzyl ester gives better protection (Prestidge *et al.*, 1976) and is removed without difficulty by HF at 0°C. The same authors propose the use of the *p*-nitrobenzyl ester for the side chains of aspartic and glutamic acid when it is desired to leave these protected during cleavage of the peptide from the resin with HBr in acetic acid, as in the preparation of protected fragments to be used further in the synthesis of larger peptides. Yamashiro (1977) recommended the *p*-bromobenzyl ester for protecting the side chain of aspartic acid in solid-phase synthesis; it is about five times more stable than the benzyl ester in 50% TFA in CH_2Cl_2 and is completely removed in liquid HF at 0°C in 10 min.

Some aspartic acid-containing peptides protected by a β-benzyl ester undergo cyclic imide formation during treatment with HF; the reaction [Eq. (7)] depends on the sequence, and seems to be minimized or prevented entirely by a free amino group on the aspartyl residue and by a free carboxyl group on the residue following aspartic acid. Thus, the sequence Boc-Glu(OBzl)-Asp(OBzl)-Gly-Thr(Bzl)-Resin produced 99% of the aspartimide (**29**) after 1 hr at 0°C in HF containing 10% anisole but H-Asp(OBzl)-Gly-Thr(Bzl)-Resin and H-Glu(OBzl)-Asp(OBzl)-Ala-Resin formed only 1.6% and 7.8% of the imide, respectively (Yang and Merrifield, 1976).

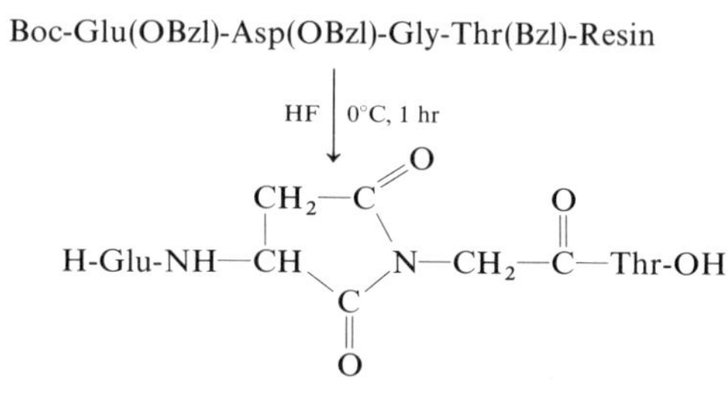

(7)

(29)

Glutamic acid γ-benzyl ester residues undergo two side reactions in HF–anisole; formation of a pyrrolidone (**30a**) and acylation of the anisole (**30b**) [Eq. (8)] (Sano and Kawanishi, 1975; Feinberg and Merrifield, 1975). Presumably, both reactions occur through an acyl carbonium ion and cannot be prevented by leaving the γ-carboxyl unprotected. Both reactions can be minimized by limiting the time and temperature of the HF cleavage reaction, e.g., Bpoc-Leu-Lys(Z)-Glu(OBzl)-Ala-Val-Gly-Resin gave only 1.5% of the by-products when it was treated at 0°C for ½ hr, but gave 90% of the undesired products when treated at 24° for 2 hr.

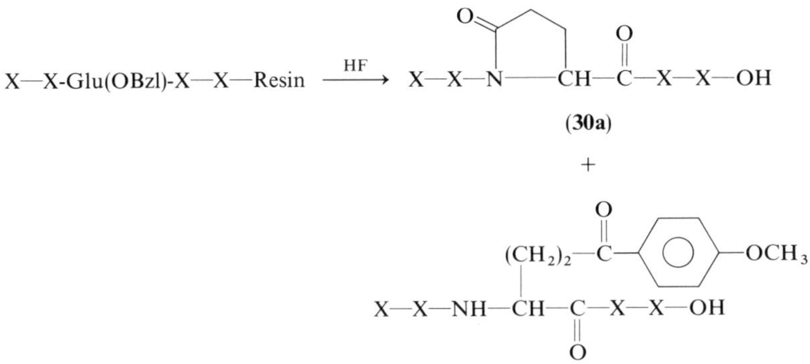

(30a)

(30b)

(8)

V. PROTECTING GROUPS CLEAVED BY HYDROGENOLYSIS

A. General Remarks

Hydrogenolysis is still an important method of removing protecting groups, particularly from small peptides. The currently used carboxyl protecting groups that can be cleaved by this method are the benzyl ester and its various derivatives, such as the *p*-nitro (ONb), the 2,4,6-trimethylbenzyl, the diphenylmethyl, and the 4-picolyl (OPic) esters. It is highly unlikely that any of these groups can be cleaved selectively by catalytic hydrogenolysis in the presence of any of the others—we are unaware of any published attempts to do this. The benzyloxycarbonyl group is also removed under the same conditions; Häussler (1960) showed that during the hydrogenation of various benzyloxycarbonyl di- and tripeptide benzyl and *p*-nitrobenzyl esters, there was no break in the curve of hydrogen uptake vs time—it climbed steadily from start to finish. The amount of catalyst, its concentration on the carrier, and the kind of carrier had no influence on the lack of selectivity. However, the *N*-triphenylmethyl group is more resistant than the benzyl ester; tritylamino acids can be prepared by hydrogenolysis of tritylamino acid benzyl esters (Stelakatos *et al.*, 1959).

Hydrogenolysis is usually carried out between 20°C and 45°C in the presence of palladium black or palladium on carbon, under 2 to 3 atm of hydrogen pressure in a shaker or by passing a stream of hydrogen gas through a stirred solution. The use of platinum oxide is avoided because of its greater tendency to catalyze the hydrogenation of aromatic rings. Schröder and Lübke (1965) have prepared an excellent summary of the applications of hydrogenation in peptide chemistry.

B. Benzyl and *p*-Nitrobenzyl Esters

A benzyl ester can generally be cleaved completely from a small peptide in less than an hour, but the same reaction might require a considerably longer time when the substrate is an octa- or decapeptide ester. In fact, one of the disadvantages of hydrogenolysis is its unpredictable course with larger peptides, where the peptide conformation might restrict the association of catalyst and substrate.

Before the introduction of the *tert*-butyloxycarbonyl amine protecting group, benzyl esters were often used in combination with a benzyloxycarbonyl group, the latter being removed selectively by acidolysis or both groups being removed in the final deprotection by hydrogenolysis. The use of the *p*-nitrobenzyl ester was suggested (Schwyzer and Sieber, 1959; Schwarz and Arakawa, 1959) in response to the finding that the benzyl ester was not completely stable under the conditions required to remove a benzyloxycarbonyl group, for example, 2 *N* HBr in AcOH for 30 to 60 min at room temperature. The *p*-nitrobenzyl ester is much more stable to acidolysis, and is normally removed by hydrogenolysis or, in uncomplicated cases, by saponification.

Hydrogenolysis in the usual solvents (alcohols, dioxane, dimethylformamide) often is not successful with peptides that contain sulfur in the form of sulfides, disulfides, or thiols because the catalyst is inactivated. However, Kuromizu and Meienhofer (1974) showed that when the reaction is carried out in liquid ammonia at its boiling point, $-33°C$, benzyloxycarbonyl groups could be cleaved completely in 6 to 8 hours from peptides containing an *S*-benzylcysteine residue, and they illustrated the method with a synthesis of oxytocin using four such hydrogenolyses. This work has been extended by Felix *et al.* (1978a) to the synthesis of somatostatin, which required the repeated removal of the benzyloxycarbonyl group in the presence of a COOH-terminal *S*-*tert*-butylcysteine residue. The average yield for the twelve stages of catalytic hydrogenolysis of the benzyloxycarbonyl group was 87%. This promising work may extend the scope of hydrogenation considerably.

Benzyl esters have been the most popular protecting groups for the side chains of aspartic and glutamic acid, both in solution synthesis and in solid-phase synthesis. But the problem of imide formation, discussed in the section on *tert*-butyl esters, remains only partly resolved. A study of the sequence-dependence of the amine-catalyzed imide formation during coupling reactions [Eq. (9)], carried out by Bodanszky and Kwei (1978), indicates that acidic or bulky side chains (R) slow the formation of the aspartimide derivative (**31**) and that serine and threonine side chains accelerate the reaction.

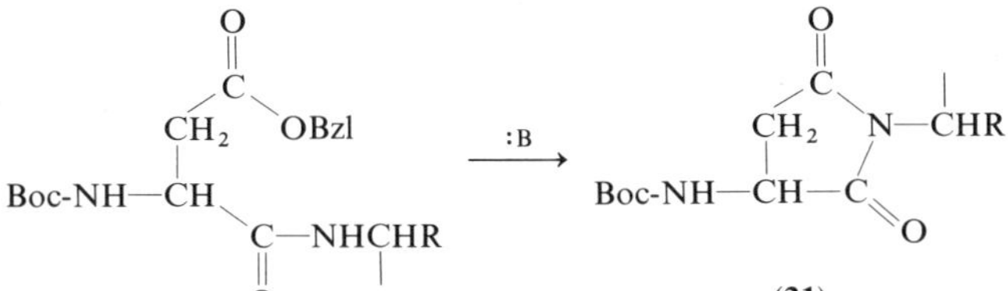

(**31**) (9)

When the coupling reaction is fast, as in the dicyclohexylcarbodiimide or mixed anhydride reactions, the amount of imide (**31**) produced is usually not significant, but when active esters are used, the coupling is slower and should be catalyzed, e.g., with hydroxybenzotriazole. This catalyst does not accelerate the formation of the aspartimide.

C. Reduction with Zinc in Acetic Acid

There is a recent report of the selective cleavage of *p*-nitrobenzyl esters in the presence of benzyl esters by reduction with powered zinc in 90% acetic acid (Suzuki *et al.*, 1976). As illustrated in Eq. (10), the procedure works well

$$\text{Boc-Glu(OBzl)-Gly-ONb} \xrightarrow[\text{0°C, 1 1/2 hr}]{\text{Zn, 90\% HOAc}} \underset{92\%}{\text{Boc-Glu(OBzl)-Gly-OH}} \qquad (10)$$

with a series of small peptides including a methionine-containing peptide, but it has not been tried on larger substrates. The benzyloxycarbonyl group was not cleaved under these conditions.

D. Catalytic Transfer Hydrogenolysis

Catalytic transfer hydrogenation is becoming a useful alternative to the use of gaseous hydrogen for the removal of various protecting groups from peptides. The procedure involves stirring a protected peptide in acetic acid or ethanol with palladium catalyst and a five- to tenfold excess of a hydrogen donor, such as cyclohexene (Jackson and Johnstone, 1976), cyclohexadiene (Felix *et al.*, 1978b), hydrazine (Anwer *et al.*, 1978), or formic acid (El Amin *et al.*, 1979; Sivanandaiah and Gurusiddappa, 1979). Benzyloxycarbonyl groups and benzyl esters are cleaved in five to ten minutes at 20–25°C (reaction 11); benzyl ethers require a somewhat longer time and the nitro group on arginine requires about five hours.

$$\underset{200\ \text{mg}}{\text{Z-Met-Gly-OH}} \xrightarrow[\substack{200\ \text{mg Pd black,}\\ 10\ \text{min}}]{\substack{\text{MeOH,}\\ 4.4\%\ \text{HCOOH}}} \underset{92\%}{\text{HCOOH}\cdot\text{H-Met-Gly-OH}} \qquad (11)$$

Generally, a 1:1 ratio of catalyst to reducible group is required for a fast reaction. Recently a more convenient procedure using palladium black or palladium on charcoal packed in a column has been described (El Amin *et al.*, 1979). The same column can be used repeatedly.

A useful review of earlier work on catalytic transfer hydrogenation has appeared (Brieger and Nestrick, 1974).

E. 4-Picolyl Esters

This group provides an attractive approach to carboxyl protection in that it combines the advantages of working in solution with one of the advantages of solid-phase peptide synthesis—that of using an excess of acylating agent (Camble *et al.*, 1969). After the coupling reaction is complete, the mixture is adsorbed on an ion exchange resin (H^+ form); the resin is washed with tetrahydrofuran or dimethylformamide to remove the excess acylating agent and any other nonbasic contaminants and the product, e.g., **32** is eluted with a

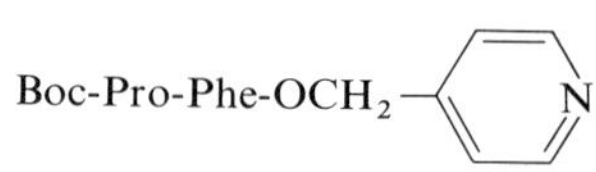

32

tertiary amine in a suitable solvent. Decomposition of a *tert*-butyloxycarbonyl group by the resin can be prevented by first saturating the resin with 3-bromopyridine, which is then displaced by the more basic picolyl group. An alternative procedure, if solubility permits, is to extract the product from an organic solvent into aqueous acid.

The method has been used effectively for the synthesis of bradykinin, where the protected peptide was obtained in 42% yield (Shafer *et al.*, 1971) and of [Val[5]]angiotensin II, to give the protected peptide in 38% yield (Garner and Young, 1971). It has the advantage of providing pure products in high yields and allows the characterization of the product at each step. A potential disadvantage of the method is the need to remove the picolyl ester by hydrogenolysis from the completed peptide. This capricious reaction may be very slow (e.g., 72 hr) and may then cause some side reactions, such as saturation of aromatic rings. However, the picolyl ester presents some interesting possibilities for selective removal—it can be cleaved by electrolytic reduction at a mercury cathode under the conditions used for removal of the nitro group from nitroarginine (Young and Kotai 1973)—conditions which do not cleave a benzyloxycarbonyl group. It can also be cleaved by treatment with sodium in liquid ammonia or by saponification.

Picolyl esters have been prepared by reaction of benzyloxycarbonyl amino acids with 4-picolyl alcohol and dicyclohexylcarbodiimide followed by acidolysis of the benzyloxycarbonyl group with HBr in AcOH. They are considerably more stable to this reagent than are benzyl esters. A more convenient method of preparation is that of Maclaren (1972) which uses ethyl acetoacetate as a transient amine protecting group.

The principle of attaching a basic "handle" to a peptide to facilitate purification has been extended to the preparation of COOH-terminal hydrazides (Macrae and Young, 1974) which are useful in the synthesis of longer peptides by fragment condensation. Equation (12) illustrates the deprotection of hydrazide.

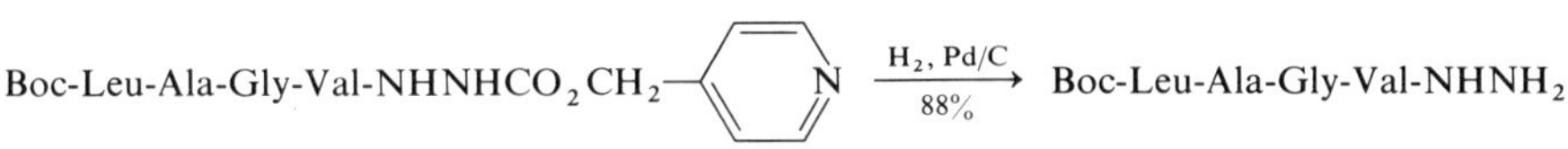

(12)

F. Hydrogenolysis in Solid-Phase Synthesis

It had been assumed until recently that hydrogenolysis could not be used in solid-phase synthesis because of the difficulty of allowing two solid phases—the catalyst and the resin-bound peptide—to interact. But Schlatter *et al.* (1977) showed that Boc-Tyr-Gly-Gly-Phe-Val-OH that was attached to 1% cross-linked polystyrene by the usual benzyl ester bond could be cleaved from the resin in excellent yield by this method. This was accomplished by equilibrating the peptide resin with about two equivalents of $Pd(OAc)_2$ in dimethylformamide for 15 min, then hydrogenating at 60 psi for 24 hr at 40°C. The peptide was obtained in 84% yield in the first such reaction, and the rest was removed in a repeat run. In an accompanying note, Jones (1977) described the preparation of Boc-Tyr-Gly-Gly-Phe-Leu-OH by the same procedure. This exciting development may well extend the usefulness of solid-phase synthesis, particularly in the preparation of N-protected fragments for the synthesis of long peptide chains.

VI. PROTECTING GROUPS REMOVED BY TREATMENT WITH ALKALI

With the exception of the *tert*-butyl ester, all of the esters used as carboxyl protecting groups can be removed by treatment with alkali. However, in most cases this is not the best method. The groups described in this section can be cleaved under mild conditions, either by $B_{AC}2$ hydrolysis or by β-elimination.

The discussion on methyl esters pertains of course to other simple alkyl esters as well, but since they are saponified more slowly there is usually no advantage in using them as protecting groups when the methyl esters would do.

A. Methyl Esters

Methyl esters have been popular protecting groups for many years because they are easy to prepare and easy to remove by saponification. But the number of known side reactions caused by the treatment of peptides with alkali has grown to the point that one is reluctant to commit a peptide to strong base. Some of these side reactions, which are discussed below, can be prevented and others can be minimized by carrying out the hydrolysis under mild conditions, e.g., by adding the aqueous base in small portions (Harris and Fruton, 1951), by monitoring the consumption of base by adding thymolphthalein (color change at pH 10.5) (Grassmann and Wünsch, 1958), or by using an automatic titrater to maintain the reaction mixture at pH 11 (Rittel *et al.*, 1957).

The finding (Kenner and Seely, 1972) that the COOH-terminal residues of Z-Ala-Phe-OMe and Z-Leu-Ala-OMe were racemized to the extent of 2.8% and 0.8%, respectively, during saponification with one equivalent of 0.25 *N* NaOH in acetone for one hour at 20°C indicates that one should be aware of that possibility in the saponification of any peptide ester. The mechanism of the racemization—whether it was the result of direct exchange of the α-hydrogen atom or whether it proceeded via the formation of a small amount of 5(4*H*)-oxazolone—was not established. However, it should be noted that McDermott and Benoiton (1973) found no racemization ($<0.1\%$) during the saponification of Z-Ala-Leu-OMe with 1.1 equivalents of NaOH in aqueous methanol for one hour.

Esters of *N*-methylamino acids are more susceptible to base-induced racemization than are the analogous unmethylated compounds. Thus Z-MeIle-OMe gave 12% of Z-DMeIle-OH under the conditions (1.1 equiv OH^-, 24 hr) that produced 1% of the D-isomer from Z-Ile-OMe (McDermott and Benoiton, 1973).

Extensive racemization (37% DL) of the cysteine residue in Z-Gly-Cys(Bzl)-OEt was observed when the peptide was treated with one equivalent of NaOH (Maclaren, 1958), but Z-Cys(Bzl)-Gly-OEt is not racemized under the same conditions. *N*-Benzyloxycarbonyl peptide esters having a glycine residue adjacent to the NH_2-terminal residue undergo cyclization to a hydantoin and its hydrolysis product (**33**) when they are treated with two equivalents of NaOH (Scheme 5). Hydantoin formation occurs also when Z-X-Gly-OR is treated with ammonia in methanol (Dekker *et al.*, 1949; Fruton and Bergmann, 1942). The cyclization may occur also with other peptide esters, but less readily (Waley and Watson, 1953).

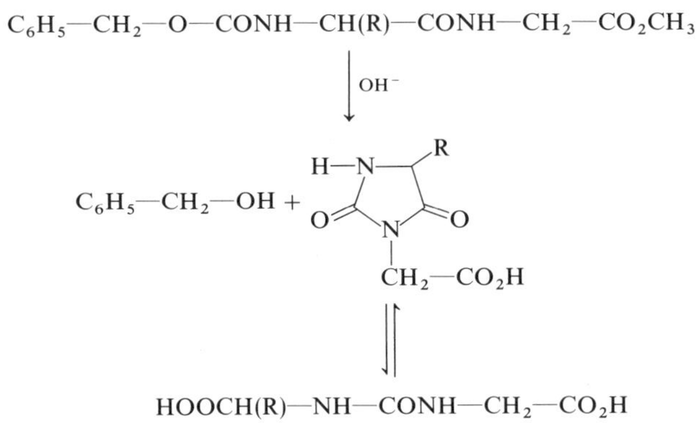

33

Scheme 5

Serine-containing peptides having an unprotected hydroxyl group are particularly susceptible to base-catalyzed side reactions. Thus, the saponification of Z-Ser-Tyr-OMe, carried out under various conditions of temperature, solvent, concentration of hydroxide ion, and time never gave more than 50% yield of Z-Ser-Tyr-OH, and free serine was detected in the saponification mixture (Guttmann and Boissonnas, 1958). The saponification of Z-Ser-Met-OMe and Z-Ser-Tyr-OEt gave the corresponding acids in yields of 67% and 50%, respectively (Hofmann *et al.*, 1957), and Z-DL-Ser-Gly-OEt gave only a 59% yield of the expected dipeptide (Zahn and Schnabel, 1957). The alkali instability of serine-containing peptides was also noticed by Harris and Fruton (1951). This problem can be avoided by protecting the serine hydroxyl with a group that is stable in base, such as the benzyl ether.

The tendency of aspartyl and, to a lesser extent, glutamyl residues having an esterified side-chain carboxyl group to cyclize in the presence of base has been discussed in the section on *tert*-butyl esters.

Several amine protecting groups are unstable under the conditions required for the saponification of methyl esters—notably the trifluoroacetyl group, which is cleaved, and the phthaloyl group, which opens to form an *N*-benzoyl derivative.

The use of nonbasic nucleophiles for cleaving methyl esters is discussed in Section VII,A.

B. Phenyl Esters

The use of phenyl esters (OPh) as carboxyl protecting groups (Kenner and Seely, 1972), appears to be an important recent contribution to the art of peptide synthesis. The key to their usefulness is their rapid reaction with peroxide anion under moderately basic conditions (pH 10.5) to give, presumably, a peracid, which spontaneously decomposes to a carboxylic acid. The reaction is carried out with an autotitrator to maintain pH 10.5 and to monitor the reaction, which is complete in 20 min or less at 20°C. No racemization could be detected when the ester groups of Z-Ala-Phe-OPh and Z-Leu-Ala-OPh were cleaved under these conditions, and Z-Asp(O*t*Bu)-Gly-OPh gave none of the $\alpha \rightarrow \beta$ aspartyl shift that heretofore had always been observed in base. Under the same conditions, but in the absence of peroxide, the rate of hydrolysis of Z-Ala-Phe-OPh and Z-Leu-Ala-OPh was an order of magnitude less and extensive racemization of the COOH-terminal residue was observed. When tryptophan or cysteine is present, a large excess of dimethyl sulfide is added to prevent oxidation of these residues.

The method has been used extensively in the successful synthesis of a 1–75 fragment of a lysozyme analogue (Galpin *et al.*, 1976) and of human big gastrin I (Choudhury *et al.*, 1976). A variety of coupling methods was used in these syntheses, including the *N*-hydroxysuccinimide ester, which seems surprising, since phenyl esters might be expected to react slowly with amines. Because of solubility problems, phenyl ester cleavage was often conducted in a mixture of solvents, such as aqueous trifluoroethanol or aqueous dimethylformamide. Methanol and ethanol are avoided when working with phenyl esters to prevent transesterification.

C. Polyethylene Glycol

The use of polyethylene glycol of about 10,000 molecular weight as a carboxyl protecting group for the synthesis of H-Leu-Thr-Ala-Leu-Gly-OH was described by Bayer and Mutter (1972; see also Mutter and Bayer, 1980). The soluble polymeric anchoring group has some of the advantages of the solid-phase method and also allows purification at each step by a method based on differences in molecular size, such as membrane filtration. Removal of the peptide from the polymer was carried out by saponification with 0.1 *N* NaOH in dioxane–water.

D. Esters Cleaved by β-Elimination

Alkyl esters of the type RCOO—CH_2—CH_2—X, where X is one of the leaving groups (**34**–**37**) undergo base-catalyzed cleavage by β-elimination and have been proposed as protecting groups in peptide synthesis. The

—$S(CH_3)_2$

(**34**)

—SO_2CH_3

(**35**)

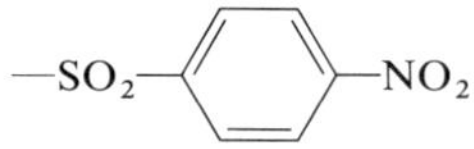

(**36**)

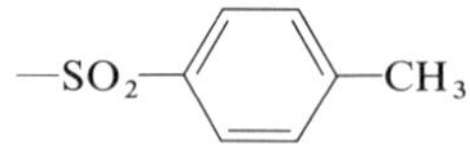

(**37**)

β-methylthioethyl ester, after conversion to the sulfonium salt (**34**), is cleaved in ten minutes at room temperature at pH 10–10.5, but formation of the sulfonium iodide is sometimes accompanied by side reactions (Amaral *et al.*, 1966). An alternative procedure starts with the same β-methylthioethyl ester but involves oxidation to the sulfone (**35**), which requires 12–24 hr for complete cleavage under the same conditions. This procedure does not prevent the intramolecular formation of aspartimides from β-esterified aspartyl residues (Hardy *et al.*, 1968). Amaral (1969) also recommended the *p*-nitro derivative (**36**); the 2-(*p*-nitrophenylthio)ethyl ester precursor can be obtained by direct esterification of amino acids with the corresponding alcohol. The only ester of this class that has been used in a complex synthesis is the β-*p*-toluenesulfonylethyl ester (**37**) which was proposed by Miller and Stirling (1968) and used by Ludescher and Schwyzer (1972) in the synthesis of a bicyclic analogue of gramicidin S, where it was cleaved in 40 min at pH 11.5 to give the desired product in 75% yield. It can be cleaved also by treatment with cyanide ion [Eq. (13)] (Miller and Stirling, 1968).

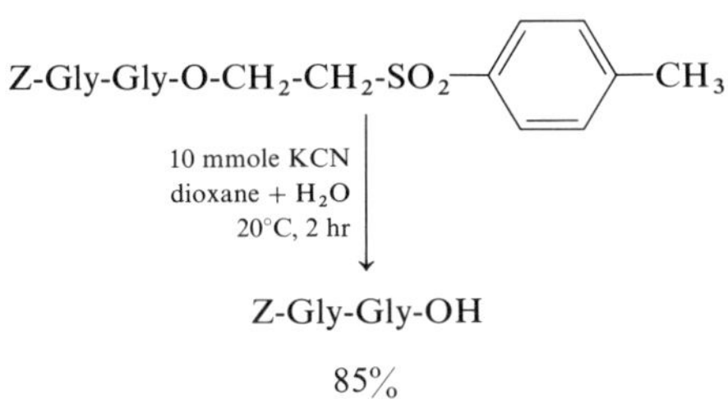

(13)

VII. PROTECTING GROUPS CLEAVED BY NUCLEOPHILES UNDER NEUTRAL CONDITIONS

There has been renewed interest recently in methods of cleaving esters under neutral conditions, and some of the methods are applicable to peptide synthesis. Most of these reactions proceed by nucleophilic attack on carbon, resulting in alkyl–oxygen cleavage of the ester [Eq. (14)].

$$\mathrm{R{-}\overset{\displaystyle O}{\overset{\|}{C}}{-}O{-}CH_2{-}R'} \xrightarrow{} \mathrm{R{-}\overset{\displaystyle O}{\overset{\|}{C}}{-}OH + R'CH_2X} \qquad (14)$$

$$\mathrm{X^-}$$

A. Methyl Esters

Taschner and Liberek (1956) showed that methyl esters and ethyl carbamates could be cleaved by treatment with excess lithium iodide in refluxing pyridine, and the method was developed further by Elsinger *et al.* (1960) and by Dean (1965) who found that the reaction proceeds faster in dimethylformamide. Lithium mercaptides (Bartlett and Johnson, 1970) are more effective than iodide, especially in hexamethylphosphoramide; e.g., methyl mesitoate is cleaved at room temperature in two hours (Kelly *et al.*, 1977). Apart from the early work of Taschner who found that phthaloyl and tosyl groups were stable but that benzyloxycarbonyl groups were cleaved by LiI in refluxing pyridine, this reaction has not been used in peptide chemistry, and seems to be worth investigating.

Iodotrimethylsilane has been proposed recently by Ho and Olah (1976) and by Jung and Lyster (1977) for the cleavage of esters (and ethers) under mild, neutral conditions [Eq. (15)]. The rate of cleavage of various esters is

$$R\text{—}\overset{\overset{\displaystyle O}{\|}}{C}\text{—}OR' + (CH_3)_3SiI \xrightarrow[50^\circ\text{–}100^\circ C,\ \sim 2hr]{\text{aprotic solvent}} R\text{—}\overset{\overset{\displaystyle O}{\|}}{C}\text{—}O\text{—}Si(CH_3)_3 \xrightarrow{H_2O} R\text{—}\overset{\overset{\displaystyle O}{\|}}{C}\text{—}OH + (CH_3)_3SiOH \qquad (15)$$

tert-butyl > benzyl > methyl esters. Amides, amines, and thioethers are not affected, but carbamates are cleaved. The reagent has not yet been used in peptide synthesis but might be useful, particularly as a last step in a synthetic scheme, when carbamate and ester protecting groups can be removed simultaneously.

B. Phenacyl Esters

Phenacyl esters (OPa), easily prepared from phenacyl bromide and a carboxylate salt, undergo alkyl–oxygen cleavage when they are treated with sodium thiophenoxide in dimethylformamide [Eq. (16)], and can also be cleaved by reduction with zinc in acetic acid (Sheehan and Daves, 1964; Ledger and Stewart, 1965; Stelekatos *et al.*, 1966; Taylor-Papadimitrou *et al.*, 1967; Hendrickson and Kendall, 1970). Yang and Merrifield (1976) studied the use of -Asp(OPa)- in solid-phase synthesis. The phenacyl ester

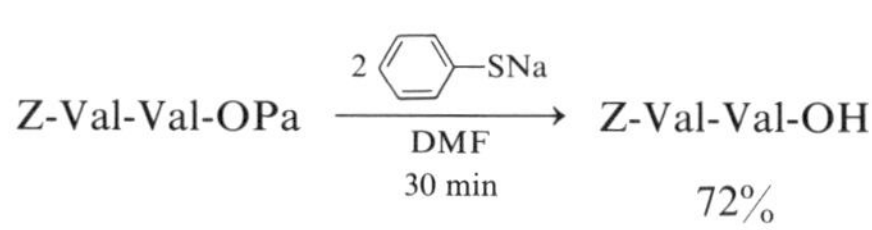

(16)

was removed from the resin-bound peptide by treatment with 1 *M* thiophenoxide in dimethylformamide at 25°C for 8 hr, before the final treatment with liquid HF. This virtually eliminated the aspartimide formation that was found to be a serious side reaction when peptides containing -Asp(OBzl)- residues were treated with HF. Although H-Asp(OPa)-OH is stable in liquid HF, an -Asp(OPa)- residue in a peptide chain will undergo aspartimide formation readily in HF, hence the need to remove the ester before liquid HF treatment. In a similar vein, Suzuki *et al.* (1977) used the γ-phenacyl ester of glutamyl residues to minimize the side reactions that occur when glutamic-containing peptides are treated with liquid HF. The OPa group was removed later by treatment with zinc powder in 90% acetic acid.

Phenacyl esters are incompletely cleaved by catalytic hydrogenation with Pd–C because of the slower but simultaneous reduction of the ketone group to give a β-phenethyl ester, which is not cleaved (Taylor-Papadimitrou *et al.*, 1967).

Phenacyl esters have been used in solid-phase synthesis to attach the first amino acid to the resin (Weygand, 1968; Mizoguchi *et al.*, 1969; Shigezane and Mizoguchi, 1972), and the completed peptide was then cleaved from the resin by thiolysis or by hydrazinolysis. An α-methylphenacyl ester link was used recently (Tam *et al.*, 1977) and removed by reaction with KCN in dimethylformamide, catalyzed by dicyclohexyl-18-crown-6 [Eq. (17)]. The α-methylphenacyl ester was cleaved also by photolysis and by

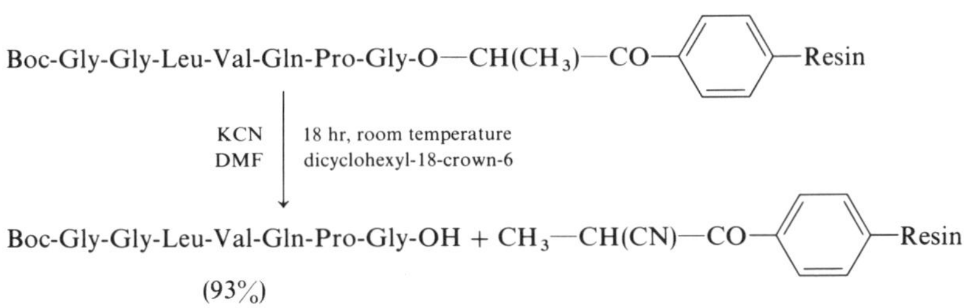

(17)

hydrazinolysis (Wang, 1976). A side reaction encountered during the use of a phenacyl ester in solid-phase synthesis is cyclization of the first amino acid residue, causing about 30% of the chain to be terminated (Birr, 1977; Tam *et al.*, 1977) [Eq. (18)].

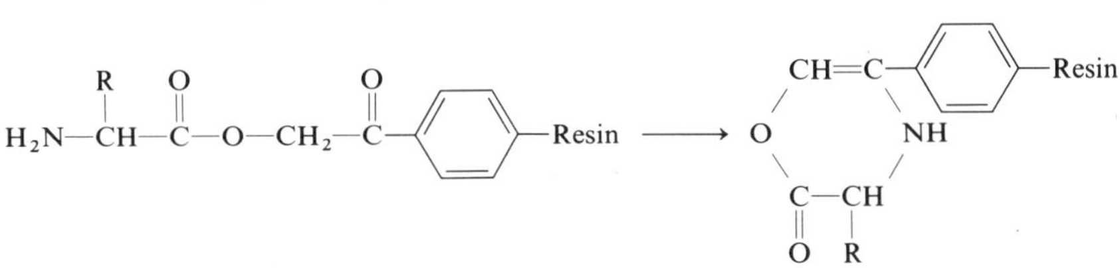

(18)

C. Phthalimidomethyl Esters

Phthalimidomethyl esters, discussed in Section IV,E, are cleaved also by reaction with sodium thiophenoxide (Sheehan and Daves, 1964) and by zinc in acetic acid but not by catalytic hydrogenation (Nefkens *et al.*, 1963).

D. 9-Anthrylmethyl Esters

The 9-anthrylmethyl ester (**38**), introduced into peptide chemistry by F. H. C. Stewart (1965) as an acid-sensitive group, was later found to be extraordinarily reactive toward thiolate anions (Kornblum and Scott, 1974) [Eq. (19)]. Thus, 9-anthrylmethyl mesitoate is deblocked completely in less than 30 sec by reaction with CH_3SNa in hexamethylphosphoramide. This aspect of 9-anthrylmethyl esters has not been put to use in peptide chemistry.

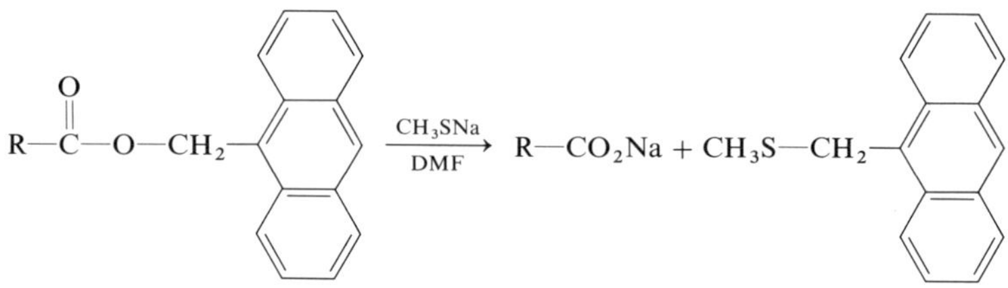

(19)

(**38**)

E. 2-Trimethylsilylethyl Esters

The 2-trimethylsilylethyl ester (OTms) was recently proposed as a carboxylic acid protecting group that can be removed by treatment with fluoride ion at room temperature (Sieber *et al.*, 1977). 2-Trimethylsilylethyl esters can be prepared [Eq. (20)] from 2-trimethysilylethanol (**39**) which is not commer-

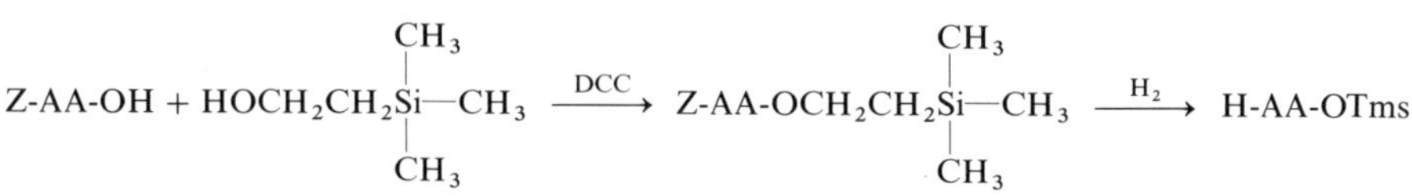

(**39**) (20)

cially available but can be synthesized in two steps from vinyl acetate and dichloromethylsilane. The OTms group is cleaved readily with a quaternary ammonium fluoride in dimethylformamide [Eq. (21)] and is complete in less than 30 min at 25°C when an excess of $[Et_4N]^+F^-$ is used. Under these conditions, the Boc, Bpoc, *N*-Trt, *S*-Trt, S-Acm and *tert*-butyl ether groups are

$$\text{Prot-AA-O—CH}_2\text{CH}_2\text{—SiMe}_3 \xrightarrow{F^-} \text{Prot-AA—OH} + \text{CH}_2\text{=CH}_2 + \text{FSiMe}_3 \qquad (21)$$

stable, but benzyloxycarbonyl groups tend to form hydantoins, -Asp(O*t*Bu)- forms the aspartimide, benzyl esters are cleaved partially, and disulfides undergo disproportionation. In spite of its limitations, the 2-trimethylsilylethyl ester is useful in some situations; it served well in the synthesis of the chain insulin B segment (1–15) [Eq. (22)].

Boc-Phe-Val-Asn-Gln-His-Leu-Cys(Acm)-Gly-Ser(O*t*Bu)-His-Leu-Val-Glu(O*t*Bu)-Ala-Leu-OTmse

$$\xrightarrow[\text{DMF, 25°C, 10 min}]{\text{10 equiv. } [\text{Et}_4\text{N}]^+\text{F}^-}$$ quantitative yield of the corresponding acid, without racemization

(22)

VIII. PROTECTING GROUPS REMOVED BY PHOTOLYSIS

The concept of removing a protecting group by photolysis is appealing, but the practice has developed slowly. In model experiments, Patchornik *et al.* (1970) removed the 2,2′-dinitrodiphenylmethyl ester (ONpm) from Boc-Ala-ONpm and Boc-Phe-ONpm by photolysis, and Sheehan *et al.* (1971) cleaved a substituted benzoin ester of phthaloylglycine [Eq. (23)].

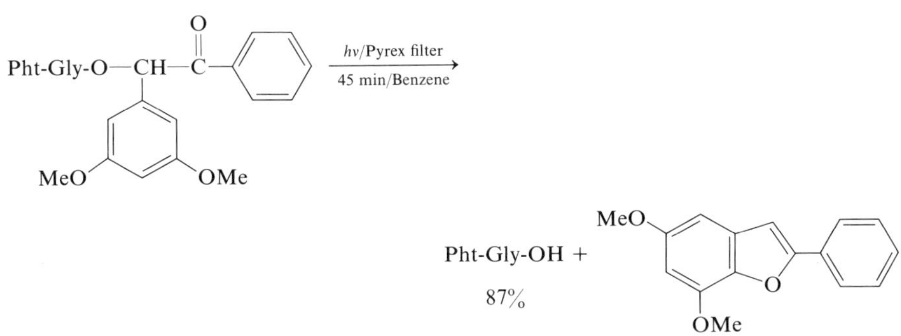

(23)

Rich and Gurwara (1975a,b), using an *o*-nitrobenzyl ester link to polystyrene, prepared a series of protected peptide acids related to luliberin (LHRH). The protected peptides were cleaved from the resin by photolysis for 18–24 hr in methanol and were obtained in excellent yields [Eq. (24)].

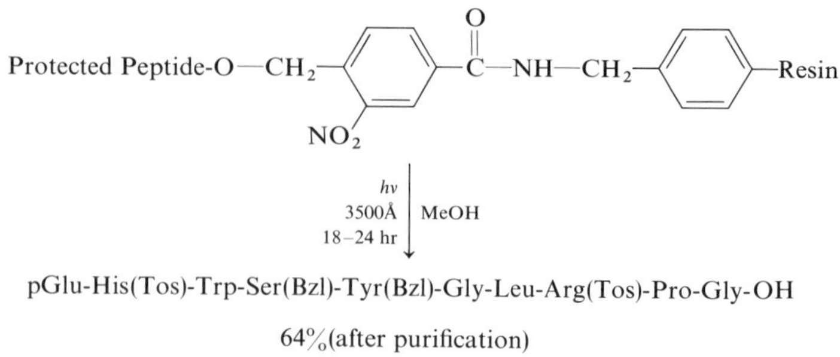

(24)

The corresponding COOH-terminal amide was prepared by photolysis of an *o*-nitrobenzyl amide link (Scheme 6).

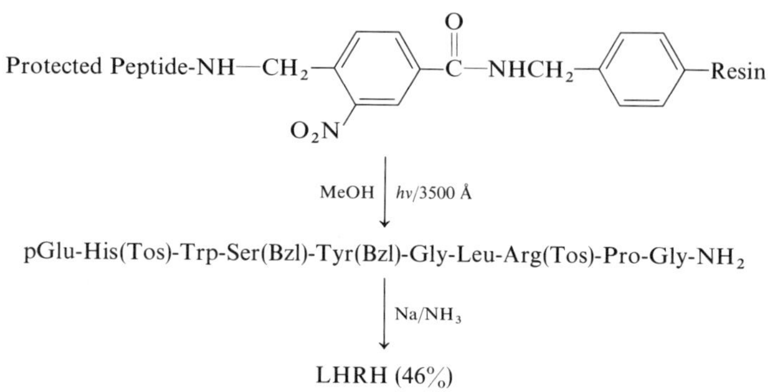

Scheme 6

Wang (1976) used the α-methylphenacyl ester, proposed by Sheehan and Umezawa (1973) as a photolabile link to polystyrene for the solid-phase synthesis of the tetrapeptide **40** [Eq. (25)]. The same peptide was obtained

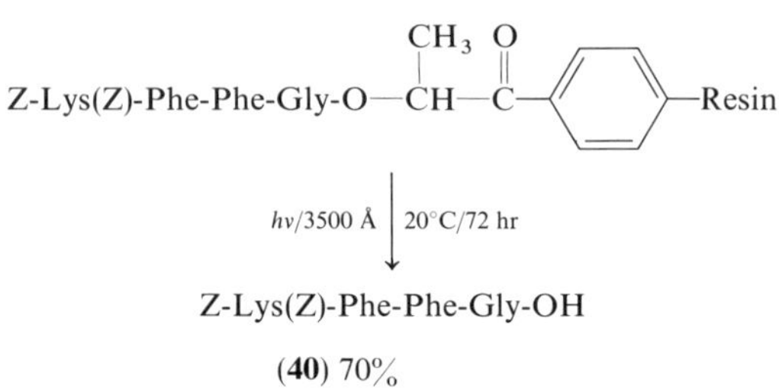

(25)

in only 40% yield when the *o*-nitrobenzyl ester link was used, presumably because photolysis of this ester is slower than that of the α-methylphenacyl ester; 72-hr photolysis was used in both cases. Preliminary experiments indicated that photolysis of the α-methylphenacyl ester of amino acids other than glycine was only $\frac{1}{2}$ to $\frac{1}{5}$ as fast as that with glycine, so that removal from a polystyrene resin might require one to two weeks under the same conditions.

IX. METHODS OF PREPARING AMINO ACID ESTERS

A. From Free Amino Acids

The direct acid-catalyzed esterification of an unprotected amino acid by reaction with an alcohol is the simplest method of preparing esters but it is applicable to only a few alcohols. The preparation of methyl and ethyl esters by this method was reviewed by Schröder and Lübke (1965). Benzyl esters of amino acids are usually prepared by refluxing a suspension of the amino acid and one equivalent of *p*-toluenesulfonic acid in a mixture of benzyl alcohol and benzene or toluene, with azeotropic distillation of the water using a Dean-Stark trap (Cipera and Nicholls, 1955). Variations on this procedure are also discussed by Schröder and Lübke (1965). The 2-(*p*-nitrophenylthio)ethyl esters of glycine and leucine were prepared by this method (Amaral, 1969).

Unprotected amino acids react with diphenyldiazomethane to give good yields of the diphenylmethyl esters (Aboderin *et al.*, 1965). The reaction is run on the *p*-toluenesulfonate or the β-naphthalenesulfonate salts, which dissolve in dimethylformamide.

Many unprotected amino acids are sufficiently soluble in a mixture of dioxane, concentrated sulfuric acid and isobutene to give the *tert*-butyl esters in good yields (Roeske, 1959, 1963). *p*-Toluenesulfonate salts can be used in place of sulfuric acid in this reaction.

B. From N-Protected Amino Acids

When esterification of the free amino acid is not feasible, esters are prepared generally via the *N*-protected amino acid, followed by a selective cleavage of the *N*-protecting group. Two general methods predominate; the alkylation of a protected amino acid carboxylate salt with an alkyl halide or tosylate and the carbodiimide-mediated coupling of a protected amino acid with an alcohol. Many variations on these general reactions have been tried, e.g., the silver salts of *o*-nitrophenylsulfenyl, trityl, or formyl amino acids react with with diphenylmethyl chloride to give the diphenylmethyl esters in good yields (Gazis *et al.*, 1966). The triethylamine salts of benzyloxycarbonylamino acids, when refluxed with *p*-nitrobenzyl chloride in ethyl acetate, form the *p*-nitrobenzyl esters (Schwyzer and Sieber, 1959; Schwarz and Arakawa, 1959). In a similar manner, the 9-anthrylmethyl (F. H. C. Stewart, 1965), phthalimidomethyl (Nefkens *et al.*, 1963) and trimethylbenzyl (F. H. C. Stewart, 1967a) esters have been made from their chloromethyl precursors.

Recently the cesium salts of various N-protected amino acids and peptides have been used to prepare esters by reaction with alkyl halides in dimethylformamide. The method gave good yields of the methyl, benzyl, *p*-methoxybenzyl, *o*-nitrobenzyl, α-methylphenacyl, and phthalimido esters and is undoubtedly applicable in a wide variety of situations (Wang *et al.*, 1977). Even *tert*-butyl bromide reacted with Z-Ala-O^-Cs^+ to give Z-Ala-O*t*Bu, but in only 17% yield. The relatively high cost of cesium salts is a deterrent to the large-scale use of this procedure, but preliminary experiments indicate that the same results can be obtained with much less expense by using the corresponding potassium salts with a catalytic amount (10%) of 18-crown-6 in dimethylformamide at room temperature (R. W. Roeske, unpublished observation, 1977).

Synthesis of esters with dicyclohexylcarbodiimide is probably less satisfactory than the alkylation method because of the formation of *N*-acylurea by-products. Some examples of its use are in the preparation of 2-sulfonylethyl esters (Miller and Stirling, 1968), of 4-picolyl esters (Camble *et al.*, 1969), and of 2-trimethylsilylethyl esters (Sieber *et al.*, 1977).

Maclaren (1972) used ethyl acetoacetate as temporary protection for the amino group; the intermediate (**41**) is alkylated at the carboxyl group by various benzyl halides, and mild acidolysis removes the amine protecting group. The reaction is convenient for the preparation of *p*-nitrobenzyl, 4-picolyl, trimethylbenzyl, and 4-methoxybenzyl esters and can be carried out without isolating the intermediates.

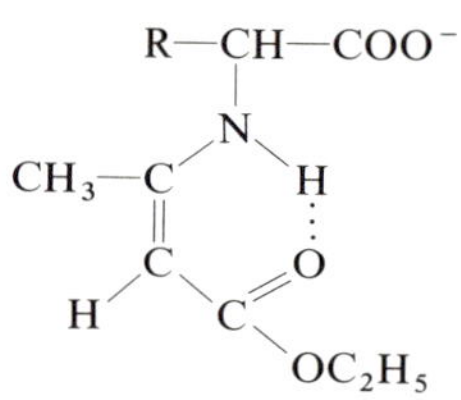

(**41**)

The ancient method of Bergmann *et al.* (1935) for the preparation of esters of N^ε-protected lysine is still useful [Eq. (26)].

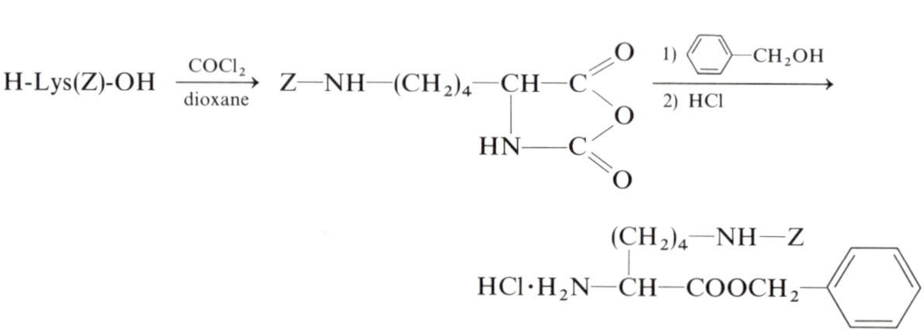

(26)

Many procedures have been developed for the synthesis of ω-esters of glutamic and aspartic acids by partial esterification of the free amino acids, and these are reviewed by Schröder and Lübke (1965). Ledger and Stewart

Table I Cleavage Conditions for Carboxyl Protecting Groups and Some Amine Protecting Group

Protecting groups	Abbrev	Reagents and Conditions[a]										
		H_2O, Room temp	75% AcOH, 1 hr, 25°C	1 *N* HCl in AcOH, 30 min, 25°C	50% TFA in CH_2Cl_2, 30 min	2 *N* HBr in AcOH, 30–60 min, 25°C	OH^- in ROH + H_2O, 25°C, 2 hr	HF, 0°C, 30 min	RS^-Na^+, DMF, 25°C	NH_2NH_2 in ROH, 25°C 2 hr	H_2–Pd–C, 25°C	Zn in 90% AcOH, 0°C, 1½ hr
Carboxyl												
Trimethylsilyl ester	OTms	+	+	+	+	+	+	+				+
Diphenylmethyl ester	ODpm	−	−	+	+	+	+	+		R	+	
tert-Butyl ester	O*t*Bu	−	−	+	+	+	−	+	−	−	−	−
p-Methoxybenzyl ester	OMob	−	−	+	+	+	+	+	−	R	+	−
Pentamethylbenzyl ester	OPmb	−	−	+	+	+	+	+	−	R	+	−
Trimethylbenzyl ester	OTmb	−	−	−+	−+	+	+	+	−	R	+	−
Phthalimidomethyl ester	OPtm	−	−	−+	−+	+	R	+	+	+	−	
9-Anthrylmethyl ester	OAm	−	−	−+	−+	+	+	+	+	+	+	
Benzyl ester	OBzl	−	−	−	−	−	+	+	−	R	+	−
p-Nitrobenzyl ester	ONb	−	−	−	−	−	+	−	−	R	+	+
Methyl ester	OMe	−	−	−	−	−	+	−	−	R	−	−
Phenyl ester	OPh	−	−	−	−	−	+		−	R	−	−
4-Picolyl ester	OPic	−	−	−	−	−	+		−	R	+	
Phenacyl ester	OPa	−	−	−	−	−	+	−	+	+	+−	+
Amine												
2-(*p*-Biphenylyl) propyl(2)oxycarbonyl	Bpoc	−	+	+	+	+	−	+	−	−	−	+
tert-Butyloxycarbonyl	Boc	−	−	+	+	+	−	+	−	−	−	−
Benzyloxycarbonyl	Z	−	−	−	−	+	−	+	−	−	+	−
Phthaloyl	Pht	−	−	−	−	−	R	−	−	+	−	−
Trityl	Trt		+	+	+	+	−	+	−	−	+ (slow)	+

[a] + indicates cleavage; − indicates no reaction; R: reaction, but not to give a carboxylic acid; + −: partial cleavage.

(1965) prepared the ω-benzyl esters of aspartic and glutamic acid by reaction of benzyl iodide with the alkali metal salts of the copper complexes of these amino acids. The yield of H-Glu(OBzl)-OH by this procedure was 27%.

Better yields of the monoesters are obtained using the method of Prestidge *et al.* (1975), which involves the cupric ion-assisted selective hydrolysis of the diester under very mild conditions. Thus, glutamic acid γ-*p*-nitrobenzyl ester was obtained in 54% yield by this procedure [Eq. (27)].

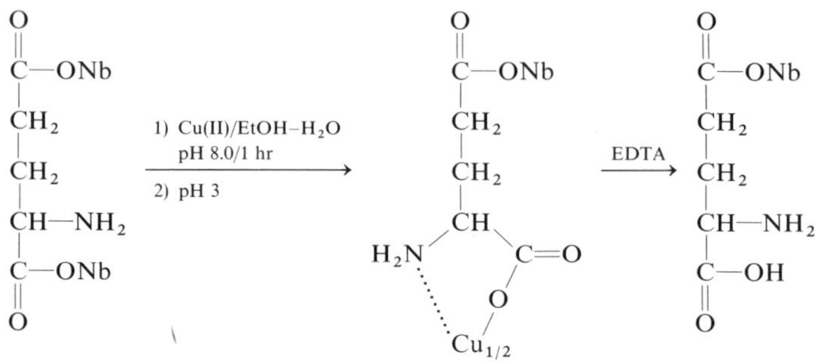

(27)

X. CONCLUDING REMARKS

Although the variety of carboxyl protecting groups now available (Table I) is adequate for the synthesis of most small peptides, there is still a definite need to develop conditions for the selective removal of a terminal carboxyl protecting group from a completely protected peptide for use in a fragment condensation scheme. The weakly basic nucleophiles such as thiolate anions and cyanide ion seem to offer promise in this regard, but they need to be explored further, particularly in combination with crown ethers. The phenyl ester, removable at pH 10.5 in the presence of peroxide ion, also appears to warrant much wider application.

A preliminary report (Royer and Anantharamaiah, 1979) of the use of carboxypeptidase Y to hydrolyze an ethyl ester without hydrolysis of peptide bonds is exciting. The pure enzyme has only esterase activity at pH 8.5, and may become quite useful in synthesis if it can be persuaded to work in mixtures of organic solvents and water.

REFERENCES

Aboderin, A. A., Delpierre, G., and Fruton, J. (1965). *J. Am. Chem. Soc.* **87**, 5469–5472.

Adamson, J. R., Bywood, R., Eastlick, D. T., Gallagher, G., Walter, D., and Wilson, E. M. (1975). *J. Chem. Soc., Perkin Trans. 1* pp. 2030–2033.

Amaral, M. (1969). *J. Chem. Soc. C* pp. 2495–2497.

Amaral, M., Barrett, G. C., Rydon, H. N., and Willett, J. E. (1966). *J. Chem. Soc. C* pp. 807–813.

Anderson, G. W., and Callahan, F. M. (1960). *J. Am. Chem. Soc.* **82**, 3359–3363.
Anwer, M. K., Khan, S. A., and Sivanandaiah, K. M. (1978). *Synthesis.* pp. 751–752.
Bartlett, P. A., and Johnson, W. S. (1970). *Tetrahedron Lett.* pp. 4459–4462.
Battersby, A. R., and Robinson, J. C. (1955). *J. Chem. Soc.* pp. 259–269.
Bayer, E., and Mutter, M. (1972). *Nature (London)* **237**, 512–513.
Beacham, J., Bentley, P. H., Kenner, G. W., MacLeod, J. K., Mendive, J. J., and Sheppard, R. C. (1967). *J. Chem. Soc. C* pp. 2520–2529.
Bergmann, M., Zervas, L., and Ross, W. F. (1935). *J. Biol. Chem.* **111**, 245–260.
Bywood, R., Gallagher, G., Sharma, G. K., and Walker, D. (1975). *J. Chem. Soc. Perkin Trans. 1* pp. 2019–2022.
Birkhofer, L., and Ritter, A. (1965). *Angew. Chem., Int. Ed. Engl.* **4**, 417–429.
Birr, C., Wengert-Müller, M., and Buku, A. (1977). *In* "Peptides, Proceedings of the Fifth American Peptide Symposium" (M. Goodman and J. Meienhofer, eds.), pp. 510–513. Wiley, New York.
Bodanszky, M. (1968). *In* "Peptides 1968" (E. Bricas, ed.), p. 150. North-Holland Publ., Amsterdam.
Bodanszky, M., and Kwei, T. (1978). *Int. J. Pept. Protein Res.* **12**, 69–74.
Brenner, M. (1968). *In* "Peptides 1968" (E. Bricas, ed.), pp. 154–155. North-Holland Publ., Amsterdam.
Brieger, G., and Nestrick, T. (1974). *Chem. Rev.* **74**, 567–580.
Camble, R., Garner, R., and Young, G. T. (1969). *J. Chem. Soc. C* pp. 1911–1916.
Chimiak, A., Kolasa, T., and Biernat, J. F. (1972). *Z. Chem.* **12**, 264–265.
Choudhury, A. M., Kenner, G. W., Moore, S., Ramage, R., Richards, P. M., and Thorpe, W. D. (1976). *In* "Peptides 1976" (A. Loffet, ed.), pp. 257–261. Editions de l'Université de Bruxelles, Belgium.
Cipera, J. D., and Nicholls, R. V. V. (1955). *Chem. Ind. (London)* pp. 16–17.
Clayton, D. W., Kenner, G. W., and Sheppard, R. C. (1956). *J. Chem. Soc.* pp. 371–380.
Davies, J. S., Hassall, C. H., and Hopkins, K. H. (1973). *J. Chem. Soc., Perkin Trans. 1* pp. 2614–2618.
Dean, P. D. G. (1965) *J. Chem. Soc.* p. 6655.
Dekker, C. A., Taylor, S. P., and Fruton, J. S. (1949). *J. Biol. Chem.* **180**, 155–173.
El Amin, B., Anantharamaiah, G. M., Royer, G. P., and Means, G. E. (1979). *J. Org. Chem.* **44**, 3442–3444.
Elsinger, F., Schreiber, J., and Eschenmoser, A. (1960). *Helv. Chim. Acta* **43**, 113–118.
Erickson, B. W., and Merrifield, R. B. (1973). *J. Am. Chem. Soc.* **95**, 3750–3756.
Feinberg, R. S., and Merrifield, R. B (1975). *J. Am. Chem. Soc.* **97**, 3485–3496.
Felix, A. M., Jimenez, M., Mowles, T., and Meienhofer, J. (1978a). *Int. J. Peptide Protein Res.* **11**, 329–339.
Felix, A. M., Heimer, E. P., Lambros, T. J., Tzougraki, C. and Meienhofer, J., (1978). *J. Org. Chem.* **43** (4194–4196).
Fruton, J. S., and Bergmann, M. (1942). *J. Biol. Chem.* **145**, 253–265.
Galpin, I. J., Handa, B. K., Hudson, D., Jackson, A. G., Kenner, G. W., Ohlsen, S. R., Ramage R., Singh, B., and Tyson, R. G. (1976). *In* "Peptides 1976" (A. Loffet, ed.), pp. 247–256. Editions de l'Université de Bruxelles, Belgium.
Garner, R., and Young, G. T. (1971). *J. Chem. Soc. C* pp. 50–53.
Gazis, E., Borovas, D., Hamalidis, C., Stelakatos, G. C., and Zervas, L. (1966). *In* "Peptides 1963" (L. Zervas, ed.), pp. 107–113. Pergamon, Oxford.
Goodacre, J., Ponsford, R. J., and Stirling, I. (1975). *Tetrahedron Lett.* pp. 3609–3612.
Grassmann, W., and Wünsch, E. (1958). *Chem. Ber.* **91**, 449–455.
Gray, C. J., and Khoujah, A. M. (1969). *Tetrahedron Lett.* pp. 2647–2650.
Guttmann, S., and Boissonnas, R. A. (1958). *Helv. Chim. Acta* **41**, 1852–1867.

Hardegger, E., El Heweihi, Z., and Rabinet, F. G. (1948). *Helv. Chim. Acta* **31**, 439–445.
Hardy, P. M., Rydon, H. N., and Thompson, R. C. (1968). *Tetrahdedron Lett.* pp. 2525–2526.
Harris, J. I., and Fruton, J. S. (1951). *J. Biol. Chem.* **191**, 143–151.
Häussler, R. (1960). *Chimia* **14**, 369–370.
Hendrickson, J. B., and Kendall, C. (1970). *Tetrahedron Lett.* pp. 343–344.
Hiskey, R. G., and Adams, J. B. (1965). *J. Am. Chem. Soc.* **87**, 3969–3973.
Hiskey, R. G., Staples, J. T., and Smith, R. L. (1967). *J. Org. Chem.* **32**, 2772–2777.
Hiskey, R. G., Beacham, L., and Matl, V. G. (1972a). *J. Org. Chem.* **37**, 2472–2478.
Hiskey, R. G., Wolters, E. T., Ülkü, G., and Rao, V. R. (1972b). *J. Org. Chem.* **37**, 2479–2483.
Ho, T. L., and Olah, G. A. (1976). *Angew. Chem., Int. Ed. Engl.* **15**, 774–775.
Hofmann, K., Jöhl, A., Furlenmeier, A. E., and Kappeler, H. (1957). *J. Am. Chem. Soc.* **79**, 1636–1641.
Jackson, A. E., and Johnstone, R. A. W. (1976). *Synthesis* pp. 685–687.
Jones, D. A. (1977). *Tetrahedron Lett.* pp. 2853–2856.
Jung, M. E., and Lyster, M. A. (1977). *J. Am. Chem. Soc.* **99**, 968–969.
Kamber, B., Riniker, B., Sieber, P., and Rittel, W. (1976). *Helv. Chim. Acta* **59**, 2830–2840.
Kelly, T. R., Dali, H., and Tsang, W.-G. (1977). *Tetrahedron Lett.* pp. 3859–3860.
Kenner, G. W., and Seely, J. H. (1972). *J. Am. Chem. Soc.* **94**, 3259–3260.
Kopple, K. D., Saito, T., and Ohnishi, M. (1969). *J. Org. Chem.* **34**, 1631–1635.
Kornblum, N., and Scott, A. (1974). *J. Am. Chem. Soc.* **96**, 590–591.
Kricheldorf, H. R. (1972). *Justus Liebigs Ann. Chem.* **763**, 17–38.
Kricheldorf, H. R., and Fehrle, M. (1974). *Chem. Ber.* **107**, 3533–3547.
Kuromizu, K., and Meienhofer, J. (1974). *J. Am. Chem. Soc.* **96**, 4978–4981.
Ledger, R., and Stewart, F. H. C. (1965). *Aus. J. Chem.* **18**, 1477–1484.
Ledger, R., and Stewart, F. H. C. (1968). *Aust. J. Chem.* **21**, 1101–1105.
Ludescher, U., and Schwyzer, R. (1972). *Helv. Chim. Acta* **55**, 2052–2060.
McDermott, J. R., and Benoiton, N. L. (1973). *Can. J. Chem.* **51**, 1915–1919.
Maclaren, J. A. (1958). *Aust. J. Chem.* **11**, 360–365.
Maclaren, J. A. (1972). *Aust. J. Chem.* **25**, 1293–1299.
Macrae, R., and Young, G. T. (1974). *J. Chem. Soc., Chem. Commun.* pp. 446–447.
Miller, A. W., and Stirling, C. J. M. (1968). *J. Chem. Soc. C* pp. 2612–2617.
Mitchell, A. R., Erickson, B. W., Ryabtsev, M. N., Hodges, R. S., and Merrifield, R. B. (1976). *J. Am. Chem. Soc.* **98**, 7357–7362.
Mizoguchi, T., Shigezane, K., and Takamura, N. (1969). *Chem. Pharm. Bull.* **17**, 411–412.
Mutter, M., and Bayer, E. (1980) *In* "The Peptides" (E. Gross and J. Meienhofer, eds.), vol. 2, pp. 285–332. Academic Press, New York.
Natarajan, S., and Bodanszky, M. (1976). *J. Org. Chem.* **41**, 1269–1272.
Nefkens, G. H. L. (1962). *Nature* (*London*) **193**, 974–975.
Nefkens, G. H. L., Tesser, G. I., and Nivard, R. J. F. (1963). *Recl. Trav. Chim. Pays-Bas* **82**, 941–953.
Patchornik, A., Amit, B., and Woodward, R. B. (1970). *J. Am. Chem. Soc.* **92**, 6333–6335.
Prestidge, R. L., Harding, D. R. K., Battersby, J. E., and Hancock, W. S. (1975). *J. Org. Chem.* **40**, 3287–3288.
Prestidge, R. L., Harding, D. R., and Hancock, W. S. (1976). *J. Org. Chem.* **41**, 2579–2583.
Rich, D. H., and Gurwara, S. K. (1975a). *J. Am. Chem. Soc.* **97**, 1575–1579.
Rich, D. H., and Gurwara, S. K. (1975b). *Tetrahedron Lett.* pp. 301–304.
Riniker, B., Kamber, B., and Sieber, P. (1975). *Helv. Chim. Acta* **58**, 1086–1094.
Rittel, W., Iselin, B., Kappeler, H., Riniker, B., and Schwyzer, R. (1957). *Helv. Chim. Acta* **40**, 614–624.
Roeske, R. W. (1959). *Chem. Ind.* (*London*) pp. 1121–1122.

Roeske, R. W. (1963). *J. Org. Chem.* **28**, 1251–1253.
Roeske, R. W. (1977). Unpublished observation.
Royer, G. P. and Anantharamaiah, G. M. (1979). *J. Am. Chem. Soc.* **101**, 3394–3396.
Rosamund, J. D., and Ferger, M. F. (1976). *J. Med. Chem.* **19**, 873–876.
Sano, S., and Kawanishi, S. (1975). *J. Am. Chem. Soc.* **97**, 3480–3484.
Schlatter, J., Mazur, R. H., and Goodmonson, O. (1977). *Tetrahedron Lett.* pp. 2851–2852.
Schnabel, E., and Oberdorf, A. (1968). *In* "Peptides 1968" (E. Bricas, ed.), pp. 261–265. North-Holland Publ., Amsterdam.
Schröder, E., and Lübke, K. (1965). "The Peptides," Vol. 1. Academic Press, New York.
Schwarz, H., and Arakawa, K. (1959). *J. Am. Chem. Soc.* **81**, 5691–5695.
Schwyzer, R., and Sieber, P. (1959). *Helv. Chim. Acta* **42**, 972–977.
Schwyzer, R., Iselin, B., Kappeler, H., Riniker, B., Rittel, W., and Zuber, H. (1963). *Helv. Chim. Acta* **46**, 1975–1996.
Shafer, D. J., Young, G. T., Elliott, D. F., and Wade, R. (1971). *J. Chem. Soc. C* pp. 46–49.
Sheehan, J. C., and Daves, G. D. (1964). *J. Org. Chem.* **29**, 2006–2008.
Sheehan, J. C., and Umezawa, K. (1973). *J. Org.* Chem. **38**, 3771–3774.
Sheehan, J. C., Wilson, R. M., and Oxford, A. W. (1971). *J. Am. Chem. Soc.* **93**, 7222–7228.
Shigezane, K., and Mizoguchi, T. (1972). *Chem. Pharm. Bull.* **21**, 972–980.
Sieber, P., Riniker, B., Brugger, M., Kamber, B., and Rittel, W. (1970). *Helv. Chim. Acta* **53**, 2135–2150.
Sieber, P., Kamber, B., Hartmann, A., Jöhl, A., Riniker, B., and Rittel, W. (1974). *Helv. Chim. Acta* **57**, 2617–2621.
Sieber, P., Kamber, B., Eisler, K., Hartmann, A., Riniker, B., and Rittel, W. (1976). *Helv. Chim. Acta* **59**, 1489–1497.
Sieber, P., Andreatta, R., Eisler, K., Kamber, B., Riniker, B., and Rink, H. (1977). *In* "Peptides, Proceedings of the Fifth American Peptide Symposinm" (M. Goodman and J. Meienhofer, eds.), pp. 543–548. Wiley, New York.
Sivanandaiah, K. M., and Gurusiddappa, S. (1979). *J. Chem. Res.* (S), p. 108.
Smithwick, E. L., and Shuman, R. T. (1974). *J. Org. Chem.* **39**, 3441–3442.
Sondheimer, E., and Holley, R. W. (1954). *J. Am. Chem. Soc.* **76**, 2467–2470.
Sparrow, J. T. (1976). *J. Org. Chem.* **41**, 1350–1353.
Stelakatos, G. C., and Argyropoulos, N. (1970). *J. Chem. Soc. C* pp. 967–970.
Stelakatos, G. C., Theodoropoulos, D. M., and Zervas, L. (1959). *J. Am. Chem. Soc.* **81**, 2884–2887.
Stelakatos, G. C., Paganou, A., and Zervas, L. (1966). *J. Chem. Soc. C* pp. 1191–1199.
Stewart, F. H. C. (1965). *Aust. J. Chem.* **18**, 1699–1703.
Stewart, F. H. C. (1966). *Aust. J. Chem.* **19**, 1067–1083.
Stewart, F. H. C. (1967a). *Aust. J. Chem.* **20**, 365–373.
Stewart, F. H. C. (1967b). *Aust. J. Chem.* **20**, 1991–2002.
Stewart, F. H. C. (1968). *Aust.J. Chem.* **21**, 2543–2550.
Suzuki, K., Endo, N., Nitta, K., and Sasaki, Y. (1976). *In* "Peptide Chemistry 1976" (T. Nakajima, ed.), pp. 45–48.
Suzuki, K., Endo, N., and Sasaki, Y. (1977). *Chem. Pharm. Bull.* **25**, 2613–2616.
Tam, J. P., Cunningham-Rundles, W. F., Erickson, B. W., and Merrifield, R. B. (1977). *Tetrahedron Lett.* pp. 4001–4004.
Taschner, E., and Liberek, B. (1956). *Rocz. Chem.* **30**, 323; *Chem. Abstr.* **51**, 1039 (1957).
Taschner, E., Biernat, J. F., Rzeszotarska, B., and Wasielewski, C. (1961a). *Justus Liebigs Ann. Chem.* **646**, 123–124.
Taschner, E., Chimiak, A., Bator, B., and Sokolowska, T. (1961b). *Justus Liebigs Ann. Chem.* **646**, 134–136.

Taylor-Papadimitrou, J., Yovanidis, C., Paganou, A., and Zervas, L. (1967). *J. Chem. Soc. C* pp. 1830–1836.
Vorbrüggen, H., and Krolikiewicz, K. (1975). *Angew. Chem., Int. Ed. Engl.* **14**, 818–819.
Waley, S. G., and Watson, J. (1953). *J. Chem. Soc.* pp. 475–479.
Wang, S. S. (1976). *J. Org. Chem.* **41**, 3258–3261.
Wang, S. S., Gisin, B. F., Winter, D. P., Makofske, R., Kulesha, I. D., Tzougraki, C., and Meienhofer, J. (1977). *J. Org. Chem.* **42**, 1286–1290.
Weygand, R. (1968). *In* "Peptides 1968" (E. Bricas, ed.), pp. 183–184. North-Holland Publ., Amsterdam.
Wünsch, E. and Zwick, A. (1963). Z. physiol. Chem. **333**, 108–113.
Yamashiro, D. (1977). *J. Org Chem.* **42**, 523–525.
Yang, C. C., and Merrifield, R. B. (1976). *J. Org. Chem.* **41**, 1032–1041.
Young, G. T., and Kotai, A. (1973). *In* "Peptides 1972" (H. Hanson and H-D. Jakubke, eds.) p. 128. North-Holland Publ., Amsterdam.
Zahn, H., and Schnabel, E. (1957). *Justus Liebigs Ann. Chem.* **605**, 212–232.

Chapter 3

Sulfhydryl Group Protection in Peptide Synthesis

RICHARD G. HISKEY

I. INTRODUCTION

The organization of this chapter utilizes the framework developed in an earlier review on this subject which surveyed literature extending into early 1972 (Hiskey *et al.*, 1973). The present review covers pertinent examples and

THE PEPTIDES, VOLUME 3

ISBN 0-12-304203-8

leading references from 1972 to early 1978. Several reviews on sulfur protection have appeared during this period. These have included monographs on the sulfhydryl group (Friedman, 1973; Patai, 1974); chapters on thiol protection (Photaki, 1973; Wolman, 1974); discussions of peptide synthesis in solution (Katsoyannis and Schwartz, 1977) and solid phase (Doscher, 1977), novel sulfhydryl reagents (Kenyon and Bruice, 1977), and excellent reviews covering peptide and protein literature for the period 1972–1975 (Jones and Ridge, 1972; Ridge and Jones, 1975; Schafer, 1976; Atherton and Sheppard, 1976). A review of the reactions of sulfenic acids in proteins (Allison, 1976) is also pertinent to this chapter.

II. THIOETHERS AND RELATED PROTECTIVE GROUPS

A. Preparation

As noted in the previous review (Hiskey *et al.*, 1973), thioether derivatives of cysteine are generally prepared by nucleophilic substitution reactions in which the sulfur functions as the nucleophile. These reactions require either a basic catalyst to generate the more nucleophilic mercaptide ion or an acidic catalyst for the generation of a cation which then reacts with the thiol. Examples of the former method include generation of *S*-4-picolyl-L-cysteine by the action of 4-picolyl chloride on the solution obtained by reduction of cystine by sodium in liquid ammonia (Gosden *et al.*, 1977), the preparation of 9-anthrylmethyl thioethers by reaction of the mercaptide ion with 9-chloromethylanthracene in dimethylformamide at 25°C (Kornblum and Scott, 1974), the alkylation of cysteine with *p*-nitrobenzyl chloride in aqueous sodium hydroxide–dioxane solution at 0°–3°C (Bachi and Ross-Petersen, 1972), or the preparation of *S*-(4-methylbenzyl)-L-cysteine [Cys(Meb)] using triethylamine and 4-methylbenzyl bromide in aqueous ethanol (Erickson and Merrifield, 1973). The general synthetic route is shown in Eq. (1).

$$\mathrm{RSH} + \mathrm{B}^- \rightleftharpoons \mathrm{RS}^- + \mathrm{BH} \xrightarrow{\mathrm{R'CH_2L}} \mathrm{RSCH_2R'} + \mathrm{L}^- \qquad (1)$$

The preparation of *S*-alkylcysteine derivatives via acid catalysis may utilize carbinols, alkenes etc. as potential cations for the alkylation of sulfur. For example, *S*-dibenzosuberyl-L-cysteine (**3**) was prepared by reaction of 10,11-dihydro-5*H*-dibenzo[*a,d*]cycloheptan-5-ol (**1**) with cysteine in trifluoroacetic acid (Pless, 1976) [Eq. (2)]. The *S*-diphenyl-4-pyridylmethyl group has been introduced in similar manner (Coyle and Young, 1976). Treatment of diphenyl-4-pyridylmethanol (**4**) with L-cysteine hydrochloride

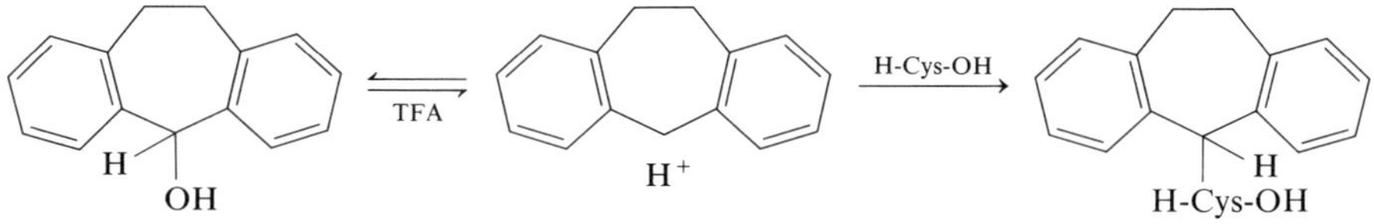

(1) (2) (3)

(2)

in acetic acid containing boron trifluoride etherate complex provided the *S*-alkylated cysteine (**5**) [Eq. (3)]. *S*-Indenyl- and fluorenyl-L-cysteine derivatives, *S*-dialkylphenanthrylmethyl- and *S*-α-methyl-α-phenylphenanthryl-

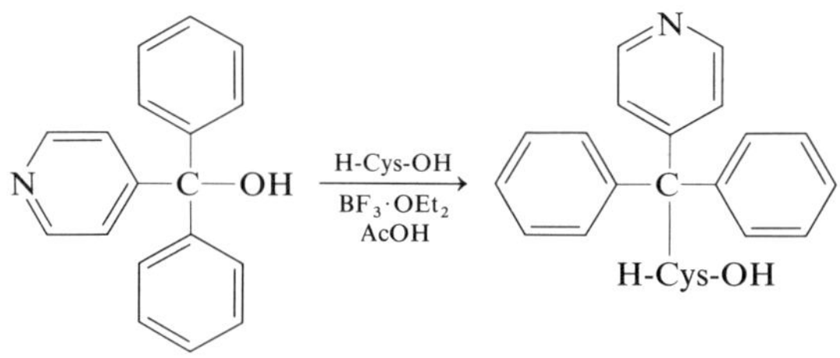

(3)

(**4**) (**5**)

methyl thioethers of cysteine have been prepared by similar methods (Mladenova-Orlinova and Tsvetkova, 1973; Mladenova-Orlinova, 1973). *S*-*tert*-Butyl-L-cysteine (**7**) has been prepared via *tert*-butyl acetate (Chimiak, 1963) or using isobutylene (Beyerman, 1963) [Eq. (4)]. In either case the initial product was the *tert*-butyl ester derivative (**6**) which was selectively cleaved to the *S*-alkyl derivative **7**.

$$\text{H-Cys-OH} \xrightarrow[(CH_3)_2C{=}CH_2,\ H^+]{CH_3COO t\text{Bu–}HClO_4 \text{ or}} \underset{(\mathbf{6})}{\text{H-}\overset{t\text{Bu}}{\overset{|}{\text{Cys}}}\text{-O}t\text{Bu}} \xrightarrow[\text{HBr–AcOH}]{\text{TFA or}} \underset{(\mathbf{7})}{\text{H-}\overset{t\text{Bu}}{\overset{|}{\text{Cys}}}\text{-OH}} \quad (4)$$

The ability of the thiol group to function in conjugate additions has been utilized. For example *S*-3,4-dimethylbenzyl-DL-[α-^{2}H]cysteine (**8**) was prepared as shown in Eq. (5) (Hruby *et al.*, 1977). The synthesis of *S*-benzyl-DL-[α-^{2}H]cysteine, *S*-benzyl-DL-[*β*,*β*-$^2H_2$]-cysteine, and *S*-benzyl-DL-[α,*β*,*β*-$^2H_3$]cysteine, together with the *N*-*tert*-butyloxycarbonyl derivatives has also been reported (Upson and Hruby, 1976). The DL-cysteine derivatives were incorporated into several oxytocin analogues; optical purity was maintained by partition chromatography of the diastereomers.

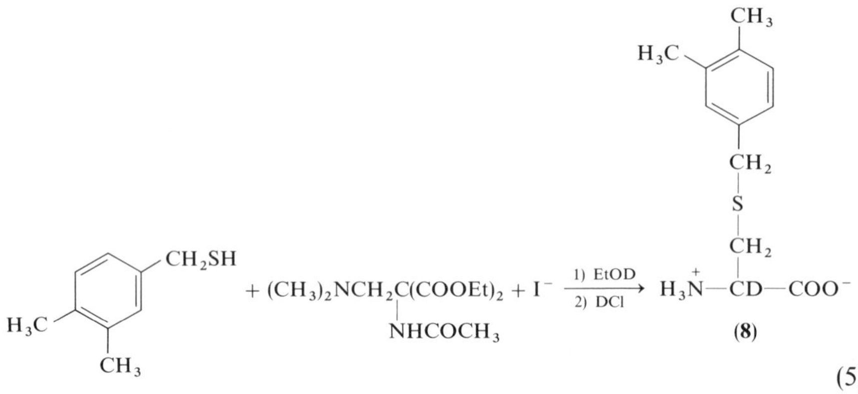

(5)

B. Removal of Thioether Protective Groups

1. Removal with Metal—Ammonia Combinations

The use of sodium in liquid ammonia to remove the two *S*-benzyl groups from various deuterated derivatives of the protected nonapeptide **9** has been reported (Yamamoto *et al.*, 1977). Several interesting observations resulted from this study including: the stability of the *S*-3,4-dimethylbenzyl group

H-Cys(Bzl)-Tyr-Phe-Gln-Asn-Cys(Dmb)-Pro-Arg(Tos)-Gly-NH_2

(9)

(Dmb) to the repetitive use of trifluoroacetic acid in solid-phase synthesis (Yamashiro *et al.*, 1973; Erickson and Merrifield, 1973), the lack of deuterium exchange during the sodium in liquid ammonia reduction, and the separation of the various oxytocin diastereoisomers by partition chromatography. The use of a nitrogen atmosphere before and during the sulfhydryl oxidation step (potassium ferricyanide, pH 8.5) was found to suppress dimer and oligomer formation as suggested by Wälti and Hope (1973).

The use of protective groups such as the 9-anthrylmethyl thioethers that are cleaved by an ion-radical mechanism (sodium methylmercaptide in hexamethylphosphoramide at 25°C) has been proposed (Kornblum and Scott, 1974).

The reductive cleavage of *S*-4-picolyl-L-cysteine has been reported (Coyle and Young, 1976) using zinc and acetic acid (20°C, 15 min) or electrolytic reduction at a mercury cathode (0°C, 45 min). The substance was not affected by trifluoroacetic acid (48 hr, 20°C) or 45% hydrogen bromide in acetic acid (20°C).

2. Removal with Acid

The search for an *S*-alkyl thioether derivative of cysteine that could be cleaved with hydrogen fluoride but was stable to trifluoroacetic acid or trifluoroacetic acid–methylene chloride mixtures has led to several interesting investigations. Sakakibara *et al.* (1976) noted that removal of the *S*-benzyl group with anhydrous hydrogen fluoride did not proceed to completion, even in the presence of anisole or tyrosine. The *S*-4-methoxybenzyl (Mob) group is predicted to be too labile for the solid-phase synthesis of large peptides (Erickson and Merrifield, 1973; Yamashiro *et al.*, 1973) and *S*-4-methylbenzyl (stable to TFA–CH_2Cl_2; cleaved with HF–anisole within 60 min at 0°C) and *S*-3,4-dimethylbenzyl (Dmb) were suggested. The synthesis of the COOH-terminal cyclic 12-peptide of human growth hormone (Yamashiro *et al.*, 1973) using the latter group (Dmb) indicated complete cleavage with hydrogen fluoride (0°C, 15 min) but less than 0.2% cleavage with trifluoroacetic acid–methylene chloride mixture in 23 hr. A careful study of the relative merits of the *S*-benzyl, *S*-4-methoxybenzyl, and *S*-4-methylbenzyl protective groups in the solid-phase synthesis of oxytocin and arginine vasopressin has recently been conducted (Live *et al.*, 1977). The yield of oxytocin using *S*-benzyl protection was about half as great (25%) after treatment with hydrogen fluoride, oxidation, and chromatography as that obtained from similar treatment of the *S*-4-methoxybenzyl and *S*-4-methylbenzyl peptides (55% yield). The lower yield with *S*-benzyl is attributed to the more vigorous conditions required for removal of the group (60 min, 20°C). Hruby *et al.* (1977) also noted reduced yields of oxytocin using hydrogen fluoride to remove *S*-benzyl groups and the peptide from the resin; improved yields were obtained with the *S*-3,4-dimethylbenzyl group. Tan and Kaiser (1976, 1977) have employed the *S*-4-methoxybenzyl protective group for the solid-phase synthesis of pancreatic trypsin inhibitor and inhibitor homologs. An important aspect of these studies was the careful attention given to the hydrogen fluoride deprotection steps. Rivier (1974) also utilized *S*-4-methoxybenzyl protection in the solid-phase synthesis of somatostatin.

The synthesis of the neurotoxic component of bee venom, apamin, **(10)** was recently critically evaluated (Van Rietschoten *et al.*, 1977). In earlier studies (Van Rietschoten *et al.*, 1975) the *S*-*tert*-butylsulfenyl group was used for protection of the four cysteine residues during solid-phase synthesis.

H-Cys-Asn-Cys-Lys-Ala-Pro-Glu-Thr-Ala-Leu-Cys-Ala-Arg-Arg-Cys-Gln-Gln-His-NH_2

(10)

This protective group (S*t*Bu) was sensitive to hydrogen fluoride; thus, three other *S*-protective groups were evaluated using the same synthetic scheme. The *S*-4-methoxybenzyl system provided the easiest manipulation of the peptide (crude yield 57%) but the lowest yield of purified apamin (7%). The *S*-ethylsulfenyl system provided 63% of crude peptide after HF cleavage and 20% of purified **10**; the *S*-acetamidomethyl system provided 66% of crude peptide and a 12% yield of purified peptide. Merrifield *et al.* (1977) report the preparation of [Har^4]apamin using the *S*-4-methylbenzyl protective group; the overall yield in a comparable synthetic sequence was 14%.

Yajima *et al.* (1975) have studied the use of trifluoromethylsulfonic acid and methanesulfonic acid on various protective groups. Yajima *et al.* (1977) found the *S*-4-methoxybenzyl (Mob) group was stable to the action of ethanesulfonic acid in acetic acid solution while the *N*-*tert*-butyloxycarbonyl and/or *N*-*p*-methoxybenzyloxycarbonyl (Moz) groups were lost under these conditions. Fujii and Yajima (1975) have found that trifluoromethylsulfonic acid will remove all protective groups from **11** to yield peptides with somatostatin activity.

Z-Ala-Gly-Cys(Mob)-Lys-(Z)-Asn-Phe-Phe-Trp-Lys-(Z)-Thr-Phe-Thr-Ser-Cys(Mob)-OH

(**11**)

Stability studies on *S*-dibenzosuberyl-L-cysteine (**3**) (Pless, 1976) indicate that this protective group is similar to the *S*-(4,4-dimethoxy)diphenylmethyl (Ddm) group (Hanson and Law, 1965). The thioether is stable toward dry hydrogen chloride in organic solvents and trifluoroacetic acid although the group is partially cleaved by hydrogen bromide in acetic acid. Similar acid lability is observed for *S*-triphenylmethyl thioethers. A practical problem with the use of acid to remove these groups is the equilibrium nature of the reaction [Eq. (6)]. Studies by Zervas *et al.* (1968) and Photaki *et al.* (1970);

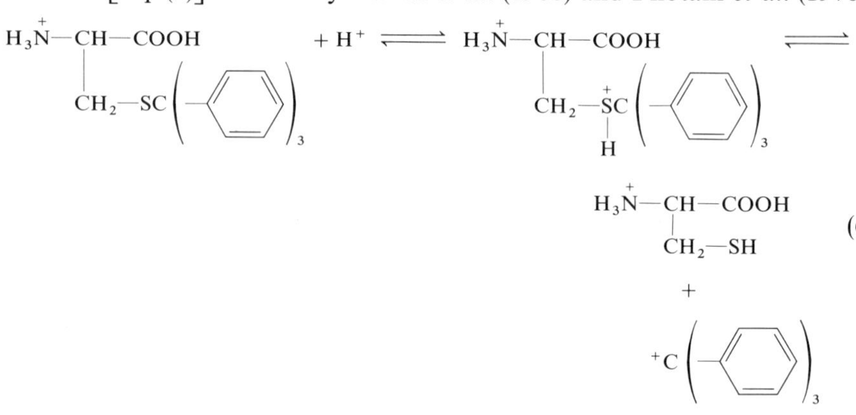

(6)

see also Hiskey *et al.*, 1973; Photaki, 1976) clearly indicate the incomplete removal of *S*-trityl (or other *S*-alkyl protective groups of similar pK_{R^+} value) with trifluoroacetic acid or hydrogen bromide or chloride in acetic acid.

Removal of *S*-trityl can in fact be almost completely suppressed by addition of triphenylcarbinol to the reaction mixture. Treatment of *S*-diphenylmethyl-L-cysteine (*S*-Dpm)peptides with refluxing trifluoroacetic acid in the presence of phenol provides complete cleavage; cleavage can also by accomplished at 20°–30°C with these reagents (Photaki *et al.*, 1970). Several other *S*-arylalkyl thioethers have been evaluated (König *et al.*, 1968) including *S*-phenylcyclohexyl-L-cysteine, which can be converted to cysteine with trifluoroacetic acid at 70°C. A detailed discussion of the behavior of various *S*-alkyl thioether protective groups in acid and the utility of the pK_{R^+} values in the design of *S*-protective groups has been presented (Hiskey *et al.*, 1973). Clearly, the more highly stabilized cations (those with the more positive pK_{R^+} values) are also the most sensitive to the nucleophilicity of the other reactants. Thus removal of, for example, the *S*-triphenylmethyl group with acid is complicated by incomplete reaction due to the selectivity factor of the trityl cation–thiol interaction. To avoid these problems one must either use alkyl groups of low stability (and selectivity) or utilize cleavage methods that are essentially irreversible (the use of scavengers for the alkyl cations produced in the cleavage reaction or heavy metals to irreversibly bind the thiol that is released).

3. Removal with Heavy Metals

The use of thiol protective groups that are removed by the action of mercury(II) ion or silver ion offers one of the presently most attractive routes to complex peptides containing cysteine or cystine. For example *S*-protective groups are available which are stable to strong acid but which are readily cleaved with heavy metal ions. Used in conjunction with amine protective groups labile to acid or other means, this combination of protective groups provides great selectivity and specificity. An illustration of the potential of this strategy is the recent synthesis of somatostatin by Felix *et al.* (1977). The stepwise addition of *N*-benzyloxycarbonylamino acid active esters provided the tridecapeptide derivative **12**. The *N*-benzyloxycarbonyl group was removed at each stage by hydrogenolysis using palladium-on-charcoal in liquid ammonia (average yield 87% for the twelve reductions). Addition of the NH_2-terminal alanine residue gave the completely protected 14-peptide **13** [Eq. (7)]. Removal of the blocking groups at oxygen and nitrogen was accomplished with anhydrous trifluoroacetic acid. The highly nucleophilic sulfur atoms still remain blocked by the bulky *tert*-butyl thioethers which are quite stable under these conditions (Hiskey *et al.*, 1973). Thus, the majority of *tert*-butyl cations are removed before cleavage of the S-protective groups. The use of mercury(II) acetate (pH 4.0) was found to provide quantitative release of the thiol groups; oxidation with potassium ferricyanide provided the hormone **14** [Eq. (8)]. In a separate study Fujino and Nishimura (1976) report that mercury(II) trifluoroacetate can also be utilized for removal of the *S*-*tert*-butyl protective group.

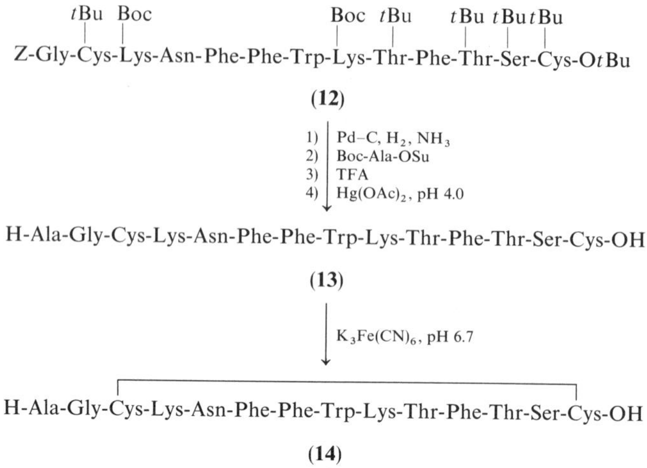

(7)

(8)

The use of heavy metals for the selective cleavage of *S*-triphenylmethyl thioethers in the presence of *S*-diphenylmethyl thioethers has been beneficial for the stepwise synthesis of bis-cystine peptides (Hiskey *et al*., 1975b). Thus treatment of the protected heptapeptide **15** with silver ion in pyridine pro-

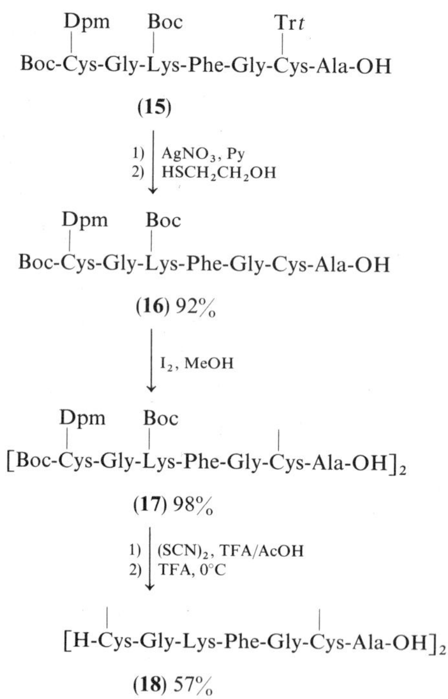

(9)

(10)

(11)

vided good yields of the thiol **16** [Eq. (9)]. Iodine oxidation of **16** provided the monocystine peptide derivative **17** [Eq. (10)]. The parallel bis-cystine peptide (**18**) was produced from **17** by oxidative removal of the *S*-diphenylmethyl groups with thiocyanogen in a trifluoroacetic acid–acetic acid solvent followed by complete removal of the remaining amine protective groups [Eq. (11)].

4. Direct Conversion to Disulfides

The conversion of a thioether of cysteine to an intermediate capable of direct conversion to cystine eliminates isolation of the thiol. More importantly, however, this strategy permits the stepwise formation of two or more disulfide bonds without the posibility of thiol–disulfide interchange (Hiskey *et al.*, 1973, 1975b). The essence of this approach is to utilize a sulfur protective group that will provide a highly stabilized cation (L^+) upon interaction of the sulfur atom with an electrophile [Eq. (12)]. The intermediate sulfenyl compound may then function as the electrophile and react with another molecule of substrate or another added thioether or thiol to provide the disulfide, **20**.

$$R\text{—}S\text{—}L + E^+ \rightleftharpoons R\text{—}\overset{+}{S}\begin{matrix} L \\ E \end{matrix} \xrightleftharpoons{-L^+} \underset{(\mathbf{19})}{R\text{—}S\text{—}E} \longrightarrow \underset{(\mathbf{20})}{R\text{—}S\text{—}S\text{—}R} \quad (12)$$

Table I "Half-Times (t_H" for Iodine Oxidations of R-S-Trt- and R-S-Acm[a]

Solvent	t_{HT} (R-S-Trt)	t_{HA} (R-S-Acm)	t_{HA}/t_{HT}
MeOH	4 sec	1 min	15
MeOH–HOH (8:2)	<1 sec	5 sec	10
AcOH	75 sec	45 min	35
AcOH–HOH (8:2)	2 sec	50 sec	25
Nitromethane	1 sec	5 sec	5
1,2-Dichloroethane	1.5 min	200 min	130
CCl_4	2.5 min	300 min	120
Dioxane	1 min	100 min	100
$CHCl_3$	1 sec	90 min	5,400
CH_2Cl_2	1 sec	60 min	3,600
TFE–CH_2Cl_2 (3:1)	7 sec	>4 hr	>2,500
HFIP–CH_2Cl_2 (3:1)	1 sec	≫3 hr	>10,000
MeOH–$CHCl_3$ (1:1)	1 sec	15 min	900

[a] R: Boc-Cys(X)-Gly-Glu(O*t*Bu)$_2$; concentration of peptide: 5×10^{-3} *M*; concentration of iodine: 15×10^{-3} *M*. From Kamber (private communication, 1978).

This approach has been elegantly employed by the Ciba group in the total synthesis of human insulin (Sieber *et al.*, 1974, 1976, 1977; Kamber *et al.*, 1976) via a route involving the stepwise formation of the three disulfide bonds. The selective formation of the 6,11 loop in the presence of Cys^7(Acm) illustrates the potential of the method (Sieber *et al.*, 1976). Oxidation of **21** [Eq. (13)] provided the octapeptide loop **22**; acylation with the protected A(1–5) azide provided the protected A(1–13) sequence **23** in 40% yield. The protected 13-peptide **23** could also be prepared by iodine oxidation of **24** [Eq. (15)]. Hexafluoroisopropanol (HFIP) was found to be a suitable solvent for the peptide. Mixtures of HFIP and methylene chloride were found to influence profoundly the selectivity of the oxidation process (Table I) (B. Kamber, private communication, 1978).

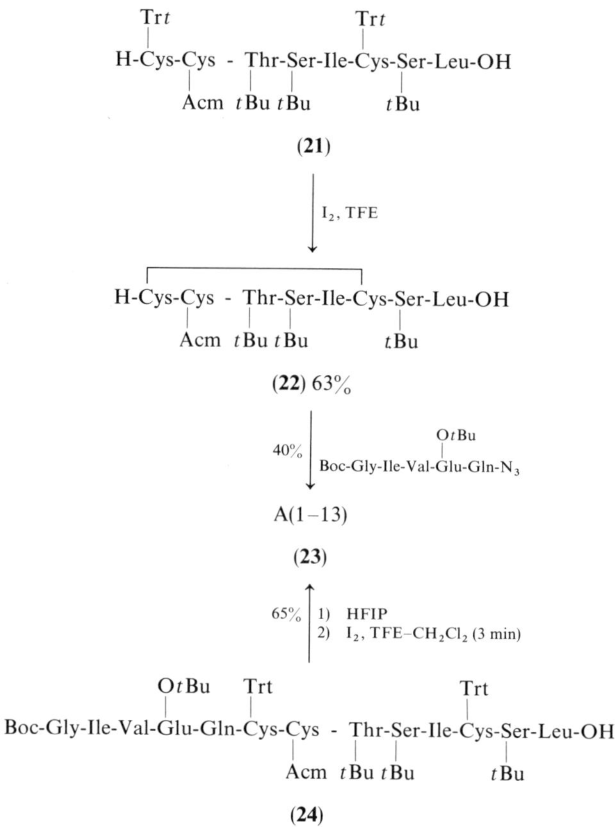

(13)

(14)

(15)

Similar oxidations of *S*-triphenylmethylcysteine peptides with iodine have been conducted on the protected calcitonin M(1–10) decapeptide (**25**) [Eq.

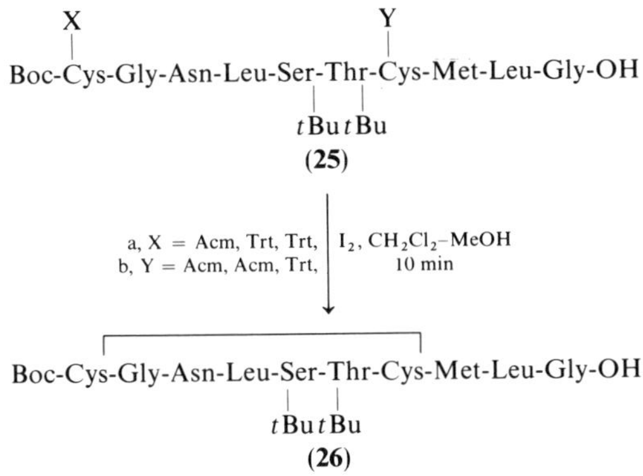

(16)

(16)] (Kamber and Rittel, 1968). In addition the protected arginine vasopressin (**28**) has been prepared by iodine oxidation of the di-*S*-triphenylmethyl cysteine precursor, **27** (Jones *et al.*, 1973) [Eq. (17)].

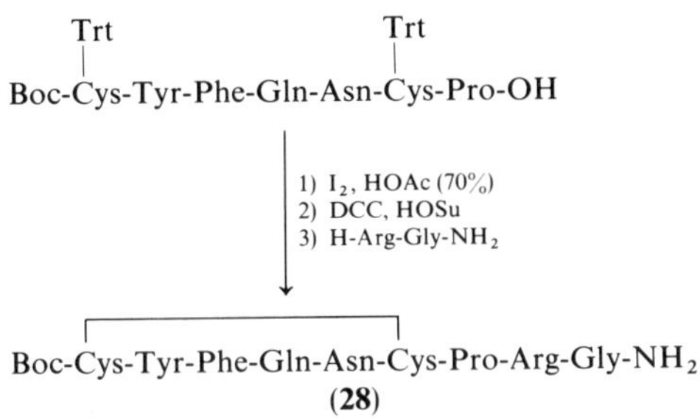

(17)

Thiocyanogen in various solvents also exhibits similar behavior toward thioethers. For example, treatment of the thiol **16** with thiocyanogen followed by addition of the fully protected heptapeptide derivative **29** provided the unsymmetrical cystine peptide, **30** (Hiskey *et al.*, 1975a,b). Oxidation of **30** with thiocyanogen in a more polar solvent followed by complete removal of the amine protective groups provided the antiparallel bis-cystine peptide **31** [Eq. (19)]. This substance was shown to be isomeric with the parallel isomer, **18**. The two bis-cystine peptides (**18** and **31**) could be distinguished by the products of tryptic or chymotryptic digestion; **31** provides only **32** on treatment with trypsin whereas **18** provides two symmetrical cystine peptides. The absence of **18** in **31** preparations indicated that the thiocyanogen oxidation of **16** → **30** and **30** → **31** proceeded without acid-catalyzed disulfide interchange and constituted a formal proof of structure of **31**.

Thioethers of cysteine are also cleaved by carbomethoxysulfenyl chloride (Brois *et al.*, 1970; Kamber, 1973; Hiskey *et al.*, 1975a). In addition the *S*-acetamidomethyl group is also removed with this reagent. Acid-catalyzed

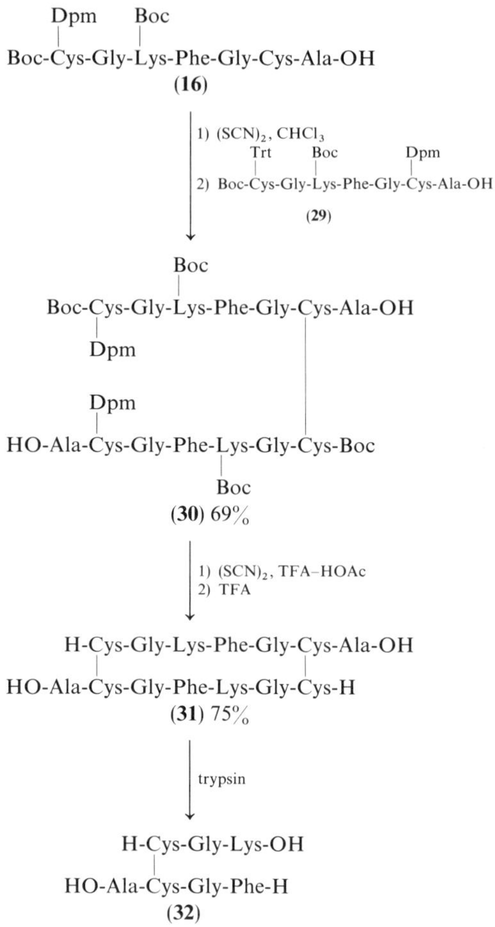

(18)

(19)

(20)

esterification that occurred in the conversion of **33** → **34** [Eq. (21)] could be suppressed (Hiskey *et al.*, 1975a). The *S*-carbomethoxysulfenyl group (Scm) is an example of a stable sulfenyl intermediate that can lead to symmetrical or

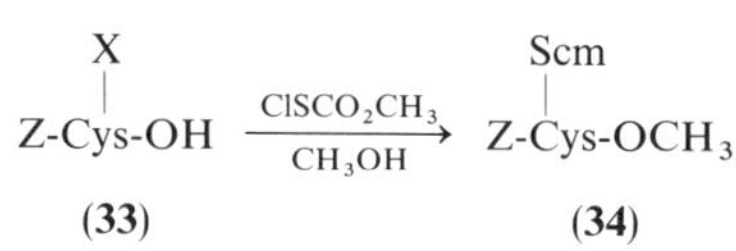

(21)

X	% *Yield*
Dpm	77
Trt	80
Bzl	25
H	85
Acm	80

unsymmetrical disulfides (Brois *et al.*, 1970). Thus, the conversion **33** → **34** is an oxidation method in which, in contrast to the use of the thiocyanogen or iodine, the sulfenyl intermediate can be isolated and characterized. Sulfenyl thiocyanates of cysteine have not been isolated; sulfenyl iodides of small peptides can generally not be isolated although stable sulfenyl iodide derivatives of proteins are known (Allison, 1976; Aune and Thomas, 1978).

III. MONOTHIOACETALS, DITHIOACETALS, AND RELATED PROTECTIVE GROUPS

A. Preparation

Protective groups in this category include the *S*-tetrahydropyranyl, *S*-isobutyloxymethyl, and the *S*-benzylthiomethyl derivatives of cysteine (**35**–**37**). The preparation of these derivatives has been reviewed (Hiskey *et al.*,

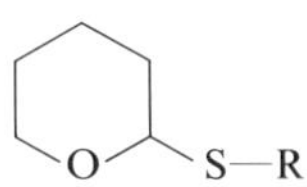

(35)

$(CH_3)_2CHCH_2OCH_2SR$

(36)

$C_6H_5CH_2SCH_2SR$

(37)

1973) and few major improvements have subsequently appeared. Schwertner *et al.* (1975a) have developed procedures for the preparation of *N*-biphenylylisopropyloxycarbonyl-*S*-tetrahydropyranylcysteine and several other Bpoc-cysteine derivatives. Wittinghofer (1974) utilized the *S*-tetrahydropyranyl (Thp) group in conjunction with the *S*-ethylcarbamoyl (Ec) group to prepare the peptide derivative **38** [Eq. (22)]. Segment coupling of **38** with the protected

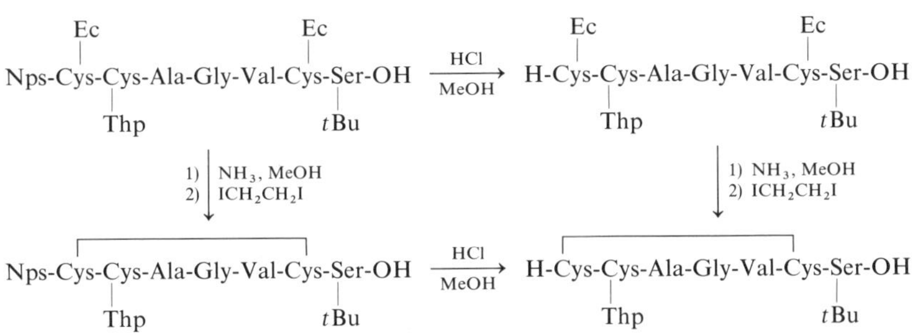

(22)

(40) **(38)**

A(1–5) derivative provided the sheep insulin sequence A(1–12), **39** [Eq. (23)]. Similar segment coupling of **40** with the A(13–21) derivative, **41**, afforded **42**

[Eq. (24)]. The *S*-Thp nonapeptide, **41**, was obtained via selective removal of the *N*-triphenylmethyl group with 75% aqueous acetic acid.

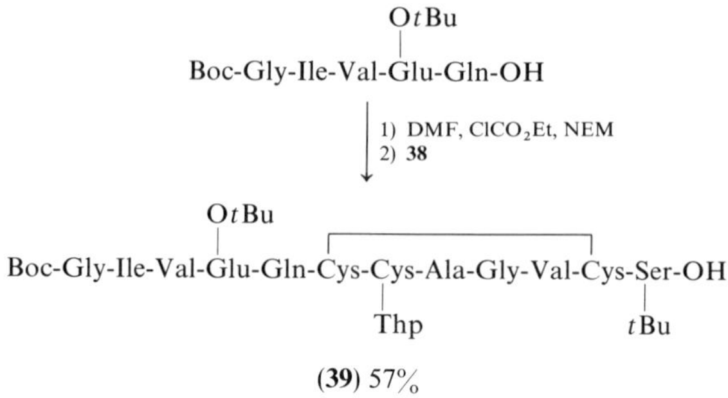

(23)

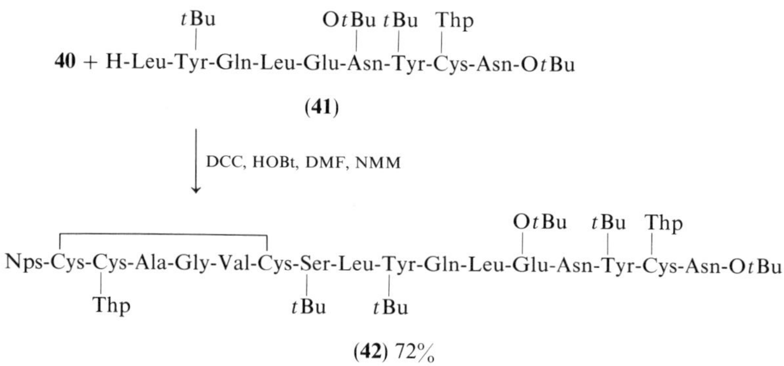

(24)

B. Removal of Thioacetal Protective Groups

Although thioacetal-type protective groups are sensitive to various acidic conditions, heavy metals, and electrophiles such as thiocyanogen (Hiskey *et al.*, 1973) relatively little recent work on cleavage conditions has appeared. The group of Zahn and collaborators has studied conditions for removal of the *S*-tetrahydropyranyl group from various synthetic insulin fragments; hydrogen bromide in trifluoroacetic acid has been successfully used (Hammerström *et al.*, 1970). Recently the *S*-tetrahydropyranyl group was removed from a protected synthetic sheep insulin A-chain, **43**, followed by direct conversion to a symmetrical disulfide (or sulfenylthiocyanate) with thiocyanogen. Oxidative sulfitolysis of the disulfide (or sulfenylthiocyanate) without isolation provided a good yield of tetra-*S*-sulfonate, **44**, in essentially a single step (Schwertner *et al.*, 1975b) [Eq. (25)].

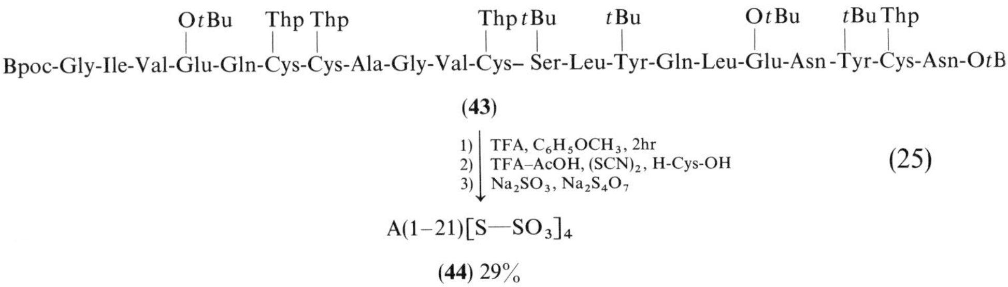

(25)

IV. THIAZOLIDINES AND RELATED PROTECTIVE GROUPS

A. Preparation

The use of carbinolamine-type reagents for sulfur protection has developed substantially during the six year period since the *S*-acetamidomethyl group was introduced (Veber *et al.*, 1972). The preparation of *S*-acetamidomethylcysteine (**47**) involves the reaction of the thiol with the stabilized cation, **46**, derived from the carbinolamine, **45**. The reaction is conducted in water at a pH $\cong$ 0.5 [Eq. (26)]. The hydrochloride may be isolated by crystallization; purification may be achieved by conversion to the free base and crystallization from aqueous alcohol (Hermann and Schreier, 1973). The *S*-isobutyrylamidomethyl and benzoylmethylaminomethyl derivatives of cysteine have also been reported (Arold and Eule, 1973).

Recently, Chakravarty and Olsen (1978) found that the attempted use of the acetamidomethyl group for protection of the thiol group of *N*-methylcysteine gave noncrystalline material that was a mixture of products. The *S*-benzamidomethyl (Bam) group was found to be a convenient alternative.

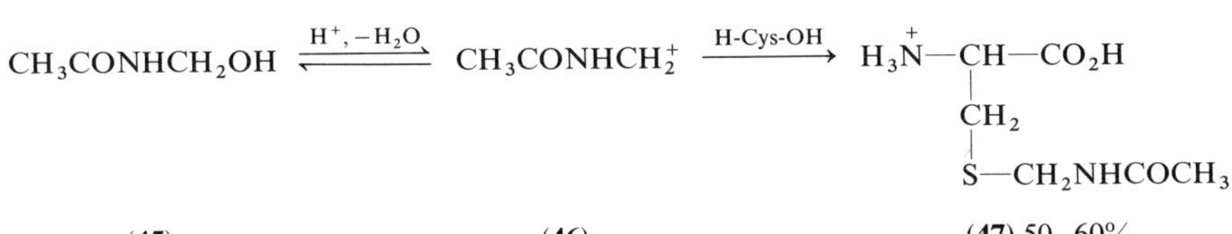

(26)

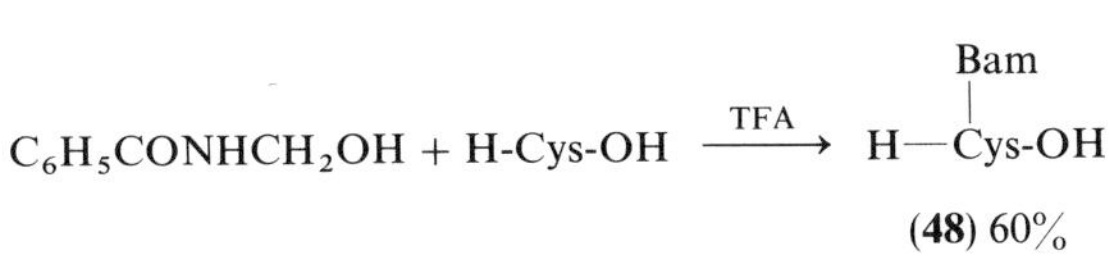

(27)

The preparation is conducted in dry trifluoroacetic acid; the product, the trifluoroacetate salt of **48** is obtained by evaporation of the solvent *in vacuo* [Eq. (27)]. Similar results were obtained with *N*-methylcysteine (88%).

Numerous cysteine peptides have been prepared using *S*-acetamidomethyl protection; the stability to acidic conditions makes the group, like the *S-tert*-butyl group, compatible with solid-phase synthetic methodology. For example, the careful synthesis of [7-(3,4-dehydroproline)]oxytocin utilized the bis-*S*-acetamidomethyl nonapeptide, **49** (Moore *et al.*, 1977). The synthesis

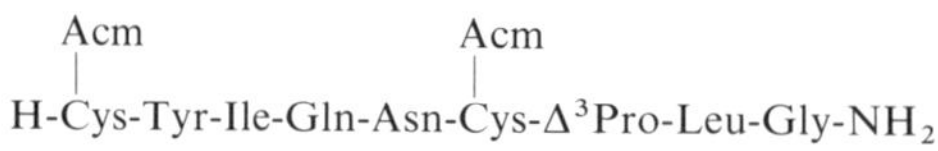

(**49**)

of apamin (**10**) using *S*-Acm was recently evaluated (Van Rietschoten *et al.*, 1977) and the utility is compared with the *S*-4-methoxybenzyl and *S-tert*-butylsulfenyl groups. Des(Ala[1], Gly[2])-[Cys[5,10], DTrp[8]]-desamino[Cys[3]]-descarboxy[Cys[14]]dicarba[3,14]-somatostatin (**50**) was obtained using *S*-Acm protection (Veber *et al.*, 1978) [Eq. (28)]. The use of the *S*-acetamidomethyl

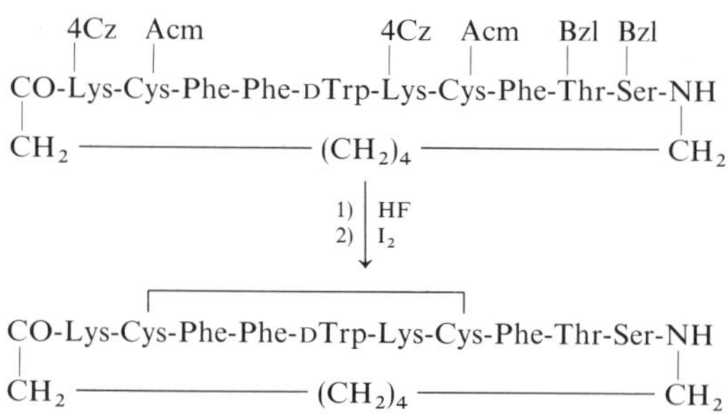

(28)

(**50**)

group to protect the six cysteine residues in the synthesis of a linear peptide corresponding to urogastrone has also been reported (Camble and Petter, 1976). Tomatis *et al.* (1977) have prepared portions of porcine pancreatic secretory trypsin inhibitor II using *S*-Acm protection for cysteine.

B. Removal of Thiazolidine and Related Derivatives

1. Removal with Heavy Metals

Removal of the *S*-acetamidomethyl protective group from peptide-bound cysteine with mercury(II) ion has not commonly been utilized except in the case of large polypeptides. Hirschmann *et al.* (1969) and Veber *et al.* (1972) established that treatment of water-soluble peptides with mercury(II) ion at pH 4 followed by hydrogen sulfide cleanly provided the thiol. Polypeptides that were insoluble in water at pH 4 could be deblocked using mercury acetate in 50% aqueous acetic acid; the reaction is somewhat slower under the latter conditions (2 hr). The use of mercury(II) acetate in 8 *M* urea is also satisfactory. The deprotection of octaS-[^{14}C]Acm-octahydro-S-protein was 98.8% complete after treatment with 9 equiv of mercury(II) acetate in 50% acetic acid for 70 min at 25°C. Air oxidation in the presence of S-peptide provided ribonuclease S with 90% of the enzymatic activity of the native enzyme. However the yield of product was about half that obtained from the oxidation of pure S-protein. Similar problems were encountered by Rocchi *et al.* (1976) in studies on bovine pancreatic trypsin inhibitor. Treatment of the reduced 58-residue polypeptide with acetamidomethanol in anhydrous hydrogen fluoride provided a polypeptide preparation S-alkylated at the six cysteine residues and C-alkylated at the 2-position of each tyrosine residue. The latter side reaction could be suppressed with anisole but removal of the *S*-acetamidomethyl protective groups with mercury(II) acetate followed by oxidation provided a preparation with only about 50% of the inhibitory power of the natural material. Thus whether the problem is the action of hydrogen fluoride on the reduced protein or removal of the *S*-acetamidomethyl group with mercury(II) ion followed by treatment with hydrogen sulfide or both is not clear. Certainly, the removal of the *S*-acetamidomethyl group from smaller polypeptides using these conditions appears to be a smooth process. Indeed Kenner (1977) has utilized the *S*-acetamidomethyl group for protection of the cysteine residues in their program aimed at the synthesis of hen egg-white lysozyme. The rational for this choice was the acid stability of the *S*-acetamidomethyl group which would permit purification of the open-chain polypeptide protected at only the eight cysteine residues; the formation of the disulfide bridges by random oxidation is anticipated as the last step of the operation.

2. Direct Conversion to Disulfide

The utility of the *S*-acetamidomethyl group lies in the stability toward acid and the lability toward iodine, thiocyanogen, sulfenyl halides, sulfenyl thiocyanates, etc. The majority of recent synthetic efforts employing the *S*-acetamidomethyl group for the protection of cysteine have employed this method for removal of the group. The procedure was first developed by Kamber (1971) who observed that the cyclo-L-cystine derivative **52** was obtained by the action of iodine on the bis-*S*-acetamidomethyl derivative **51** [Eq. (29)]. The Ciba group exploited this observation in a thorough study of

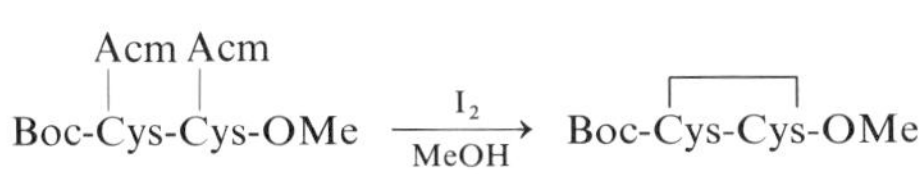

(29)

(**51**) (**52**) 84%

the selectivity of the process (Table I). Thus, Sieber *et al.* (1976) established that iodine oxidation of the two *S*-triphenylmethylcysteine residues of the A(6–13) portion of human insulin could be accomplished in the presence of the A-7 *S*-acetamidomethylcysteine residue [Eq. (13) or (15)]. Model studies (B. Kamber, private communication, 1978) showed that iodine oxidation of the two *S*-acetamidomethylcysteine residues of **53** could be accomplished without cleavage of a preformed disulfide bond [Eq. (30)]. These studies together with the careful experiments establishing the conditions for the

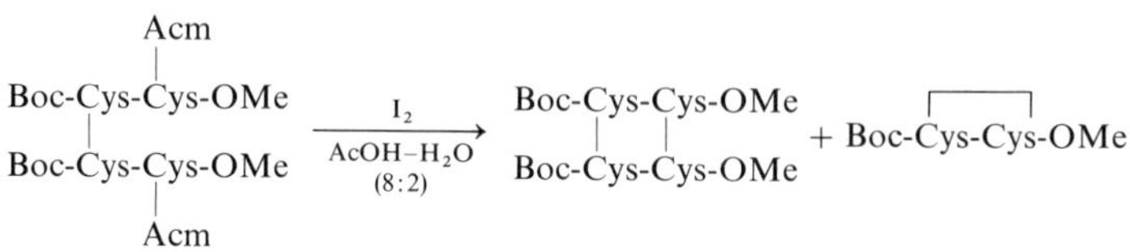

(30)

(53) **(54)** 80% 0%

selective removal of N^{α}-triphenylmethyl in the presence of N^{α}-2-(*p*-biphenylyl)propyl(2)oxycarbonyl (Bpoc) by Riniker *et al.* (1975) were the key elements in the classic total synthesis of human insulin by the stepwise method of disulfide bond formation. The starting point is the A_{20}–B_{20} disulfide bridge in **57** [Eq. (31)]. Acylation of the NH_2-terminus of cysteine A-20 with the azide derived from **58**, A(14–19), provided **59** [Eq. (32)].

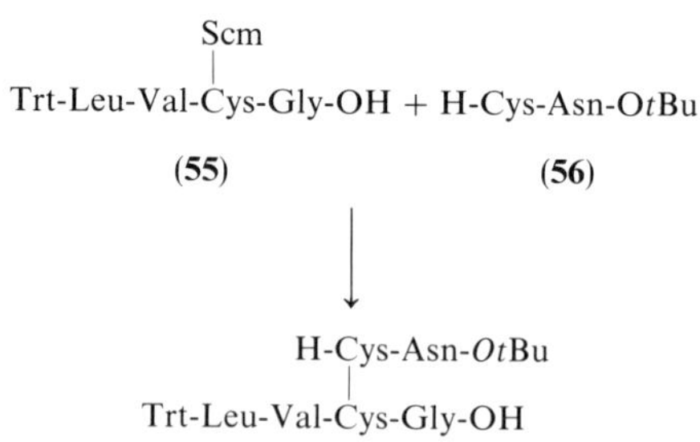

(31)

(57)

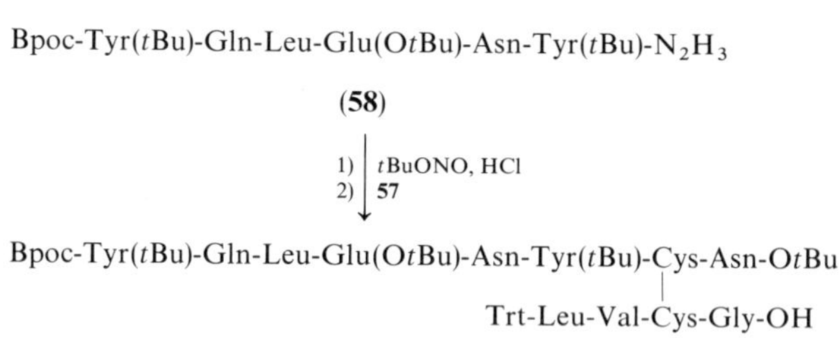

(32)

(59)

Acylation of the protected decapeptide **60** corresponding to the sequence B(21–30) provided the unsymmetrical cystine peptide **61a** [Eq. (33)]. The tendency for disulfide interchange complicated isolation and purity studies (Kamber *et al.*, 1976). The synthetic plan now required the selective removal of the N^{α}-triphenylmethyl group from **61a** (Sieber *et al.*, 1977); this was achieved using hydrochloric acid in 90% trifluoroethanol (pH 3.5, 5 min.). The resulting free base (**61b**) was coupled with the peptide representing the

59 + H-Glu(O*t*Bu)-Arg-Gly-Phe-Phe-Tyr(*t*Bu)-Pro-Lys(Boc)-Thr(*t*Bu)-O*t*Bu

(**60**)

NMN, DMF ↓ HObt, DCC

Bpoc-Tyr(*t*Bu)-Gln-Leu-Glu(O*t*Bu)-Asn-Tyr(*t*Bu)-Cys-Asn-O*t*Bu

R-Leu-Val-Cys-Glu(O*t*Bu)-Arg-Gly-Phe-Phe-Tyr(*t*Bu)-Thr(*t*Bu)-Pro-Lys(Boc)-Thr(*t*Bu)-O*t*Bu

(**61a**) R = Trt
(**61b**) R = H

(33)

remainder of the B chain to provide **62** (Scheme 1). Alcoholysis of the 2-(*p*-biphenylyl)propyl(2)oxycarbonyl group of **62** resulted when **62** was warmed in 90% trifluoroethanol; acylation of the resulting amine with the active ester derived from **23** provided the fully protected hormone lacking only the A_7–B_7 disulfide bridge **63**. Model experiments such as the conversion of **53** → **54** established that dilute acetic acid was a useful solvent for rapid oxidative conversion of the bis-*S*-acetamidomethylcysteine derivative **63** to human insulin (Scheme 1). Treatment of the peptide with 95% trifluoroacetic acid followed by iodine oxidation in 60% acetic acid provided a 70% yield of human insulin (50% as the natural isomer and 25% as the [D-tyrosine B-16]-stereoisomer). Model studies by B. Kamber (private communication, 1978) indicate that under the conditions used for the oxidative removal of the *S*-acetamidomethyl and/or *S*-trityl groups iodination of other side chains may also occur (Table II).

Sieber *et al.* (1978) have utilized this scheme to prepare two isomeric human insulin structures, [A_7–A_{11}, B_6–B_7 cystine]insulin (ii) and [A_6–A_7, A_{11}–B_7 cystine]insulin (iii). The isomeric insulins exhibit chemical stability

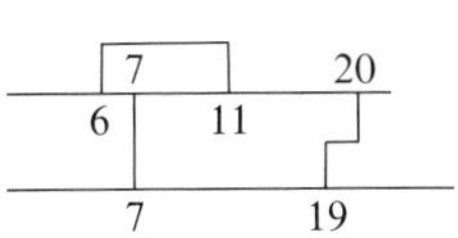

i

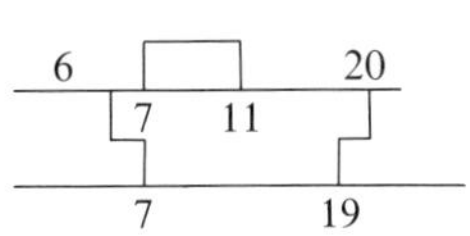

ii

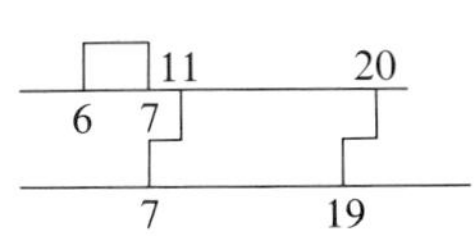

iii

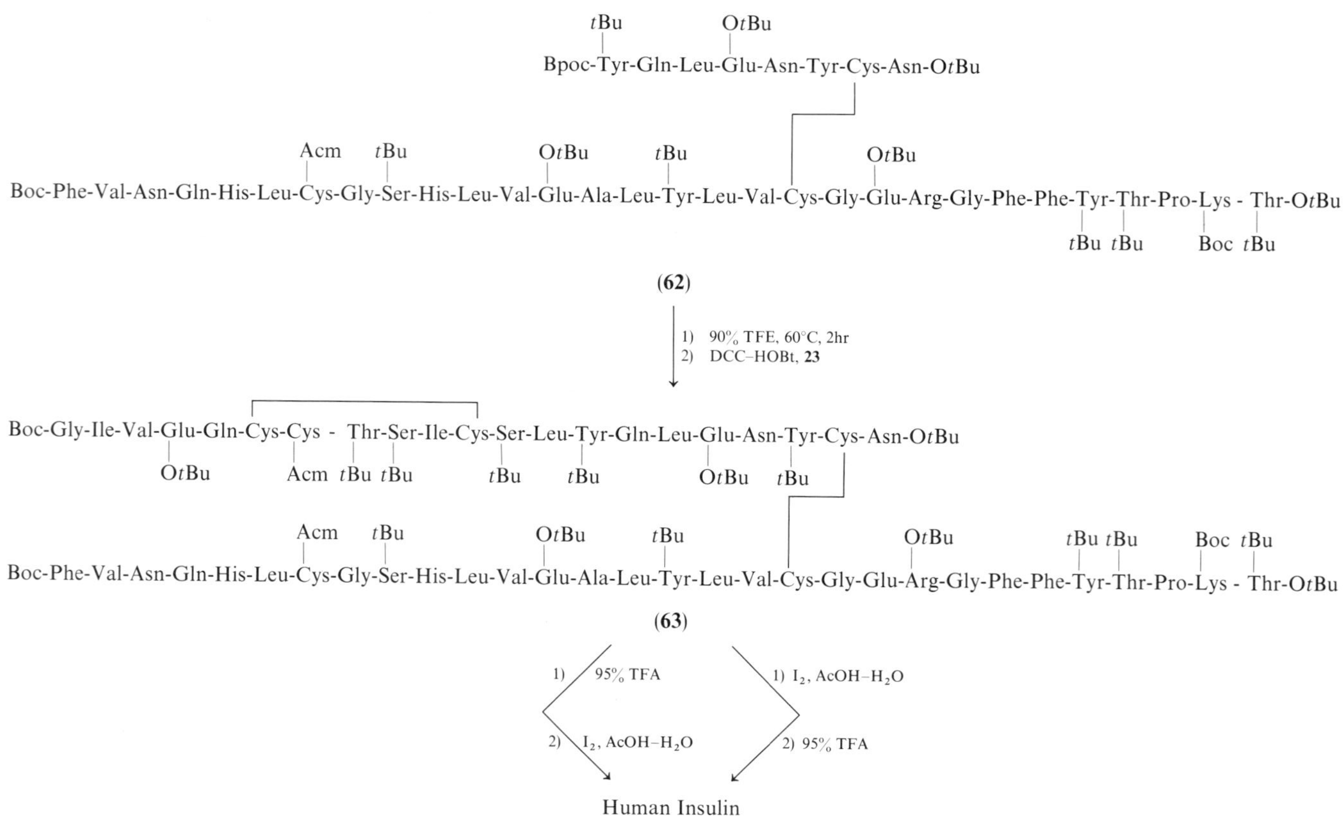
tBu OtBu
Bpoc-Tyr-Gln-Leu-Glu-Asn-Tyr-Cys-Asn-OtBu
Acm tBu OtBu tBu OtBu
Boc-Phe-Val-Asn-Gln-His-Leu-Cys-Gly-Ser-His-Leu-Val-Glu-Ala-Leu-Tyr-Leu-Val-Cys-Gly-Glu-Arg-Gly-Phe-Phe-Tyr-Thr-Pro-Lys - Thr-OtBu
tBu tBu Boc tBu
(62)
1) 90% TFE, 60°C, 2hr
2) DCC–HOBt, 23
Boc-Gly-Ile-Val-Glu-Gln-Cys-Cys - Thr-Ser-Ile-Cys-Ser-Leu-Tyr-Gln-Leu-Glu-Asn-Tyr-Cys-Asn-OtBu
OtBu Acm tBu tBu tBu tBu OtBu tBu
Acm tBu OtBu tBu OtBu tBu tBu Boc tBu
Boc-Phe-Val-Asn-Gln-His-Leu-Cys-Gly-Ser-His-Leu-Val-Glu-Ala-Leu-Tyr-Leu-Val-Cys-Gly-Glu-Arg-Gly-Phe-Phe-Tyr-Thr-Pro-Lys - Thr-OtBu
(63)
1) 95% TFA
2) I2, AcOH–H2O
1) I2, AcOH–H2O
2) 95% TFA
Human Insulin

Scheme 1

Table II Stability of Amino Acid Residues during Iodine Oxidation in Different Solvents

Residue	Stable	Unstable
Tyrosine	MeOH AcOH–HOH HFIP–CH_2Cl_2 TFE–CH_2Cl_2	MeOH–HOH MeOH–HOH
Tyrosine *tert*-butyl ether	Usual solvents	
Histidine	Imidazole protonated	Imidazole not protonated
Tryptophan	$CHCl_3$ AcOH Dioxane HFIP–CH_2Cl_2 TFE–CH_2Cl_2	MeOH AcOH–HOH DMF

[a] From B. Kamber (private communication, 1978).

at acid or neutral pH comparible to human insulin (i). In 0.1 *N* ammonia or with β-mercaptoethanol the isomers are much less stable and rearrange to primarily polymeric products. Both isomers exhibit surprisingly high levels of insulin activity and displace labeled insulin from binding to anti-insulin serum *in vitro* with potencies of 0.20 and 0.13 (insulin = 1.0).

Several bicyclic peptide derivatives have been prepared using the Kamber conditions. These include the preparation of the somatostatin analogue **50**; the conversion of **64** to the dibenzyloxycarbonyl derivative of des-*N*-tetramethyl triostin A **65** (Ciardelli and Olsen, 1977) [Eq. (34)]; and the

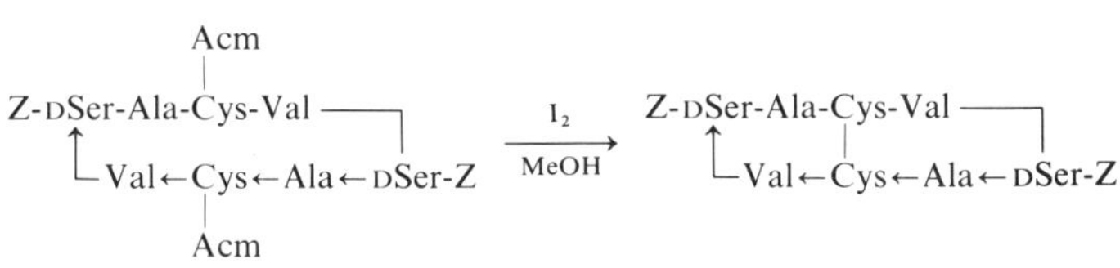

(34)

(**64**) (**65**) 89%

synthesis of disulfide bridge analogues of antamanid **68** and **71** (Scheme 2) by Wieland *et al.* (1977). It is of interest that both the iodine oxidation of the dithiol, **67**, and the bis-*S*-acetamidomethyl derivative **66b** gave similar mixtures of monomer **68** and dimer **69**. The effect of conformation on the ease of cyclization is contrasted by a comparison of the oxidation of **66b** with **70b**.

Indications that the iodine oxidation method may not always result in complete removal of *S*-acetamidomethyl groups in large molecules is given by the results of Trudelle and Caille (1977). These investigators prepared the

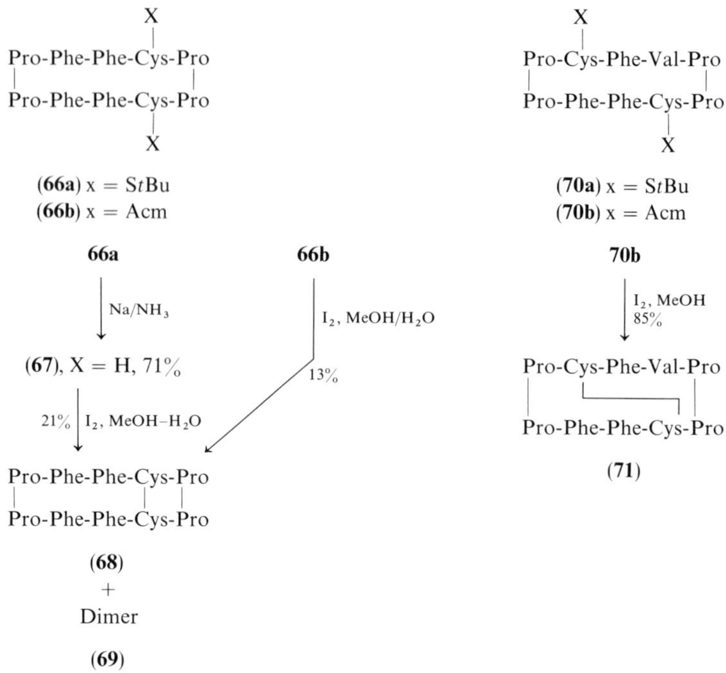

Scheme 2

pentapeptide polymer **72**, oxidatively converted **72** to the disulfide with iodine in methanol, and reduced the disulfide with β-mercaptoethanol [Eq. (35)] to the polythiol **73**. Determination of the amount of thiol per

$$\underset{(\mathbf{72})}{[\text{Cys(Acm)-Ser-Phe-Glu-Glu}]} \xrightarrow[\text{MeOH} \\ \text{2) } HSCH_2CH_2OH]{\text{1) } I_2} \underset{(\mathbf{73})}{[\text{Cys-Ser-Phe-Glu-Glu}]_n} \quad (35)$$

pentapeptide unit indicated a value of 0.32 suggesting that a substantial number of unreacted *S*-acetamidomethyl groups remain. In this connection it should be noted that the methyl hydrogen atoms of the *S*-acetamidomethyl group can be readily detected by nuclear magnetic resonance spectroscopy, thus providing a convenient assay for some *S*-acetamidomethyl group containing peptides (Veber *et al.*, 1972; Sieber *et al.*, 1976).

Fontana (1975) has suggested an alternative procedure for the removal of *S*-acetamidomethyl, *S*-triphenylmethyl, *S*-[4,4′-dimethoxydiphenylmethyl], and *S*-diphenylmethyl protective groups (and others with appropriate pK_{R^+} values). The cysteine derivatives, **74** are reported to be quantitatively converted to the unsymmetrical disulfide **75** when treated with 2-nitrophenylsulfenyl chloride. Reduction of **75** with β-mercaptoethanol, dithiothreitol (DTT), or sodium borohydride yields the thiol **76** (Scheme 3). This procedure

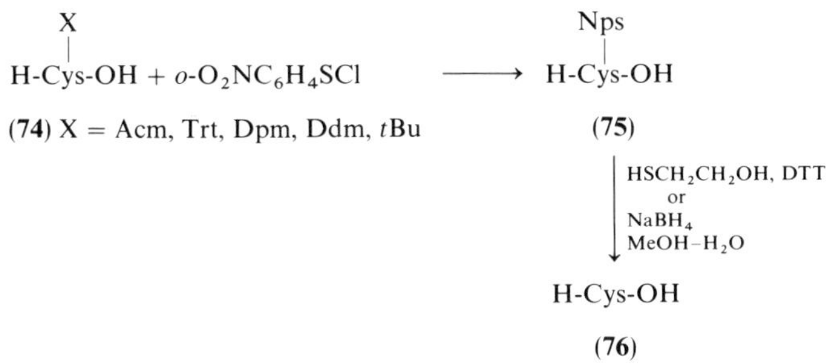

Scheme 3

is of course not applicable to the stepwise synthesis of peptides containing more than one cystine residue. Tryptophan appears to be only amino acid other than cysteine whose side chain would react with 2-nitrophenyl sulfenyl chloride. The method offers a good alternative to the use of heavy metals in the cleavage of this protective group. Pastuszak and Chimiak (1977) have also reported the cleavage of *S-tert*-butylcysteine by this route.

V. THIOESTERS AND RELATED DERIVATIVES

A. Preparation

Relatively little recent use has been made of the *S*-acyl protecting group. The tendency to undergo S → N acyl migration reduces the utility of protective groups of this sort (Hiskey *et al.*, 1973; Photaki, 1976). Heller *et al.* (1977) studied the reaction of peptides containing cysteine and histidine with *p*-nitrophenyl acetate. They concluded that there is a rapid reversible transfer of the acetyl group between the thiol and imidazole side chains. Trudelle and Caille (1977) found that the S → N acyl migration of *S*-acylcysteine can occur at pH values as low as 3. The literature of S → N and S → O acyl transfers has been recently reviewed (Sheradsky, 1974).

S-Carbamoyl and *S*-alkyloxycarbonyl derivatives of cysteine do not undergo S → N acyl migration and thus are of greater practical value. Storey *et al.* (1972) prepared *N-tert*-butyloxycarbonyl-*S*-ethylcarbamoylcysteine peptides via the 2,4,5-trichlorophenyl ester **78** [Eq. (36)]. Subse-

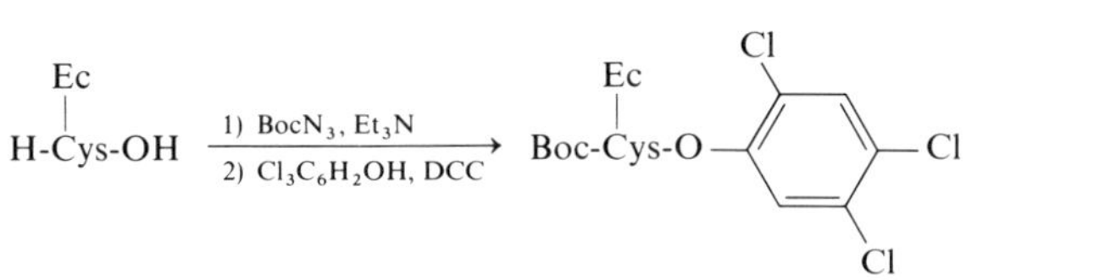

(36)

(77) **(78)** 43%

quently a modified *S*-ethylcarbamoyl group which incorporates the "safety-catch" principle was reported (Jäger and Geiger, 1973). The reagent, isocyanate **79**, is prepared as shown in Scheme 4. Reaction of cysteine with **79**,

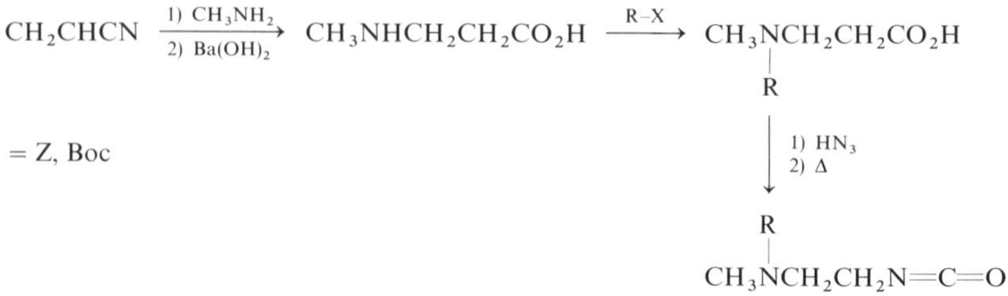

R = Z, Boc

(**79**)

Scheme 4

provides the β-(*N*-acyl-*N*-methylaminoethyl)carbamoylcysteine derivative **80** (Scheme 5). Deprotection of the cysteine residue is achieved by removal of either the benzyloxycarbonyl or the *tert*-butyloxycarbonyl group from **80** with acid and neutralization of the salt **81**. The resulting β-aminoethylcarbamoyl derivative (**82**) cyclizes to produce 1-methyl-2-imidazolidone (**83**)

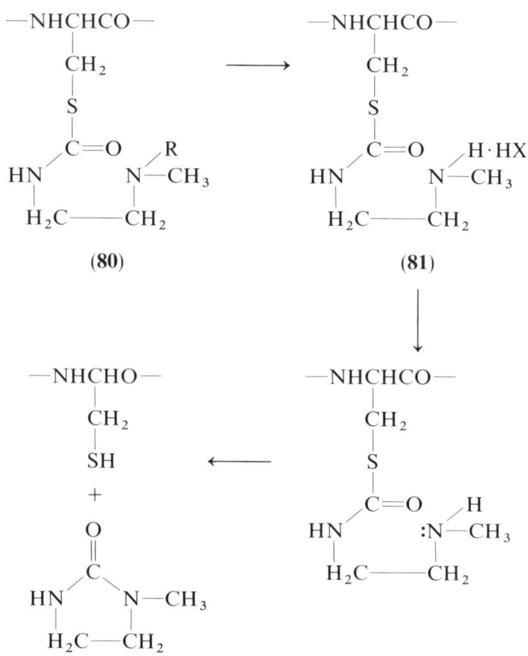

(**80**) (**81**)

R = Z, Boc

(**83**) (**82**)

Scheme 5

and the thiol containing peptide. The procedure was tested in a synthesis of oxytocin. Tschesche and Jering (1973) have reported the use of methoxymethylisocyanate for cysteine protection.

B. Removal of Thioesters and Related Derivatives

Thioesters are generally removed by alkaline hydrolysis (Hiskey *et al.*, 1973). Storey *et al.* (1972) studied conditions for removal of the *S*-ethylcarbamoyl group and found the group was stable toward trifluoroacetic acid and hydrogen bromide in trifluoroacetic acid but quite sensitive toward alkali. Wittinghofer (1974) utilized the *S*-ethylcarbamoyl group in the synthesis of the A(6–13)fragment of ovine insulin [Eq. (22)]. Removal of the groups was smoothly achieved by use of liquid ammonia. Lability to alkali may in fact be an operational problem of the *S*-ethylcarbamoyl group. Kawasaki *et al.* (1973) note that the synthetic carboxamido-terminal tetrapeptide (**84**) of ribonuclease T_1 could not be elongated by the active ester method presumably because of the lability of the S-protecting group to the added base.

```
              Ec
              |
H-Val-Glu-Cys-Thr-NH2
```

(**84**)

Studies by Storey *et al.* (1972) indicate that the *S*-ethylcarbamoyl group is cleanly removed by the action of silver ion in dilute base and mercury(II) ion including *p*-chloromercuribenzoate at pH 7.0 but not pH 4.6.

VI. SULFENYL DERIVATIVES (PREPARATION AND REMOVAL)

The sulfenyl derivatives that have been utilized as protective groups include symmetrical cystine peptides, unsymmetrical disulfides of cysteine, the *S*-sulfonate and the *S*-carbomethoxysulfenyl derivatives of cysteine, and various sulfenamides.

The use of symmetrical cystine peptides was articulated by Zahn and Schmidt (1970) who found enhanced solubility, stability, and ease of manipulation. Thus the "H-peptide" could be built up from symmetrical cystine derivatives and cleaved at the last stage to generate either the thiol or *S*-sulfonate derivative. Cyclic cystine "loops" can also be utilized for this purpose. Berndt and Zahn (1975) have described the synthesis of **85** and the conversion of this substance to the tetrathiol form using *β*-mercaptoethanol. Symmetrical disulfides have been used by others including Amaral *et al.* (1977), who employed cystine bis-*tert*-butyl ester for synthesis, and the Ciba

OtBu (Glu); tBu (Ser); tBu (Tyr); OtBu (Glu); tBu (Tyr)

Bpoc-Gly-Ile-Val-Glu-Gln-Cys-Cys-Ala-Gly-Val-Cys-Ser-Leu-Tyr-Gln-Leu-Glu-Asn-Tyr-Cys-Asn-O*t*Bu

(85)

group who employed an unsymmetrical cystine derivative **62** as a key intermediate in the synthesis of human insulin.

Other unsymmetrical disulfides of cysteine peptides have been utilized as sulfur protective groups. For example *S*-*tert*-butylsulfenylcysteine peptides are stable to hydrogen chloride in dioxane (Weber and Hartter, 1970) and 30% trifluoroacetic acid in methylene chloride but unstable toward hydrogen fluoride (1 hr, 0°C) (Van Rietschoten *et al.*, 1975). In the latter case with synthetic apamin (**10**), a disulfide polymer resulted upon cleavage of the peptide from the resin; reduction of the polymer using the MacLaren–Sweetman (1966) procedure (see also Ruegg and Gattner, 1975) with tri-*n*-butylphosphine provided the tetrathiol which was reoxidized to **10** (Van Rietschoten *et al.*, 1975). The *S*-*tert*-butylsulfenyl (S*t*Bu) group is stable toward 10% triethylamine in methylene chloride (Van Rietschoten *et al.*, 1975) but cleaved by sodium in liquid ammonia (Wieland *et al.*, 1977) and thiols (Hartter and Weber, 1973).

Hartter (1976) studied the conditions for the conversion of **86** → **87** [Eq. (37)]. He found that hydrogen chloride in acetic acid caused disulfide interchange but cleavage with boron trifluoride etherate in acetic acid gave **87** in 77% yield. These results suggest that the *S*-isopropylsulfenyl derivative

S*i*Pr (Cys), Z (Lys)

Boc-Ile-Cys-Gly-Lys-OH ⟶ H-Ile-Cys-Gly-Lys-OH (37)

(86) **(87)**

is somewhat less stable than the *S*-*tert*-butyl analogue. The *S*-ethylsulfenylcysteines have also been employed with limited success. Losse *et al.* (1977) found that the disulfide interchange encountered during removal of the *tert*-butyloxycarbonyl group from **88** with acid could be suppressed by

SEt (Cys)

Boc-Leu-Cys-Gly-O*t*Bu

(88)

conducting the reaction in the presence of ethylmercaptan. Kunzek *et al.* (1975a) used 4 *N* HCl in dioxane to remove the *tert*-butyloxycarbonyl group from **89**; high-molecular-weight products were encountered with 1*N* HCl

SEt
|
Boc-Cys-Ala-Phe-OMe

(89)

in acetic acid or trifluoroacetic acid in methylene chloride. Treatment of **90** with hydrazine also led to disulfide interchange (Kunzek *et al.*, 1975b). The *S*-ethylsulfenyl group has been shown to be cleaved by thiols and sulfite (Inukai *et al.*, 1968; Weber *et al.* 1969) without effect on other portions of the peptide chain.

SEt Dpm
| |
Boc-Cys-Cys-Ala-Gly-OMe

(90)

The *S*-sulfonate derivatives of synthetic peptides continue to receive use as vehicles for purification as well as intermediates for the formation of unsymmetrical cystine derivatives. For example, Schwartz and Katsoyannis (1977a,b) prepared the protected derivative of the [Leu^{10}]B chain of human insulin (**91**); treatment of **91** with liquid hydrogen fluoride [Eq. (38)] followed by oxidative sulfitolysis of the resulting dithiol provided the crude synthetic

Boc-Phe-Val-Asn-Gln-His-Leu-Cys(Dpm)-Gly-Ser(Bzl)-Leu-Leu-Val-Glu(OBzl)-Ala-Leu-Tyr(Bzl)-Leu-Val-Cys(Dpm)-Gly-Glu(OBzl)-Arg(NO_2)-Gly-Phe-Phe-Tyr(Bzl)-Thr-Pro-Lys(Z)-Thr(Bzl)-OBzl

(91)

↓ 1) HF 2) Sulfitolysis 3) Chromatography (38)

[Leu^{10}]B(SSO_3^-)$_2$

(92)

B chain di-*S*-sulfonate. Purification of the B-chain derivative was accomplished by chromatography on a carboxymethyl cellulose column equilibrated and eluted with acetate buffer in the presence of urea. The combination of **92** with the thiol form of porcine A chain, illustrates the activating effect of the sulfonate group on the sulfenyl sulfur atom. Treatment of the tetra-*S*-sulfonate form of A chain with β-mercaptoethanol (Schwartz and Katsoyannis, 1976) followed by precipitation from a 2-propanol–ethyl acetate mixture provided the tetrathiol (**93**) [Eq. (39)]. Reaction of **93** with **92** in a 2:1 ratio followed by purification provided the [Leu^{10}]insulin analogue [Eq. (40)]. Many insulin analogues have been prepared by this method including [Hcy^{6,11}-A]insulin in which the cysteine residues at A-6,11 were replaced by homocysteine units (providing an insulin analogue containing a 22- rather than a 20-membered loop) (Cosmatos and Katsoyannis, 1973)

$$A(S{-}SO_3^-)_4 \xrightarrow[\text{2) Precipitation}]{\text{1) } HSCH_2CH_2OH} A(SH)_4 \quad (39)$$

(93)

$$A(SH)_4 \xrightarrow[\text{2) Chromatography}]{\text{1) [Leu}^{10}\text{]B(S-SO}_3^-)_2} [\text{Leu}^{10}]\text{insulin} \quad (40)$$

and a similar modification at positions A-7, 20 (Cosmatos and Katsoyannis, 1975).

The *S*-carbomethoxysulfenyl derivatives of cysteine are more reactive toward thiols than the *S*-sulfonate derivatives. As indicated by the conversion of **33** → **34** [Eq. (21)] and **55** → **57** [Eq. (31)], the intermediate *S*-carbomethoxysulfenyl derivatives are comparable in reactivity toward thiols to the sulfenyl thiocyanates or sulfenyl iodides but offer the advantage of substantially increased stability over these sulfenyl derivatives. The reaction of *S*-carbomethoxysulfenylcysteine with thiol is irreversible [Eq. (41)]. In contrast to the sulfenyl iodides or sulfenylthiocyanates, the *S*-carbomethoxysulfenyl (Scm) derivatives do not react with *S*-triphenyl- or diphenylmethyl thioethers or the *S*-acetamidomethyl derivatives of cysteine (Hiskey *et al.*, 1975a). The *N-tert*-butyloxycarbonyl group may be removed from **94** without cleavage of the Scm group [Eq. (42)]; in effect the *S*-carbomethoxysulfenyl derivative of cysteine serves not only as an *S*-protective group but also as a derivative that may be cleanly converted into a disulfide. Harpp *et al.* (1978) have continued their earlier studies on novel sulfenyl systems that function as similar "sulfur-transfer" reagents.

$$\text{H-}\underset{}{\overset{\text{Scm}}{\text{Cys}}}\text{-OH} + \text{RSH} \longrightarrow \text{H-}\overset{\text{SR}}{\text{Cys}}\text{-OH} + \text{COS} + \text{CH}_3\text{OH} \quad (41)$$

$$\text{Boc-}\overset{\text{Scm}}{\text{Cys}}\text{-OM} \xrightarrow[\text{2) Boc-Phe-OSu, NMM}]{\text{1) TFA, 0°C}} \text{Boc-Phe-}\overset{\text{Scm}}{\text{Cys}}\text{-OMe} \quad (42)$$

(94) **(95)** 65%

REFERENCES

Allison, W. S. (1976). *Acc. Chem. Res.* **9**, 293–299.

Amaral, J. S. A., Macedo, A., and Oliveira, I. A. (1977). *J. Chem. Soc., Perkin Trans. 1* p. 205

Arold, H., and Eule, M. (1973). *In* "Peptides 1972" (H. Hanson and H.-D. Jakubke, eds.), p. 78. North-Holland Publ., Amsterdam.

Atherton, E., and Sheppard, R. C. (1976). *Amino-Acids, Pept, Proteins* **8**, 261.
Aune, T. M., and Thomas, E .L. (1978). *Biochemistry* **17**, 1005–1010.
Bachi, M. D., and Ross-Petersen, K. J. (1972). *J. Org. Chem.* **37**, 3550–3551.
Berndt, H., and Zahn, H. (1975). *Justus Liebigs Ann. Chem.* pp. 1601–1612.
Beyerman, H. C. (1963). *In* "Peptides, 1962" (G. T. Young, ed.), p. 53. Pergamon, Oxford.
Brois, S. J., Pilot, J. F., and Barnum, H. W. (1970). *J. Am. Chem. Soc.* **92**, 7629.
Camble, R., and Petter, N. N. (1976). *In* "Peptides 1976" (A. Loffet, ed.), p. 299. Editions de l' Université de Bruxelles, Belgium.
Chakravarty, P. K., and Olsen, R. K. (1978). *J. Org. Chem.* **43**, 1270–1271.
Chimiak, A. (1963). *In* "Peptides 1962" (G. T. Young, ed.), p. 37. Pergamon, Oxford.
Ciardelli, T. L., and Olsen, R. K. (1977). *J. Am. Chem. Soc.* **99**, 2806–2807.
Cosmatos, A., and Katsoyannis, P. G. (1973). *J. Biol. Chem.* **248**, 7304–7309.
Cosmatos, A., and Katsoyannis, P. G. (1975). *J. Biol. Chem.* **250**, 5315–5321.
Coyle, S., and Young, G. T. (1976). *J. Chem. Soc., Chem. Commun.* pp. 980–981.
Doscher, M. S. (1977). *In* "Methods in Enzymology" (C. H. W. Hirs and S. N. Timasheff, eds.), Vol. 47, Part E, pp. 578–617. Academic Press, New York.
Erickson, B. W., and Merrifield, R. B. (1973). *J. Am. Chem. Soc.* **95**, 3750–3756.
Felix, A. M., Jimenez, M. H., and Meienhofer, J. (1977). *In* "Peptides, Proceedings of the Fifth American Peptide Symposium" (M. Goodman and J. Meienhofer, eds.), pp. 532–535. Wiley, New York.
Fontana, A. (1975). *J. Chem. Soc., Chem. Commun.* pp. 976–977.
Friedman, M. (1973). "The Chemistry and Biochemistry of the Sulfhydryl Group in Amino Acids, Peptides and Proteins." Pergamon, Oxford.
Fujii, H., and Yajima, H. (1975). *Chem. Pharm. Bull.* **23**, 1596–1603.
Fujino, M., and Nishimura, O. (1976). *J. Chem. Soc., Chem. Commun.* p. 998.
Gosden, A., Macrae, R., and Young, G. T. (1977). *J. Chem. Res.* (*S*) **1**, 0317–0326.
Hammerstrom, K., Lunkenheimer, W., and Zahn, H. (1970). *Makromol. Chem.* **133**, 41–51.
Hanson, R. W., and Law, H. D. (1965). *J. Chem. Soc.* pp. 7285–7297.
Harpp, D. N., Steliou, K., and Chan, T. H. (1978). *J. Am. Chem. Soc.* **100**, 1222–1228.
Hartter, P. (1976). *Hoppe-Seyler's Z. Physiol. Chem.* **357**, 1683–1693.
Hartter, P., and Weber, U. (1973). *Hoppe-Seyler's Z. Physiol. Chem.* **354**, 365–370.
Heller, M. J., Walder, J. A., and Klotz, I. M. (1977). *J. Am. Chem. Soc.* **99**, 2780–2785.
Hermann, P., and Schreier, E. (1973). *In* "Peptides 1972" (H. Hanson and H. -D. Jakubke, eds.), p. 126. North-Holland Publ., Amsterdam.
Hirschmann, R., Nutt, R. F., Veber, D. F., Vitali, R. A., Varga, S. L., Jacob, T. A., Holly, F. W., and Denkewalter, R. G. (1969). *J. Am. Chem. Soc.* **91**, 507–508.
Hiskey, R. G., Rao, V. R., and Rhodes, W. G. (1973). *In* "Protective Groups in Organic Chemistry" (J. F. W. McOmie, ed.), pp. 235–308. Plenum, New York.
Hiskey, R. G., Muthukumaraswamy, N., and Vunnam, R. R. (1975a). *J. Org. Chem.* **40**, 950–953.
Hiskey, R. G., Li, C.-d., and Vunnam, R. R. (1975b). *J. Org. Chem.* **40**, 3697–3703.
Hruby, V. J., Upson, D. A., and Agarwal, N. S. (1977). *J. Org. Chem.* **42**, 3552–3556.
Inukai, N., Nakano, K., and Murakami, M. (1968). *Bull. Chem. Soc. Jpn.* **41**, 182–186.
Jäger, G. and Geiger, R. (1973). *In* "Peptides 1972" (H. Hanson and H.-D., Jakubke, eds.), pp. 90–92. North-Holland Publ., Amsterdam.
Jones, D. A., Jr., Mikulec, R. A., and Mazur, R. H. (1973). *J. Org. Chem.* **38**, 2865–2869.
Jones, J. H., and Ridge, B. (1974). *Amino-Acids, Pept, Proteins* **5**, 277–283.
Kamber, B. (1971). *Helv. Chim. Acta* **54**, 927–930.
Kamber, B. (1973). *Helv. Chim. Acta* **56**, 1370–1381.
Kamber, B., and Rittel, W. R. (1968). *Helv. Chim. Acta* **51**, 2061–2064.
Kamber, B., Riniker, B., Sieber, P., and Rittel, W. (1976). *Helv. Chim. Acta* **59**, 2830–2840.

Katsoyannis, P.G., and Schwartz, G. P. (1977). *In* "Methods in Enzymology" (C. H. W. Hirs and S. N. Timasheff, eds.), Vol. 47, Part E, pp. 501–578. Academic Press, New York.

Kawakasaki, K., Camble, R., Dupuis, G., Romovacek, H., Storey, H. T., Yanaihara, C., and Hofmann, K. (1973). *J. Am. Chem. Soc. Ser.* **95**, 6815–6824.

Kenner, G. W. (1977). *Proc. R. Soc. London, A* **353**, 441–457.

Kenyon, G. L., and Bruice, T. W. (1977). *In* "Methods in Enzymology" (C. H. W. Hirs and S. N. Timasheff, eds.), Vol. 47, Part E, pp. 407–430. Academic Press, New York.

König, W., Geiger, R., and Siedel, W. (1968). *Chem. Ber.* **101**, 681–693.

Kornblum, N., and Scott, A. (1974). *J. Am. Chem. Soc.* **96**, 590–591.

Kunzek, H., Seidel, C., Klauschenz, E., and Kaufmann, K.-D. (1975a). *Z. Chem.* **15**, 20–21.

Kunzek, H., Seidel, C., and Frache, G., (1975b). *Z. Chem.* **15**, 58–59.

Live, D. H., Agosta, W. C., and Cowburn, D. (1977). *J. Org. Chem.* **42**, 3556–3561.

Losse, G., Mauck, H., and Stange, H. (1977). *Tetrahedron* **33**, 1813–1816.

MacLaren, J.A., and Sweetman, B. J. (1966). *Aust. J. Chem.* **19**, 2347–2354.

Merrifield, R. B., Barany, G., Cosand, W. L., Engelhard, M., and Mojsov, S. (1977). *In* "Peptides, Proceedings of the Fifth American Peptide Symposium" (M. Goodman and J. Meienhofer, eds.), pp. 488–502. Wiley, New York.

Mladenova-Orlinova, L. (1973). *Dokl. Bolg. Akad. Nauk* **26**, 497–499 and 643–645; *Chem. Abstr.* **79**, 79163 and 105563 (1973).

Mladenova-Orlinova, L., and Tsvetkova, M. A. (1973). *Dokl. Bolg. Akad. Nauk* **26**, 493–495. *Chem. Abstr.* **79**, 79149 (1973).

Moore, S., Felix, A. M., Meienhofer, J., Smith, C. W., and Walter, R. (1977). *J. Med. Chem.* **20**, 495–500.

Pastuszak, J. J., and Chimiak, A. (1977). *Rocz. Chem.* **51**, 1567–1569.

Patai, S., ed. (1974). "The Chemistry of the Thiol Group," Parts 1 and 2. Wiley, New York.

Photaki, I. (1973). *In* "The Chemistry of Polypeptides" (P. G. Katsoyannis, ed.), pp. 59–85. Plenum, New York.

Photaki, I. (1976). *Top. Sulfur Chem.* **1**, 113–183.

Photaki, I., Taylor-Papadimitriou, J., Sakarellos, C., Mazarakis, P., and Zervas, L. (1970). *J. Chem. Soc. C* pp. 2683–2687.

Pless, J. (1976). *Helv. Chim. Acta* **59**, 499–512.

Ridge, B., and Jones, J. H. (1975). *Amino-Acids, Pept., Proteins* **6**, 286–297.

Riniker, B., Kamber, B. and Sieber, P. (1975). *Helv. Chim. Acta* **58**, 1086–1094.

Rivier, J. E. F. (1974). *J. Chem. Soc.* **96**, 2986–2992.

Rocchi, R., Benassi, C. A., Tomatis, R., Ferroni, R., and Menegatti, E. (1976). *Int. J. Pept. Protein Res.* **8**, 167–175.

Ruegg, U. T., and Gattner, H.-G. (1975). *Hoppe-Seyler's Z. Physiol. Chem.* **356**, 1527–1533.

Sakakibara, S., Shimonishi, Y., Kishida, Y., Okada, M., and Sugihara, H. (1976). *Bull. Chem. Soc. Jpn.* **40**, 2164–2167.

Schafer, D. J. (1976). *Amino-Acids, Pept., Proteins* **7**, 255–256.

Schwartz, G. P., and Katsoyannis, P. G. (1976). *Biochemistry* **15**, 4071–4076.

Schwartz, G. P., and Katsoyannis, P. G. (1977a). *J. Chem. Res.* (*M*) pp. 2453–2469.

Schwartz, G. P., and Katsoyannis, P. G. (1977b). *J. Chem. Res.* (*S*) pp. 220–221.

Schwertner, E., Berndt, H., Gielen, H.-G., and Zahn, H. (1975a). *Justus Liebigs Ann. Chem.* pp. 581–585.

Schwertner, H., Klostermeyer, H., and Zahn, H. (1975b). *Justus Liebigs Ann. Chem.* pp. 1092–1106.

Sheradsky, T. (1974). *In* "The Chemistry of the Thiol Group" (S. Patai, ed.), pp. 692–702. Wiley, New York.

Sieber, P., Kamber, B., Hartmann, A., Jöhl, A., Riniker, B., and Rittel, W. (1974). *Helv. Chim. Acta* **57**, 2617–2621.

Sieber, P., Kamber, B., K. Eisler, K., Hartmann, A., Riniker, B., and Rittel, W. (1976). *Helv. Chim. Acta* **59**, 1489–1497.

Sieber, P., Kamber, B., Hartmann, A., Jöhl, A., Riniker, B., and Rittel, W. (1977). *Helv. Chim. Acta* **60**, 27–37.

Sieber, P., Eisler, K., Kamber, B., Riniker, B., Rittel, W., Märki, F., and De Casparo, M. (1978). *Hoppe-Seyler's Z. Physiol. Chem.* **359**, 113–123.

Storey, H. T., Beachan, J., Cernosek, S. F., Finn, F. M., Yanaihara, C., and Hofmann, K. (1972). *J. Am. Chem. Soc.* **94**, 6710–6178.

Tan, N. H., and Kaiser, E. T. (1976). *J. Org. Chem.* **41**, 2787–2793.

Tan, N. H., and Kaiser, E. T. (1977). *Biochemistry* **16**, 1531–1541.

Tomatis, R., Guggi, A., Salvadori, S., and Rocchi, R. (1977). *Int. J. Pept. Protein Res.* **9**, 137–147.

Trudelle, Y., and Caille, A. (1977). *Int. J. Pept. Protein Res.* **10**, 291–298.

Tschesche, H., and Jering, H. (1973). *Angew. Chem., Int. Ed. Engl.* **12**, 756.

Upson, D. A., and Hruby, V. J. (1976). *J. Org. Chem.* **41**, 1353–1358.

Van Rietschoten, J., Granier, C., Rochat, H., Lissitzky, S., and Miranda, F. (1975). *Eur. J. Biochem.* **56**, 36–40.

Van Rieschoten, J., Muller, E. P., and Granier, C. (1977). In "Peptides. Proceedings of the Fifth American Peptide Symposium" (M. Goodman and J. Meienhofer, eds.), pp. 522–524. Wiley, New York.

Veber, D. F., Milkowski, J. D., Vargas, S. L., Denkewalter, R. G., and Hirschmann, R. (1972). *J. Am. Chem. Soc.* **94**, 5456–5461.

Veber, D. F., Holly, F. W., Paleveda, W. J., Nutt, R. F., Bergstand, S. J., Torchiana, M., Glitzer, M. S., Saperstein, R., and Hirschmann, R. (1978). *Proc. Natl. Acad. Sci., U.S.A.* **75**, 2636–2640.

Wälti, M., and Hope, D. B. (1973). *Experientia* **29**, 389.

Weber, U., and Hartter, P. (1970). *Hoppe-Seyler's Z. Physiol. Chem.* **351**, 1384–1388.

Weber, U., Herzog, K. H., Grossmann, H., Hörnle, S., and Weitzel, G. (1969). *Hoppe-Seyler's Z. Physiol. Chem.* **350**, 1425–1429.

Wieland, T., Abel, K. J., and Birr, C. (1977). *Justus Liebigs Ann. Chem.* pp. 371–380.

Wittinghofer, A. (1974). *Justus Liebigs Ann. Chem.* pp. 290–305.

Wolman, Y. (1974). *In* "The Chemistry of the Thiol Group" (S. Patai, ed.), Part 2, pp. 669–684. Wiley, New York.

Yajima, H., Kisco, Y., Ogawa, H., Fujii, N., and Irie, H. (1975). *Chem. Pharm. Bull.* **23**, 1164–1166.

Yajima, H., Ogawa, H., Fujii, N., and Funakoshi, S. (1977). *Chem. Pharm. Bull.* **25**, 740–747.

Yamamoto, D. M., Upson, D. A., Linn, D. K., and Hruby, V. J. (1977). *J. Am. Chem. Soc.* **99**, 1564–1570.

Yamashiro, D., Noble, R. L., and Li, C. H. (1973). *J. Org. Chem.* **38**, 3561–3565.

Zahn, H., and Schmidt, G. (1970). *Justus Liebigs Ann. Chem.* **737**, 91–105.

Zervas, L., Photaki, I., and Phocas, I. (1968). *Chem. Ber.* **101**, 3332–3333.

Chapter **4**

Protection of the Hydroxyl Group in Peptide Synthesis

JOHN M. STEWART

THE PEPTIDES, VOLUME 3

ISBN 0-12-304203-8

I. INTRODUCTION

The proper handling of hydroxyamino acids is an important aspect of effective peptide synthesis. However, syntheses of peptides containing these amino acids often lead to problems. Some of these problems arise from uncertainty as to whether a protecting group is actually needed for the hydroxyl function. Many peptides have been synthesized successfully using amino acid derivatives containing free hydroxyl groups, whereas in other situations such attempts have led to serious side reactions. This chapter will survey the field of synthesis of peptides containing serine, threonine, and tyrosine, as well as related hydroxyl-containing amino acids, and will offer suggestions as to the need for and use of protecting groups.

Recent reviews have covered this field, and the reader may refer to them both for details of procedures not described here and for more extensive literature references. Problems associated with side-chain hydroxyl groups were essentially ignored in the classic work of Greenstein and Winitz (1961), while Schröder and Lübke (1965) gave a good coverage of the problems and the solutions available at that time. The very extensive work of Wünsch (1974) presents a thorough discussion of all aspects of peptide synthesis, including protection of the hydroxyl group, and gives representative laboratory procedures for synthesis of derivatives and peptides, as well as tables of appropriate amino acid derivatives. Chapters on protection of alcoholic and phenolic hydroxyl groups are included in the work of McOmie (1973). This present chapter will review the standard methods and bring the field up to date with the significant advances described in the literature through the year 1977.

II. SYNTHETIC STRATEGY FOR HYDROXYL-CONTAINING PEPTIDES: "MINIMAL" VERSUS "GLOBAL" PROTECTION

Two different philosophies have arisen for the strategy of synthesis in solution of large peptides. The "Minimal Protection Philosophy" is advanced by those who are most impressed by the difficulty of complete removal of side-chain protecting groups from large peptides and by the damage done to peptides by the reagents needed for such blocking-group removal. These workers are willing to accept the occurrence of some degree of side reactions involving the side-chain functional groups of serine, threonine, tyrosine, histidine, and arginine, assuming that these by-products will be removed by careful purification of the intermediate peptides. It is clear that this philosophy is most useful in the segment condensation approach, where purification of products is relatively simpler than in the case of stepwise condensation methods. The opposing "global" protection philosophy states that all side-chain functional groups must be protected throughout the synthesis to avoid, insofar as possible, all side reactions arising from accessible side-chain functions. The global protection technique is obviously necessary in solid-phase synthesis, where large excesses of acylating amino acid derivatives must be used to insure as complete reaction as is possible at every step. It is also favored for stepwise condensation methods in solution, since the by-products may be very difficult to remove by ordinary purification methods once the peptide being synthesized has reached any considerable size. Global protection has also been used with impressive success in syntheses of several large peptides by the segment condensation method.

Outstanding examples of the minimal protection approach are the synthesis of ribonuclease S-protein by the Merck group (for a summary, see Hirschmann, 1971) and of ribonuclease T_1 fragments by the Hofmann laboratory (see Finn and Hofmann, 1976). In the S-protein work, the S-acetamidomethyl group for cysteine [Cys(Acm)] and the N^{ε}-benzyloxycarbonyl group for lysine [Lys(Z)] were the only protecting groups used in the later stages of the synthesis. The Acm groups were removed by treatment with iodine or mercuric ions, and the carbobenzoxy groups by anhydrous HF; these procedures were shown not to harm S-protein seriously. The azide reaction was used exclusively for segment coupling during late parts of the synthesis, and it and the *N*-carboxyanhydride method were used at earlier stages of the synthesis. These coupling reactions do not generally form esters at the hydroxyamino acid residues or acylate histidine imidazole groups. In the ribonuclease T_1 work, segments were assembled by use of the azide method, while active esters also were used at earlier stages of the work. In this case only the lysine and cysteine residues were blocked at the end of the synthesis by N^{ε}-formyl (For) and *S*-ethylcarbamoyl (Ec) groups, respectively. The critical point in the minimal protection approach is that the coupling reactions must be chosen with great care so as not to cause acylation of slightly reactive side-chain functional groups, such as hydroxyls. Coupling reactions must be of the "low activation" type, and no large excess of acylating agent can be used.

The use of the global protection approach is well-exemplified by syntheses of secretin (Bodanszky *et al.*, 1967; Wünsch and Wendlberger, 1972) and insulin (Sieber *et al.*, 1974); it has also been discussed in detail by Sakakibara (1977). The first synthesis of secretin (Bodanszky *et al.*, 1967) was by the stepwise condensation method, and used benzyl blocking groups for hydroxyl "permanent" protection and *tert*-butyloxycarbonyl (Boc) for α-amine group protection. Active esters were used exclusively for coupling and final deprotection was by means of hydrogenolysis. The Wünsch and Wendlberger (1972) secretin synthesis was by segment condensation, and used *tert*-butyl

Table I Cleavage of Hydroxyl Protecting Groups[a]

Hydroxyl blocking group	Cleavage reagent[b]								
	H_2–Pd	HBr–AcOH, HF	OH^-	H^+–H_2O	Na–liq NH_3	$h\nu$ (350 nm)	TFA	Thiol	e^-
p-Toluenesulfonyl (Tos)	−	−	−	−	+	−	−		−
Acetyl (Ac)	−	−	+	−	−	−	−		−
Benzyloxycarbonyl (Z) 2-Bromobenzyloxycarbonyl (2Bz)	+	+	+	−	+	−	−	(+)	−
Benzyl (Bzl)	+	+	−	−	+	−	(−)		−
tert-Butyl (*t*Bu)	−	+	−	(+)	−	−	+		−
Tetrahydropyranyl (Thp)	−	−	−	+	−	−	−	−	−
Trimethylsilyl (Tms)	−	+	+	+		−			−
1-Benzyloxycarbonylamino-2,2,2-trifluoroethyl (Ztf)	+	+	+	−	+	−	−		−
2-Nitrobenzyl (2 Nb) [Tyr]	+	−	−	−	+	+	−		−
2,4-Dinitrophenyl (Dnp) [Tyr]	−	−	(+)	−	+	−	−	+	−
Picolyl (Pic) [Tyr]	−	−	−	−	−	−	−	−	+

[a] +, The group is cleaved; (+), cleaved, but not a recommended procedure; −, the group is stable; (−), the group is of marginal stability. A blank indicates that no information is available.

[b] e^-, Electrolysis; $h\nu$, photolytic cleavage; TFA, trifluoroacetic acid.

[c] Benzyloxycarbonyl is not fully stable.

Table II Compatible α-Amino and Hydroxyl Blocking Groups

Hydroxyl protection	α-Amine protection	Cleavage reagent for N^{α} deblocking[a]	Cleavage reagent for hydroxyl deblocking[a]
Benzyl (Bzl) 2-Bromobenzyl (2Bz) 1-Benzyloxycarbonylamino-2,2,2-trifluoroethyl (Ztf)	*tert*-Butyloxycarbonyl (Boc)	TFA (25–50% in CH_2Cl_2)	HF, HBr
	2-(*p*-Biphenylyl)propyl(2)oxycarbonyl (Bpoc) 3,4,5-Trimethyloxybenzyloxycarbonyl (Tmz) α,α-Dimethyl-3,5-dimethyloxycarbonyl (Ddz)	TFA (1–5% in CH_2Cl_2)	
	o-Nitrophenylsulfenyl (Nps)	Thiol; HCl in EtOAc	
tert-Butyl (*t*Bu)	Benzyloxycarbonyl (Z)	H_2–Pd	TFA
	2-(*p*-Biphenylyl)propyl(2)oxycarbonyl (Bpoc) 3,4,5-Trimethyloxybenzyloxycarbonyl (Tmz) α,α-Dimethyl-3,5-dimethyloxycarbonyl (Ddz)	TFA (1–5% in CH_2Cl_2)	TFA
p-Toluenesulfonyl (Tos)	Any	As required	Na in liquid NH_3
Picolyl (Pic)	Any	As required	Electrolysis
2,4-Dinitrophenyl (Dnp)	Benzyloxycarbonyl (Z) *tert*-Butyloxycarbonyl (Boc) 2-(*p*-Biphenylyl)propyl(2)oxycarbonyl (Bpoc)	As required (see above)	Thiol

[a] TFA, trifluoroacetic acid.

hydroxyl protection with temporary N^{α}-protection by *o*-nitrophenyl (Nps) and benzyloxycarbonyl groups which could be removed without serious damage to the side chain blocking groups. Coupling reactions were mediated by dicyclohexylcarbodiimide (DCC) with addition of hydroxybenzotriazole (HOBt) or *N*-hydroxysuccinimide, and final deprotection was by means of anhydrous trifluoroacetic acid (TFA). Even though relatively labile side-chain blocking groups were used, some difficulty was encountered in complete removal of these groups by TFA. The synthetic strategy used in the insulin synthesis cited (Sieber *et al.*, 1974) resembled that of the Wünsch and Wendlberger secretin synthesis in that *tert*-butyl hydroxyl protection was used. The outstanding feature of this synthesis was the selective and specific formation of the three disulfide bridges. In contrast to these syntheses using relatively labile side-chain protecting groups, Sakakibara (1977) favors more stable groups removable finally by anhydrous HF. In this respect his approach resembles that of solid-phase synthesis. Clearly, this latter approach is limited to those peptides and proteins not harmed severely by HF, and must be modified for more labile molecules. The trend will most likely be toward ever more labile blocking groups. Global protection allows use of more vigorous reaction conditions for coupling (higher activation and larger excesses of acylating agents) which can be expected to give higher yields and shorter reaction times. See Tables I and II for compatible systems.

III. SERINE IN PEPTIDE SYNTHESIS

Serine is a common and very important amino acid in many peptides and proteins of biological origin. It occurs, as is well known, in the active center of the serine proteases. It is also frequently found as a constituent of peptide hormones, although in these molecules it does not appear generally to play a critical role, but is rather usually a "spacer" to hold other more critical groups in the appropriate steric relationships. Several peptides and proteins contain serine with glycosyl or phosphoryl residues attached to the hydroxyl group.

Serine contains a primary hydroxyl group which is readily accessible to acylating reagents (activated peptides and amino acid derivatives). The hydroxyl group can also be acylated by N → O shift of acyl groups under anhydrous acid conditions. Moreover, under dehydrating conditions, a serine (or threonine) hydroxyl group may be acylated by glutamic and aspartic acid side-chain carboxyl groups in neighboring positions within a peptide. *O*-Acylserines may also undergo β-elimination with production of dehydroalanine peptides. Whereas some of these problems also arise during removal of serine protecting groups, they are more widespread and serious when serine is unprotected in peptide synthesis. Nevertheless, if these problems are recognized, and conditions which avoid them are used, serine may often be used in synthesis without protection of the hydroxyl group.

A. Synthesis of Serine Peptides without Protection of the Hydroxyl Group

1. Serine with a Free Hydroxyl Group in the Amine Component of Coupling Reactions

Serine esters and serine-containing peptides with free hydroxyl groups have been used as amine components in a wide variety of coupling reactions. Successful syntheses have used azide, mixed anhydride, and active ester couplings as well as dicyclohexycarbodiimide-mediated reactions. In all such condensations, the most important consideration is to be sure that the free serine hydroxyl group is not exposed to an excess of active acylating agent for any length of time. If it is, the serine hydroxyl will likely be acylated by the activated amino acid or peptide derivative. Such acylation is favored if the coupling reaction involves a sterically hindered amino acid at the site of coupling. In that case, the coupling is likely to be slow and the serine hydroxyl may prove to be a better and more accessible nucleophile than the exposed amino group. Sakakibara (1977) reported acylation of the serine hydroxyl of H-Phe-Ser-Pro-Phe-Arg(NO_2)-OBzl by Aoc-Pro-Pro-Gly-ONp in the synthesis of bradykinin. This result indicates that problems can be encountered even when "over activation" in the coupling reaction is avoided; presumably, the side reaction could have been even worse if dicyclohexylcarbodiimide had been used for the coupling reaction.

Bodanszky *et al.* (1977) reported that the presence of an imidazole-unprotected histidine residue in the amine component of a coupling reaction promoted O-acylation of serine residues by *tert*-butyloxycarbonylamino acid *p*-nitrophenyl or *N*-hydroxysuccinimide esters. O-acylation was inhibited when nitrophenyl esters were used in the presence of 1-hydroxybenzotriazole (HOBt), although this was not effective for *N*-hydroxysuccinimide esters. O-acylation is also favored by presence of excess tertiary base, excess active ester, or by addition of HOBt to active ester reaction mixtures for synthesis of peptides not containing histidine.

Problems can also be anticipated when an NH_2-terminal serine peptide is acylated by a derivative in which the activated carboxyl is that of aspartic acid or asparagine. In these cases, formation of cyclic aspartimide-containing peptides (**1**) and acylation of the serine hydroxyl by the aspartic carboxyl (**2**) can be anticipated as undesirable side reactions (Scheme 1). Although cyclic aspartimide-containing peptides (**1**) can be readily opened by treatment

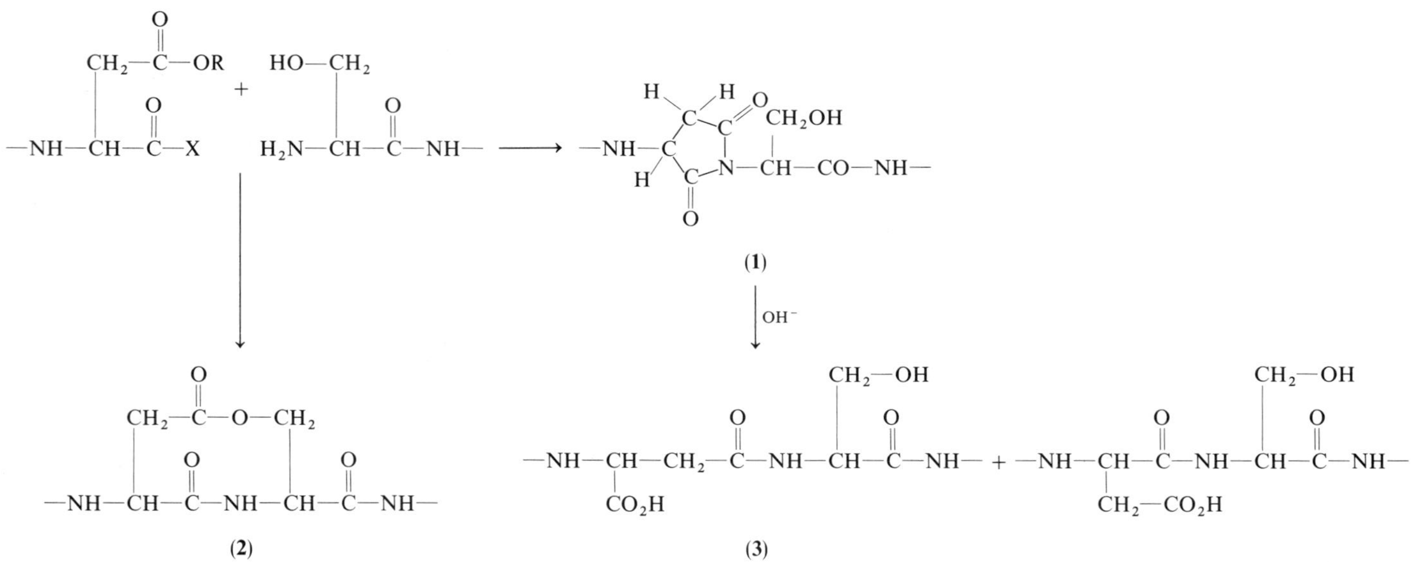

Scheme 1

with mild base, they unfortunately open in predominantly the wrong direction to yield β-aspartyl peptides (**3**).

2. Use of Serine with a Free Hydroxyl Group as the Carboxyl Component in Coupling Reactions

When an unprotected serine hydroxyl is present (either as a serine derivative or as a peptide derivative) in the carboxyl component of a coupling reaction, the method of carboxyl activation for coupling must generally be restricted to those methods that do not cause very high activation and consequent ester formation. For example, mixed anhydrides and Woodward's reagent, as well as some active esters, might be expected to give satisfactory results. Self-acylation problems will be aggravated if the activated compound stands for long periods of time in a solvent which promotes nucleophilic reactions (such as dimethylformamide), which may happen if an excess of carboxyl component is used or if the coupling reaction is slow due to steric hindrance in the amine component.

Several N-protected serine derivatives have been successfully used for addition of this amino acid to the amino terminus of peptides. Benzyloxycarbonylserine dinitrophenyl and pentachlorophenyl active esters have been used with success, although synthesis of N-protected serine *p*-nitrophenyl ester has been reported to be difficult (Bodanszky and Ondetti, 1966). Similar reactions have also been carried out with *tert*-butyloxycarbonyl serine derivatives. *tert*-Butyl- and benzyloxycarbonylserine have been coupled successfully by the mixed anhydride and azide methods, although in the latter case, if the reaction does not go rapidly to completion, the Curtius rearrangement may occur, leading to isocyanate formation and loss of the serine as a cyclic derivative [Eq. (1)]. Problems may also be anticipated in dicyclohexylcarbo-

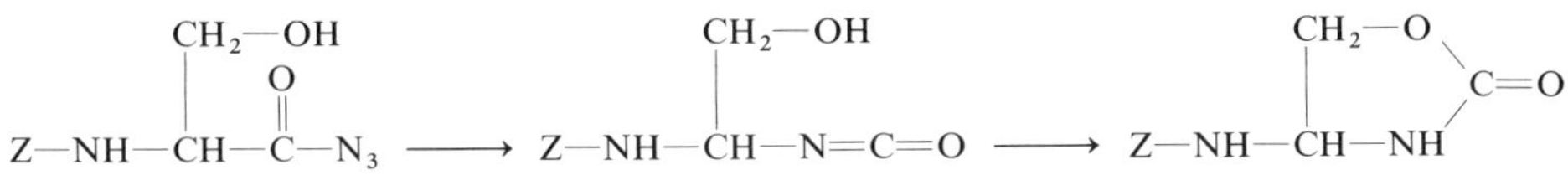

(1)

diimide-mediated coupling reactions, since these involve a very high degree of activation of the carboxyl function. Difficulties may often be minimized in this case by conducting these reactions in the presence of 1- hydroxybenzotriazole (although, see Bodanszky *et al*., 1977).

3. Other Problems in the Use of Unprotected Serine

Acylation of unprotected serine hydroxyls may occur not only during peptide coupling reactions, but also during removal of N-blocking groups. For example, if benzyloxycarbonyl groups are removed by HBr–AcOH (which is a good acetylating agent), the hydroxyl group of serine will be acetylated by the reagent. Although this is now a well-known reaction, it caused some embarrassment to peptide chemists several years ago. It can be avoided by removal of the benzyloxycarbonyl group with HBr in trifluoroacetic acid, or by including some water in the HBr–AcOH reagent. This latter modification introduces the risk of concomitant hydrolysis. No problems are to be anticipated when the benzyloxycarbonyl group is removed by hydrogenolysis, although a little water is often included in these reaction media for safety.

If undesired 0-acylation of serine does occur, it can be corrected by base treatment. This was an accepted procedure formerly when methyl esters were often used for COOH-terminal protection, but is no longer commonly the case, since now COOH-terminal blocking groups removable by acidolysis or hydrogenolysis are normally used. Base treatment of *O*-acyl peptides to remove the undesired group may not only cause partial racemization of the serine residue, but may also lead to partial hydrolysis of the peptide at the serine residue in some sequences, since Ser-X bonds are often quite labile. Amides of asparagine and glutamine may also be hydrolyzed, especially when these residues are adjacent to arginine or lysine in the sequence (Robinson, 1974). Asparagine is less stable than glutamine. The scope of peptide synthesis has been greatly broadened by elimination of alkaline hydrolysis as a deprotection procedure, since many peptides not stable to base can thus be successfully synthesized.

The N → O acyl shift has not been thoroughly studied under the conditions of peptide synthesis. Most of the work on this reaction has been with the goal of promoting selective cleavage of peptides and proteins at serine and threonine residues as an aid to sequence determination. Anhydrous HF, which is a very strong acid, is probably the best reagent known for promoting N → O shifts in peptides and proteins (Lenard and Hess, 1964), and its use in removal of blocking groups from peptides during synthesis undoubtedly leads to some N → O shift in synthetic peptides. Removal of blocking groups with HF should occur rapidly and at low temperature, while extensive N → O rearrangement generally requires extended treatment at room temperature. In most cases this shift, if it does occur, can be readily reversed by treatment of the peptides with dilute hydroxyl-containing solutions for a brief time [Eq. (2)]. In many cases, it may not be necessary to use solutions above neutral pH. For example, Levy and Carpenter (1970) found that treatment of insulin with anhydrous 0.1 *N* HCl–MeOH at room temperature for one hour led to a quantitative N → O shift at threonine residue B27. The reverse O → N shift took place readily in water at all pH values above 2.2. This ready reversal of the N → O shift has undoubtedly improved the yields of many synthetic peptides, but if an *O*-acyl blocking group was used for an NH_2-terminal serine, the chemist should be wary of irreversible N-acylation if conditions are suitable [see Eq. (2)].

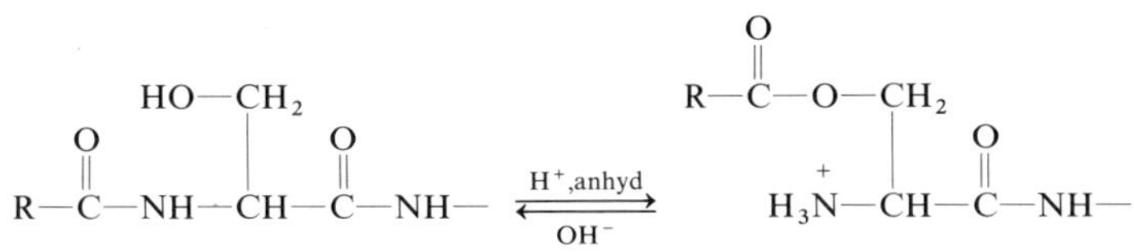

(2)

The esterification of serine hydroxyls by aspartic and glutamic side-chain carboxyls mentioned above (Scheme 1) is also promoted by the dehydrating conditions present during acidolytic removal of blocking groups, and has been observed in cleavage of a peptide containing a Glu-Thr sequence from a solid-phase resin by HBr–TFA (Young *et al.*, 1968) as well as in HF cleavage of peptide resins containing Asp-Ser sequences (J. M. Stewart, unpublished). Under these conditions aspartimide peptides (Scheme 1) may also be formed from Asp-Ser sequences in addition to esters, and may lead to significant amounts of byproducts in solid-phase procedures. Aspartimide peptides are most readily formed from Asp-Gly sequences. The cyclic aspartic and glutamic esters appear to be opened readily by very mild base treatment, even chromatography in aqueous amine systems (Young *et al.*, 1968). These side reactions in peptides synthesized by solid-phase procedures involved sequences in which the *O*-benzyl ether had been used for serine and threonine. Presumably, the side reaction occurred after the ether had been cleaved by the HBr or HF used. Aspartimide formation is favored by the presence of an ester on the aspartic side chains, and in some cases can be largely prevented by either removing the ester from the aspartic residue before other blocking groups are removed with HF (Yang and Merrifield, 1976) or by avoiding acidic dehydrating conditions. On the other hand, the esterification of a serine or threonine hydroxyl by an aspartic or glutamic side chain is favored by having the free carboxyl group present in the strong dehydrating medium, and can be prevented by using a more stable blocking group for the carboxyl function and removing this ester later under non-dehydrating conditions (Young *et al.*, 1968).

In no case should serine (or threonine) be used in solid-phase peptide synthesis without protection of the hydroxyl group. Since an excess of the activated amino acid derivative is always necessary to ensure completion of the coupling reaction, side-chain acylation is to be expected. As hydroxyl-protected derivatives of serine were already available at the time solid-phase synthesis was developed, they have been used exclusively in this method. Initial attempts to use threonine in solid-phase synthesis without protection of the hydroxyl group led to formation of branched-chain peptides, where acylation of the threonine hydroxyl led to generation of a new peptide chain (see below). This type of problem has subsequently been avoided by general use of O-blocked threonine in all solid-phase synthesis.

B. Protection of the Serine Hydroxyl Group

1. Protection by Acylation

In the earlier years of peptide synthesis some work was done with *O*-acetyl, *O*-benzyloxycarbonyl, and *O*-toluenesulfonyl derivatives of serine. However, since suitable ethers have become available for protection of serine, these groups are generally no longer used because of the serious problems encountered. The use of a derivative such as *O,N*-bis-benzyloxycarbonyl serine is reasonable when the serine is the last residue to be introduced at the NH_2-terminal position of a peptide. In this case both blocking groups are removed simultaneously and there is no opportunity for an undesired O → N shift, to cause irreversible blockade of the amine group. In contrast, if *O*-acetyl,*O*-benzoyl, or *O*-toluenesulfonyl groups are used and the synthesis is continued after removal of the N-protection, O → N migration can lead to permanent termination of the chain. Facile β-elimination from *O*-acylserine derivatives and peptides in the presence of anhydrous base can lead to incorporation of dehydroalanine residues into peptides. Although this has been used as a synthetic device for preparation of peptide amides, it is generally not a desirable reaction.

2. Ethers for Protection of the Serine Hydroxyl Group

a. Benzyl Ethers. Much peptide synthesis in recent years has been done with *O*-benzylserine derivatives. Such ethers are readily synthesized, are compatible with the use of the *tert*-butyloxycarbonyl and more labile amine blocking groups, and are readily removed at the end of the synthesis by HBr or HF. Boc-Ser(Bzl)-OH is generally synthesized by alkylation of Boc-Ser-OH with benzyl bromide. The procedure of Hruby and Ehler (1970), using sodium in liquid ammonia, yields Boc-Ser(Bzl)-OH in 45% yield, while that of Sugano and Miyoshi (1976), using sodium hydride in dimethylformamide, gives improved yields of 57%. The latter method has also been used by Turan *et al.* (1977) for synthesis of *N*-*tert*-butyloxycarbonyl-*O*-benzylhomoserine.

Boc-Ser(Bzl)-OH is the standard derivative of serine for use in solid-phase synthesis, and many peptides have been successfully synthesized using this derivative. Recently, several investigators have paid very careful attention to stability of side-chain blocking groups in order to develop techniques suitable for solid-phase synthesis of long peptides and proteins. This examination has shown that several of the commonly used side-chain blocking groups are not adequately stable for long syntheses, and some new derivatives have been introduced. Yamashiro (1977) showed that *O*-benzylserine possessed marginal stability, and introduced *O*-(4-bromobenzyl)serine as an improved derivative. He synthesized this new derivative by the procedure of Hruby and Ehler (1970) and found that, while treatment of Ser(Bzl) with trifluoroacetic acid–dichloromethane (1:1, v/v) for 23 hours caused 3% loss of the benzyl group, Ser(4Bb) lost only 1.3% protection in 71 hours treatment with the same reagent.

The benzyl ether of serine is readily cleaved by all reagents that normally remove benzyl groups, such as HF, HBr, and H_2–Pd. Scavengers such as anisole or methylethyl sulfide must be present during acidolysis to prevent benzylation of susceptible residues such as methionine or tyrosine. While anhydrous HF has been most widely used because it can cleanly remove all blocking groups and in solid-phase synthesis simultaneously cleave peptides from the resin, safe use of this toxic, volatile reagent requires expensive equipment (Peninsula Laboratories, San Carlos, Caliornia). For that reason investigators have sought alternate reagents for removal of benzyl-type blocking groups. Pless and Bauer (1973; Bauer and Pless, 1975) proposed boron tristrifluoroacetate (BTFA) as an alternate reagent for removal of benzyl-type blocking groups as well as for cleavage of solid-phase peptides from the resin. They report superior yields of peptides such as somatostatin if thioanisole is used as scavenger. They report that yields with BTFA are superior to those obtained by use of trifluoromethanesulfonic acid, which has been proposed by Yajima *et al.* (1975) as a substitute for HF. Some other investigators have found cleavage by boron tristrifluoroacetate to be inferior to that by HF. Felix (1974) proposed the use of a 1 *M* solution of boron tribromide in dichloromethane at room temperature for removal of benzyl, dichlorobenzyl, and *p*-nitrobenzyl esters and ethers, as well as benzyloxycarbonyl groups. BBr_3 does not cleave Arg(Tos), His(Bzl), or Cys(Bzl) groups, and partially degrades Arg(NO_2), Asn and Gln residues. Interestingly, it also cleaves methyl and ethyl esters, which are not cleaved by BTFA. Trifluoromethanesulfonic acid, on the other hand, does cleave Arg(Tos) groups, and has been used for synthesis of several peptides by solution methods. Pyridinium HF has also recently been proposed (Matsuura *et al.*, 1976) as a reagent for removal of benzyl blocking groups. The nonvolatility of this reagent causes problems in work-up of peptides following deblocking.

b. O-tert Butylserine. Recent evidence of instability of some larger peptides and many proteins to the rigorous conditions (such as HF or HBr) necessary for removal of benzyl-type blocking groups has prompted renewed interest in *tert*-butyl ethers as blocking groups for serine and threonine hydroxyl functions. These ethers have a stability similar to that of the *N*-*tert*-butyloxycarbonyl group, and are readily cleaved by 50% trifluoroacetic acid in dichloromethane. The total synthesis of glucagon by Wünsch and

Wendlberger (1968) was an excellent example of the use of *tert*-butyl ethers for serine, threonine, and tyrosine. The *O-tert*-butyl ethers are compatible with the use of several amine blocking groups such as *N*-benzyloxycarbonyl (removed by hydrogenolysis), *N*-2-(*p*-biphenylyl)propyl(2)oxycarbonyl (removed by very dilute TFA), or *N-o*-nitrophenylsulfenyl (removed by an equivalent of HCl or thiolysis).

Synthesis of *tert*-butyl ethers is accomplished by addition of the hydroxyl compound to isobutene in the presence of a strong anhydrous acid catalyst, such as H_2SO_4 or *p*-toluenesulfonic acid (Schröder, 1963). Recently, boron trifluoride etherate has been proposed as a catalyst (Smithwick, 1973). If free serine is used in such procedures, the product obtained is *O-tert*-butylserine *tert*-butyl ester; the amine group is protonated and does not react. *O-tert*-Butylserine can be prepared by reaction of serine methyl ester with isobutene and subsequent saponification of the ester. A suitable *N*-blocking group can then be introduced.

Whereas all *tert*-butyl blocking groups have roughly similar reactivity, these are not exactly the same, and some selectivity in removal of these groups is possible. Selective removal of *N-tert*-butyloxycarbonyl groups in the presence of *tert*-butyl esters has been reported by Goodacre *et al.* (1975), who used toluenesulfonic acid in ether at room temperature for 3–24 hr, and by Vorbrüggen and Krolikiewicz (1975), who used trimethylsilane perchlorate in dichloromethane at room temperature for 5 min. They did not specifically examine serine, threonine, and tyrosine *tert*-butyl ethers, and their stability under these conditions is not certain. Furthermore, these experiments were done with amino acids and small peptides and the selectivity might well be impossible in larger peptides where *tert*-butyloxycarbonyl groups have frequently been found to be resistant to removal by acidolysis. On the other hand, side-chain blocking groups often show abnormally high stability in large peptides.

c. Other Blocking Groups for the Serine Hydroxyl Function. Several other groups have been proposed for protection of serine hydroxyl functions, but have not received wide application in peptide synthesis. Trimethylsilyl (Tms) ethers of serine and threonine *N*-carboxyanhydrides have been used in peptide synthesis by Hirschmann *et al.* (1971). Serine *N*-carboxyanhydride (NCA) was converted to the trimethylsilyl ether derivative (**4**) by treatment with trimethylchlorosilane in tetrahydrofuran. This derivative was used in

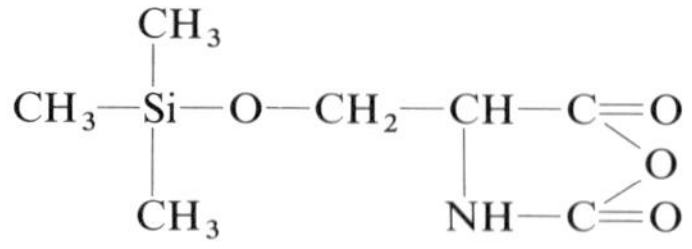

(**4**)

the rapid *N*-carboxyanhydride coupling procedure developed by the Merck group. The extreme hydrolytic lability of the trimethylsilyl group caused it to be split from the peptide in the alkaline coupling reaction medium. Hirschmann *et al.* (1971) state that yields from coupling of serine by this technique are not as good as for the other amino acid *N*-carboxyanhydrides, and that some side reactions have been observed.

Tetrahydropyranyl ethers are acetals, and are stable to base but easily cleaved by aqueous acid. Tetrahydropyranyl (Thp) derivatives of serine (**5**)

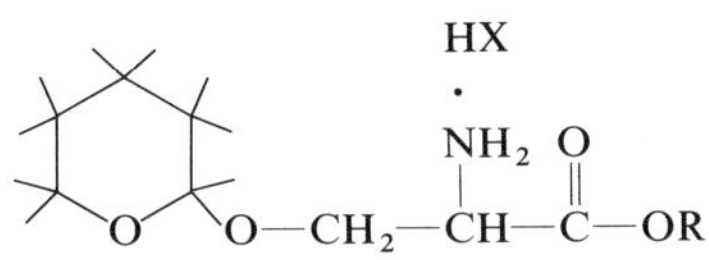

(**5**)

are prepared by reaction of a serine ester with dihydropyran in the presence of anhydrous acid (Iselin and Schwyzer, 1956). The *N*-benzyloxycarbonyl blocking group is compatible with tetrahydropyranyl ethers if it is removed by hydrogenolysis. The tetrahydropyranyl group is removed by heating in 50% aqueous AcOH for 10 min. When tetrahydropyranyl ethers are formed, a new asymmetric center is introduced into the molecule. Therefore, tetrahydropyranyl ethers of L-amino acids will be pairs of diastereomers, and will thus not have sharp melting points. Solubilities of the two forms may differ significantly. This blocking group has been relatively little used in peptide synthesis in recent years.

The benzyloxycarbonylaminotrifluoroethyl (Ztf) blocking group of Weygand *et al.* (1968) has also been used for protection of serine hydroxyl groups (**6**). However, the high cost of the derivative and the fact that it does not

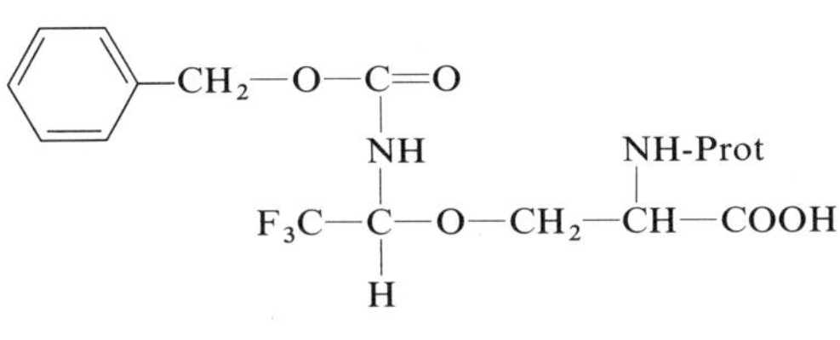

(**6**)

appear to offer significant advantages over other blocking groups have resulted in only slight use of this group in recent synthetic work. Like the tetrahydropyranyl ethers, addition of this blocking group introduces a new optical center and the derivatives of L-amino acids will be diastereomers with unsharp melting points. It is cleaved by reagents which cleave *N*-benzyloxycarbonyl groups.

IV. THREONINE IN PEPTIDE SYNTHESIS

The reactivity of the threonine hydroxyl group and the problems encountered in synthesis of threonine-containing peptides are quite similar to those of serine, and the comments made in the preceding section apply generally. Only those cases which differ significantly from serine will be discussed in this section.

A. Synthesis of Threonine Peptides without Protection of the Hydroxyl Group

The problems which arise here and the means used to overcome them are quite similar to those found with serine. Although O-acylation of unprotected threonine hydroxyls during coupling reactions would be expected to be less of a problem than with serine because of steric hindrance, it can occur. An example of such *O*-acylation comes from the early solid-phase synthesis work, before *O*-benzylthreonine was commercially available. In that case, a branched-chain by-product (7) was formed in which an Arg-Pro-Pro-Gly

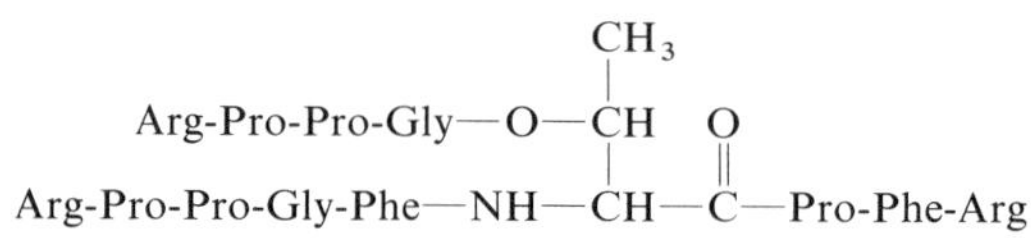

(7)

chain was attached to the threonine hydroxyl of [6-threonine]bradykinin (Stewart, cited in Stewart and Young, 1969, p. 23). This side reaction occurred in a dicyclohexylcarbodiimide-mediated coupling reaction in which a 2.5-fold amount of the *tert*-butyloxycarbonyl amino acid and dicyclohexylcarbondiimide was used. It is interesting that *O*-acylation did not occur at the first coupling reaction (addition of Boc-Phe-OH) when threonine was NH_2-terminal, but only at the following coupling, when Boc-Gly-OH was coupled. Evidently steric hindrance played a critical role here, even with the highly activated dicyclohexylisourea derivative. This result may lend support to the suggestion (Rebek and Feitler, 1974) that the actual acylating agent in dicyclohexylcarbodiimide-mediated solid-phase coupling reactions is not the isourea but the symmetrical anhydride, which would be expected to show a greater influence of steric hindrance (see Vol. I, Chapter 5, for details).

One additional concern with the use of unprotected threonine in synthesis is that an additional asymmetric center is available for racemization. Inversion of the β-center causes incorporation of *allo*-threonine into the peptide sequence. This is a particular problem in those cases where N $\rightarrow$ O shifts may take place, since the β-center may be inverted when this process is reversed.

B. Protection of the Threonine Hydroxyl Group

The same blocking groups described above for serine have been applied to threonine. In most cases results are quite similar.

The synthesis of *O*-benzylthreonine is difficult and cannot be accomplished with yields comparable to those obtained with serine. Mizoguchi *et al.* (1968) prepared H-Thr(Bzl)-OBzl by long heating of threonine with benzyl alcohol and toluenesulfonic acid. Following saponification of the benzyl ester and conversion to the N^{α}-*tert*-butyloxycarbonyl derivative, the overall yield obtained was only about 10%. The method of Hruby and Ehler (1970) gave even poorer results. The method of Sugano and Miyoshi (1976) may be slightly better, offering a 14% yield of Boc-Thr(Bzl)-OH from Boc-Thr-OH. Yamashiro (1977) used the method of Mizoguchi *et al.* (1968) to prepare the *p*-chlorobenzyl (4Cb) ether of threonine in a yield comparable to that obtained earlier with the benzyl ether. Yamashiro prepared this new derivative in order to obtain additional stability of the threonine protection during extended solid-phase synthesis runs. Thr(Bzl) lost 5% of the blocking group in 23-hr treatment with trifluoroacetic acid in dichloromethane (1:1), while (4Cb) lost only 2.5% of the blocking group in 71-hr treatment with the same reagent. The chlorobenzyl ether is readily removed by liquid HF and other similar reagents.

Threonine *tert*-butyl ethers are obtained in good yield by methods similar to those used for preparation of serine derivatives, and show similar behavior in synthesis and deprotection reactions. Hirschmann *et al.* (1971) synthesized the *O*-trimethylsilyl derivative of threonine *N*-carboxyanhydride, and used this derivative in the rapid *N*-carbonxyanhydride method of synthesis. As in the case of serine, the trimethylsilyl ether was cleaved spontaneously during the coupling step. Tetrahydropyranyl and benzyloxycarbonylaminotrifluoroethyl derivatives have also been prepared for threonine, but have received only limited use in peptide synthesis.

V. HYDROXYPROLINE IN PEPTIDE SYNTHESIS

A recently published compendium of peptides and derivatives of 4-hydroxyproline (Adams, 1977) lists the peptides and derivatives described in the literature to this time. An additional listing is to be found in Wünsch (1974). An earlier paper of Adams (1976) describes some methods used for synthesis of these peptides. That paper describes work with both the common

4- and less common 3-hydroxyprolines. A recent paper of Inouye *et al.* (1976) describes the use of *allo*-4-hydroxyproline in synthesis of collagen analogue models.

Although work on hydroxyprolines in peptide synthesis has been quite limited, it appears that similar considerations for need and selection of a hydroxyl blocking group apply as do in the cases of serine and threonine. Hydroxyproline has been used in peptide synthesis without protection of the hydroxyl group both as a carboxyl and amine component of coupling reactions. Perhaps the greatest surprise was the discovery that protection of the hydroxyl group was not necessary for satisfactory use of hydroxyproline in dicyclohexylcarbodiimide-mediated solid-phase synthesis, as illustrated by synthesis of several bradykinin analogs (Felix *et al.*, 1973; Stewart *et al.*, 1974). Hydroxyproline was used without protection of the hydroxyl group in a mixed anhydride coupling by Sivanandaiah *et al.* (1966) in the synthesis of an analogue of angiotensin II. Hydroxyl-unprotected hydroxyproline has been used successfully as the amine or carboxyl component in coupling reactions mediated by dicyclohexylcarbodiimide and active esters (Adams, 1976). In the collagen model work, tripeptides of hydroxyproline were synthesized in solution and then condensed by the Merrified method to yield repeating tripeptide sequences of defined composition. In work on 4-hydroxyproline (Sakakibara *et al.*, 1973) and *allo*-4-hydroxyproline (Inouye *et al.*, 1976) the *O*-benzyl ether was used.

In *allo*-4-hydroxyproline the hydroxyl and carboxyl groups are cis with respect to each other and can readily lactonize if the hydroxyl is not blocked before the carboxyl is activated. Although such a lactone can be used to add *allo*-4-hydroxyproline to the amino group of peptide or amino acid derivatives, it is preferable to avoid lactonization by blocking the hydroxyl group. If the lactone is opened by inversion of the β-center, the residue is converted back to the normal hydroxyproline.

VI. OTHER ALIPHATIC HYDROXY ACIDS; DEPSIPEPTIDES

Hydroxylysine has received almost no attention in peptide synthesis. The presence of two additional functional groups makes selective derivatization especially onerous, and synthesis without hydroxy protection leads to special problems. Since the hydroxyl group is in the δ-position of the chain, it is particularly prone to lactonization if the carboxyl group is activated. Such a lactone can, however, be used to add hydroxylysine to peptides (Schnabel, 1963) [Eq. (3)] by aminolysis of the lactone by the amino group of a peptide or amino acid derivative.

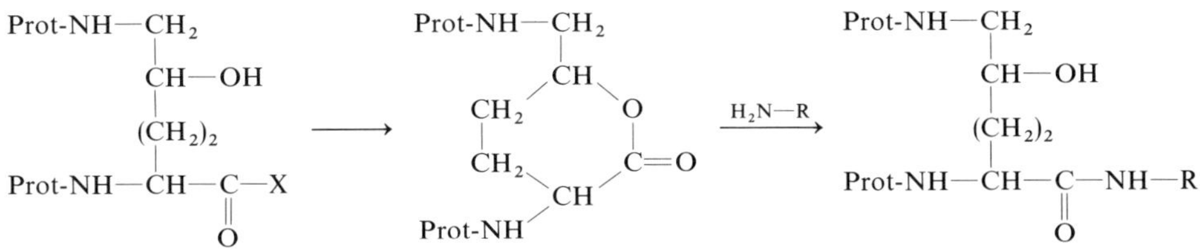

(3)

Homoserine (Hse) occurs in several natural products, but in peptides it is mainly the result of chemical scission of methionine-containing proteins by cyanogen bromide. A homoserine-containing analogue of one of the peptide toxins of the mushroom *Amanita phalloides* has biological activity, and has been synthesized by Wieland's group (Wieland, 1972; Buku *et al.*, 1976). For this synthetic work Z-Hse(Bzl)-OH was used. Synthetic procedures for benzyloxycarbonylhomoserine (Flavin, 1971) and its *p*-nitrobenzyl ester (Z-Hse-ONb) (Schnyder and Rottenberg, 1975) have been described. Z-Hse-OH and Boc-Hse-OH can be synthesized with retention of optical purity from the corresponding methionine derivatives by alkylation with methyl iodide and hydrolysis of the sulfonium compound (Okumura *et al.*, 1973). In contrast, heating methionine (L or DL) with benzyl alcohol and toluenesulfonic acid (TOS) in refluxing benzene gave only the DL-form of homoserine lactone (Morley and Smith, 1968). The latter investigators acylated the DL-homoserine lactone with benzyloxycarbonyl-L-tryptophan trichlorophenyl ester (Z-Trp-OTcp) and coupled the resulting Z-Trp-Hse-lactone with H-Asp-Phe-NH_2 to give the blocked gastrin tetrapeptide analogue in poor yield. This coupling is analogous to that shown [Eq. (3)] for hydroxylysine lactone. They also converted Z-Trp-Hse-lactone to the hydrazide and azide without blocking the hydroxyl group; the same tetrapeptide was synthesized in comparable yield by this route. Turan *et al.* (1977) synthesized Boc-Hse(Bzl)-OH by a modification of the Sugano and Miyoshi (1976) method, and used this derivative for synthesis of an oxytocin analogue.

Depsipeptides, in which hydroxy acids replace amino acids, have been of interest as analogues of peptide hormones and collagen, and as ionophores. In most of the work on depsipeptides, if a single hydroxy acid is to be incorporated, the usual technique is to acylate an ester of the hydroxy acid with an activated *N*-blocked amino acid. This dipeptide can then be further incorporated in solution (Stewart, 1969) or by solid-phase synthesis (Gisin *et al.*, 1969; see Erickson and Merrifield, 1976). In these procedures protection of the hydroxyl group is not a problem. This approach was used by Gisin *et al.* (1969) for the solid-phase synthesis of valinomycin (see Scheme 2) because of the difficulty of formation of ester bonds on the solid phase. When alternate

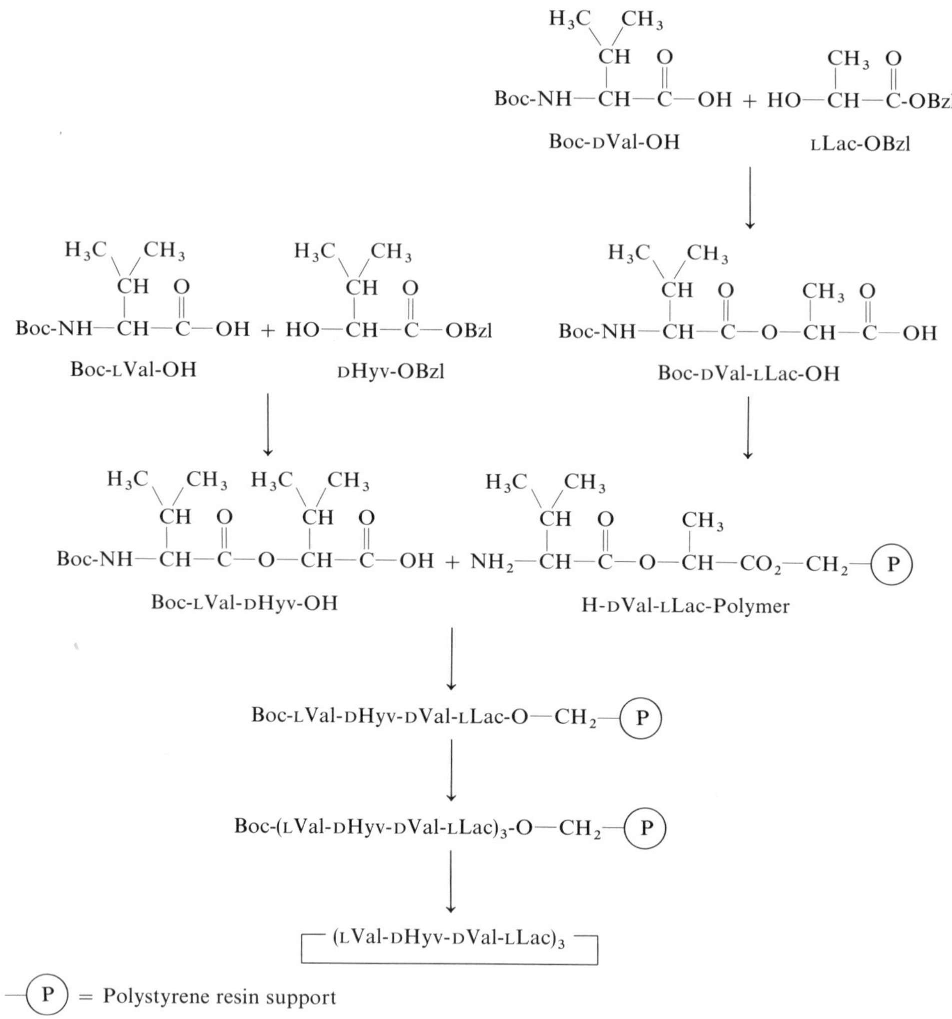

Scheme 2

approaches to synthesis are used, or when several hydroxy acids are assembled in series, hydroxyl protection is usually necessary. Andreeva *et al.* (1969) used benzyl ethers for this purpose, whereas Shchukina *et al.* (1967) used *tert*-butyl ethers for protection in the solid-phase synthesis of a tetradepsipeptide. In this case, mixed anhydrides with benzenesulfonic acid were used for coupling each residue. One interesting solid-phase depsipeptide synthesis involved coupling of the mixed anhydride of β-(4-benzyloxyphenyl)-

lactic acid with benzenesulfonic acid to a resin-bound peptide amino group (Semkin *et al.*, 1968). Although the phenolic hydroxyl was protected, the α-hydroxyl was free; success depended on preferential attack on the anhydride by the more nucleophilic amine.

Replacement of free α-amino groups in peptide hormones by hydroxyl has received some attention, since this is a device known to produce antagonists. Several hydroxyl analogues of oxytocin and vasopressin have been synthesized, and found to have interesting properties. Wälti and Hope (1975) and Hope and Wälti (1971) used 2-acetoxy-3-benzylthiopropionic acid in dicyclohexylcarbodiimide-mediated coupling reactions, while Lowbridge *et al.* (1977) have used 2-hydroxy-3-benzylthiopropionic acid in a dicyclohexylcarbodiimide-mediated coupling in the presence of hydroxybenzotriazole. In these cases the hydroxy acid is the last residue to be introduced into the peptide.

VII. TYROSINE IN PEPTIDE SYNTHESIS

Tyrosine has been much studied in peptide synthesis. It is a crucial amino acid in the biological function of many peptide hormones, and its aromatic side chain has provided a valuable tool for physicochemical studies of peptide conformation. Most of the problems encountered in peptide synthesis with aliphatic hydroxy amino acids are also found in the use of tyrosine, with several additional problems caused by the aromatic ring.

A. Synthesis of Tyrosine Peptides without Protection of the Hydroxyl Group

1. Unprotected Tyrosine in the Amine Component of Coupling Reactions

Tyrosine esters and NH_2-terminal tyrosine peptides in which the hydroxyl group is not blocked have been used successfully in synthesis, although O-acylation is a more serious problem with tyrosine than with the aliphatic hydroxyamino acids. Active ester, mixed anhydride, and dicyclohexylcarbodiimide-mediated coupling reactions have been used. If the phenolic hydroxyl ionizes, the resulting phenoxide ion is an extremely potent nucleophile, even better than an amine group. Moreover, its position on the aromatic ring makes it freely accessible to reagents without the steric hindrance present in threonine. Ionization of the tyrosine hydroxyl during coupling reactions is promoted by the presence of excess tertiary base. In a reaction of Z-Val-ONp with H-Tyr-OMe in which excess triethylamine was present (Ramachandran and Li, 1963), both Z-Val-Tyr(Z-Val)-OMe and H-Tyr(Z-Val)-OMe were isolated in addition to the desired product. However, with careful attention to detail, Bodanszky *et al.* (1976) successfully synthesized vasoactive intestinal polypeptide, a 28-peptide containing two tyrosines, using *p*- and *o*-nitrophenyl active esters. The chemist must be particularly careful to avoid excess acylating components in coupling reactions if the tyrosine hydroxyl is not blocked.

2. *Unprotected Tyrosine in the Carboxyl Component of Coupling Reactions*

Tyrosine with a free hydroxyl group has also been used in the activated component in peptide synthesis, using most of the standard coupling reactions, although not without problems. Active ester, mixed anhydride, and dicylohexylcarbodiimide-mediated coupling reactions have been used. Activation of COOH-terminal tyrosine peptides (or *N*-blocked tyrosine) by the azide method leads to special problems. The phenolic ring is subject to facile nitrosation by the reagent used to convert the hydrazide to the azide. In addition, tyrosine azides have an unusually high tendency to undergo the Curtius rearrangement to the isocyanate when the phenolic hydroxyl is not blocked. For these reasons, the azide method is not recommended with COOH-terminal unprotected tyrosine. When COOH-terminal tyrosine peptides are activated for coupling reactions, the tyrosine has a high probability of being partially racemized, with consequent loss of product and increased difficulty of purification.

In every case where acidolytic removal of blocking groups is done on peptides containing tyrosine with a free hydroxyl (this includes final removal of O-blocking groups from tyrosine, such as in HF cleavage of solid-phase peptide-resins and in final deprotection of all "globally protected" peptides), high concentrations of scavengers must be present to trap alkylcarbonium ions and prevent extensive alkylation of the tyrosine ring. Benzylation and *tert*-butylation of tyrosine are well-known reactions. Anisole and methyl ethyl sulfide are widely used for this purpose in HF and HBr deblocking reactions; thioanisole, dimethyl sulfide, ethanedithiol and indole are advocated by some investigators, as well as labile amino acids such as methionine and tryptophan. Anhydrous HF, used for cleavage of benzyl-type blocking groups (and solid-phase resins), usually contains 10% anisole or other scavengers. In this laboratory all trifluoroacetic acid reagents for removal of *tert*-butyloxycarbonyl groups contain 1 mg/ml of indole. The reagent is prepared and allowed to stand at least overnight before use; the deep burgundy color of the reagent as it is used suggests that the indole has scavenged reactive impurities (such as aldehydes) in the trifluoroacetic acid. This reagent has been used for solid-phase synthesis of tryptophan-containing peptides in which the chain was extended many residues past the tryptophan without apparent significant damage to the tryptophan (Stewart *et al.*, 1976).

In cases where O-acylation of tyrosine occurs, further complications may be anticipated by transfer of the acyl group to exposed amine groups at later stages of the synthesis. Acyl phenols are indeed active esters and acylate amine groups effectively, particularly when intramolecular transfer of the acyl group is involved, as in NH_2-terminal tyrosine.

B. Synthesis of Peptides with O-Blocked Tyrosine

1. Use of Acyl Blocking Groups

a. O-Acetyltyrosine. The acetyl group cannot be recommended for protection of tyrosine hydroxyls, due to the great activity of the ester as an acetylating agent.

b. O-Benzyloxycarbonyl tyrosine. The benzyloxycarbonyl group on the phenolic hydroxyl is considerably more labile than that on amino groups. It can be removed not only by acidolysis but also by nucleophilic displacement in the presence of tertiary amines. For this latter reason loss of benzyloxycarbonyl groups may occur when not anticipated. *O,N*-bis-benzyloxy-carbonyltyrosine can be readily prepared, and is a reasonable derivative for incorporation of an NH_2-terminal tyrosine residue. The benzyloxycarbonyl group can be introduced selectively into the side chain by acylation of the copper complex of tyrosine. After decomposition of the copper complex, the amine can be blocked with the desired group, such as *tert*-butyloxycarbonyl.

Although *O*-benzyloxycarbonyltyrosine has not been extensively used in peptide synthesis, the corresponding 2-bromobenzyloxycarbonyl (2Bz) derivative has recently been introduced as a recommended reagent for use of tyrosine in solid-phase synthesis. The *O*-2-bromobenzyloxycarbonyl group appears to have the requisite stability toward acidolysis for satisfactory synthesis of large peptides in solid-phase synthesis, and largely eliminates the serious problem of ring alkylation encountered with *O*-benzyltyrosine. Yamashiro and Li (1973a) synthesized Boc-Tyr(2Bz)OH in the standard way, and it has been used extensively since then. The superiority of this new blocking group for the tyrosine hydroxyl function was demonstrated (C. Pena and J. M. Stewart, unpublished) in the synthesis of an angiotensin analogue in which the residue on either side of tyrosine was alanine. When the peptide was synthesized by standard solid-phase methods using Boc-Tyr(Bzl)-OH and cleaved with HF, the only product isolated was the peptide containing 3-benzyltyrosine. When the synthesis was repeated using 2,6-dichlorobenzyltyrosine, benzylation still occurred during HF treatment to the extent of about 50%. However, when 2-bromobenzyloxycarbonyltyrosine was used, no significant amount of tyrosine alkylation occurred. Of crucial importance in the synthesis of peptides where any tyrosine derivatization may have occurred is the use of purification techniques adequate to separate any derivatized tyrosine peptides. Countercurrent distribution (CCD) has been found (J. M. Stewart, unpublished) to be particularly effective for separation of benzyltyrosine-containing peptides. The added aromatic group causes these by-products to have a much higher partition coefficient than the desired peptide in the usual CCD systems (e.g., BuOH:AcOH:H_2O = 4:1:5).

The use of adequate quantities of effective scavenger nucleophiles in reactions for removal of tyrosine blocking groups is essential. Anisole has been most widely used, and thioanisole has been proposed as a more effective reagent (Bauer and Pless, 1975). The failure of such scavengers to block ring alkylation indicates that much of the alkylation occurs by the well-known intramolecular Friedel–Crafts mechanism; HF is an excellent catalyst for this reaction. In the *O*-benzyloxycarbonyl derivative, the benzyl group is apparently far enough removed from the ring that the intramolecular reaction is minimized and the benzylcarbonium ion is produced at a sufficient distance from the phenolic ring that it has a good chance of being captured by the scavenger before it can attack the tyrosine ring. Although *O*-2-bromobenzyloxycarbonyltyrosine is the best derivative currently available for solid-phase synthesis, some improvement may still be in order for the synthesis of very long peptides; because 24-hr treatment with trifluoroacetic acid in dichloromethane (1:1) causes 1% loss of the blocking group, while similar treatment with 10% diisopropylethylamine in dimethylformamide caused 5% loss of the blocking group. This latter loss should be minimized by the use of dichloromethane in neutralization reactions instead of dimethylformamide.

c. Other Acyl Groups. *O-tert*-Butyloxycarbonyltyrosine has been prepared by standard methods, but cannot be recommended for use in extended synthesis. Bis-*tert*-butyloxycarbonyltyrosine is a suitable derivative for introduction of tyrosine as the last amino acid in a synthesis. *O-p*-Toluenesulfonyltyrosine has been used in peptide synthesis, particularly of oxytocin and derivatives. Its use requires sodium in liquid ammonia as a final deprotection reagent and therefore cannot be recommended for general use. The *N*-benzyloxycarbonyl group can be removed selectively from Z-Tyr(Tos)-OH by acidolysis or hydrogenolysis without affecting the *O-p*-toluenesulfonyl group.

2. Ethers for Protection of the Tyrosine Hydroxyl Group

a. O-Benzyltyrosine. The benzyl ether is probably the most widely used blocking group for the hydroxyl of tyrosine. It is readily synthesized from the copper complex of tyrosine in aqueous alkali. The phenolic benzyl ether present in tyrosine is much more labile than benzyl ethers of serine and threonine, and is readily cleaved by all reagents which normally cleave benzyl groups. Although there was some confusion on this point during the initial work with HF in peptide synthesis, it is now clear that the *O*-benzyltyrosine group is not adequately stable for successful solid-phase synthesis of long chains. It is cleaved significantly by trifluoroacetic acid reagents used for removal of N^{α}-*tert*-butyloxycarbonyl groups. The situation is much improved by introduction of negative substituents into the benzyl ring, and *O*-2,6-dichlorobenzyltyrosine appears to have the needed stability for synthesis of long peptides by the solid-phase method (Yamashiro and Li, 1973b; Erickson and Merrifield, 1973). As was noted above, however, this derivative can still give rise to ring alkylation of tyrosine during HF deprotection when the amino acid sequence of the peptide favors this migration.

Migration of an *O*-benzyl group of tyrosine to the 3-position of the ring is a problem not only with HF, but has also been observed with HBr cleavage of benzyl ethers (Trudelle and Spach, 1972; Bell *et al.*, 1975). It was found by Bell *et al.* (1975) that less ring benzylation occurred with HBr–AcOH than with HBr–TFA, although O-acetylation occurred to a significant degree. They preferred this procedure, with subsequent saponification for removal of the artifactual *O*-acetyl group. With the present availability of the 2-bromobenzyloxycarbonyl derivative, which largely avoids the problem, this procedure does not appear to be practical, particularly in light of the danger of racemization of tyrosine in alkali.

Another side reaction has recently been described in the use of HBr for deprotection of peptides containing both a tyrosine and a cysteine residue. In this case (Ito and Prota, 1977), the sulfhydryl group of the cysteine alkylated the tyrosine ring in the 3-position to give the derivative **8**.

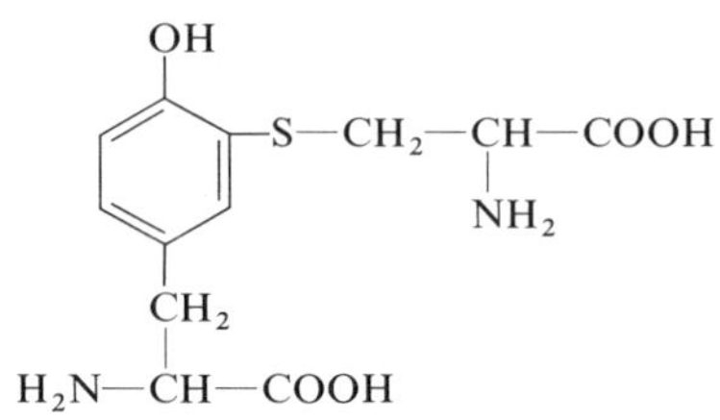

(8)

There will apparently be continued debate as to the best blocking group for tyrosine hydroxyls since the ideal group is apparently not yet available.

Both the 2,6-dichlorobenzyl (Yamashiro and Li, 1973b) and the 2-bromobenzyloxycarbonyl groups (Pena *et al.*, 1975; Stewart *et al.*, 1976) have been used for synthesis of long peptides by the solid-phase method.

The *o*-nitrobenzyl (2Nb) ether has recently been proposed (Amit *et al.*, 1977) as a tyrosine blocking group removable by irradiation. The derivative was synthesized by reaction of Z-Tyr(O^-Na^+)-OMe with *o*-nitrobenzyl chloride in dimethylformamide, followed by saponification of the ester. The blocking group was removed from Z-Tyr(2Nb)-Gly-OEt by irradiation at 350 nm in methanol solution for 12 hr at −10°C. The irradiation was done in the presence of semicarbazide or Girard's reagent T to trap the *o*-nitrosobenzaldehyde produced as the by-product of the photolysis reaction [Eq. (4)]. Some problems were reported with colored by-products which made purification difficult. At the present time, the very slow speed of this reaction would appear to limit its usefulness.

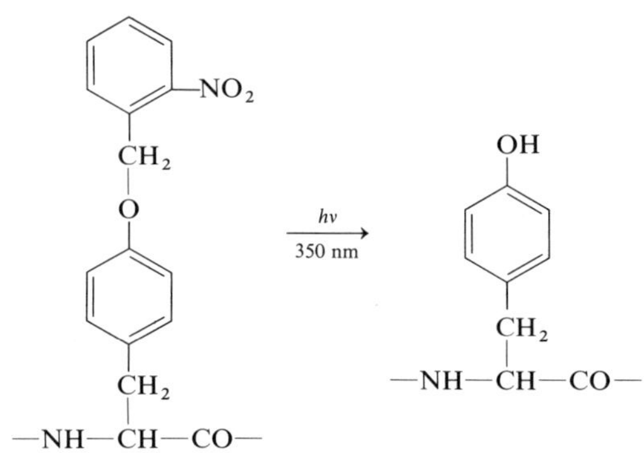

(4)

b. O-tert-Butyltyrosine. The *tert*-butyl ether has been widely used for protection of tyrosine in peptides in conjunction with groups such as N^α-benzyloxycarbonyl (removal by hydrogenolysis) or 2-(*p*-biphenylyl)propyl-(2)oxycarbonyl and *o*-nitrophenylsulfenyl (removable by very mild acidolysis). Boc-Tyr(*t*Bu)-OH has been used for introduction of a final tyrosine residue into peptides. Although selective removal of N^α-*tert*-butyloxycarbonyl groups from serine and threonine *tert*-butyl ethers has been reported, this is not likely to be successful in the case of tyrosine. The *tert*-butyl ether is readily removed from tyrosine by treatment with trifluoroacetic acid at room temperature.

c. Other Ethers. The 2,4-dinitrophenyl (Dnp) ether (**9**) has been proposed for protection of tyrosine in peptide synthesis. The derivatives are readily synthesized by treatment of Z-Tyr-OH or Boc-Tyr-OH with fluorodinitrobenzene. The *O*-dinitrophenyl group is stable to acid (HF or TFA) for

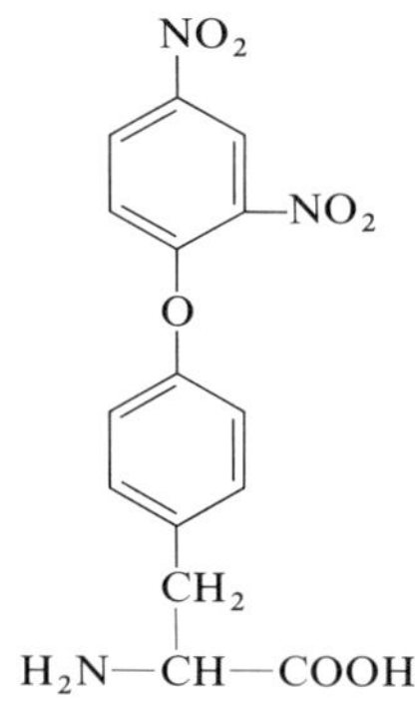

(9)

extended periods, and to triethylamine. It can be removed readily by treatment with 2-mercaptoethanol in dimethylformamide or water at pH 9; the cleavage can be conveniently followed spectrophotometrically (Fridkin *et al.*, 1977). Z-Tyr(Dnp)-OH was converted to Z-Tyr(Dnp)-OSu, which was used for peptide synthesis without complication. The use of this blocking group for tyrosine would be convenient in conjunction with N^{im}-dinitrophenyl for protection of histidine imidazole groups in peptide synthesis.

Young has recently proposed the 4-picolyl ether (**10**) as a blocking group for the phenolic hydroxyl of tyrosine. The ethers were prepared by methods similar to those used for synthesis of benzyl ethers and were used for synthesis of some small peptides (Gosden *et al.*, 1977). This ether is removable in moderate yield by electrochemical reduction.

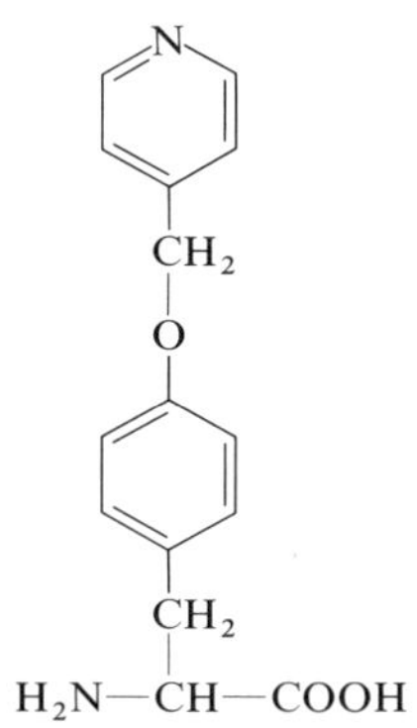

(10)

d. Tetrahydropyranyl Ethers. Acetals formed by addition of hydroxyl compounds to dihydropyran have been used for blocking tyrosine hydroxyl groups as well as the aliphatic hydroxyamino acids. Tetrahydropyranyl ethers are synthesized by anhydrous acid-catalyzed reaction of tyrosine esters or N-blocked tyrosine esters in organic solvents. The ethers are usually cleaved by dilute aqueous acid, which raises questions about their safety as general blocking groups. For this reason, they have been little used in recent years.

The most interesting recent application of this blocking group is by Hirschmann *et al.* (1971) who used the tetrahydropyranyl ether for protection of the hydroxyl group of tyrosine *N*-carboxy anhydride (**11**). Tyrosine-NCA

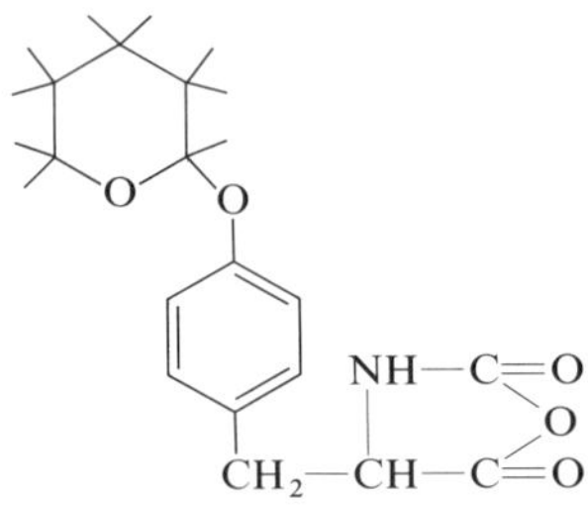

(**11**)

was treated with dihydropyran and *p*-toluenesulfonyl chloride to form the desired derivative. The interesting feature is that the ether was lost during the aqueous coupling reaction, and thus did not require an additional step for deblocking. However, this had the disadvantage of leaving the tyrosine hydroxyl unprotected in subsequent reactions. As was mentioned earlier, formation of the tetrahydropyranyl ethers introduces a second asymmetric center into the molecule with the consequence that diastereomeric derivatives are produced. Hirschmann *et al.* (1971) found that one stereoisomer of Tyr(Thp)-NCA was very soluble, and this was the derivative they used in synthesis. The other isomer was very insoluble.

VIII. PEPTIDE SYNTHESIS WITH OTHER PHENOLIC AMINO ACIDS

The biological importance of certain substituted tyrosines has begun to stimulate interest in incorporation of these residues into peptides. Initial work has been reported on peptide synthesis with 3′,5′-diiodotyrosine (Dit) and 3,4-dihydroxyphenylalanine (DOPA).

Pawelczak *et al.* (1976) used 3′,5′-diiodotyrosine with a free hydroxyl group and also as a benzyl ether in synthesis. In a synthesis conducted without hydroxyl blocking, H-Dit-OMe was acylated by Z-Gly-OSu. The blocked dipeptide was converted to the azide and this was used to acylate H-Gly-OMe. Removal of blocking groups gave Gly-DIT-Gly (**12**).

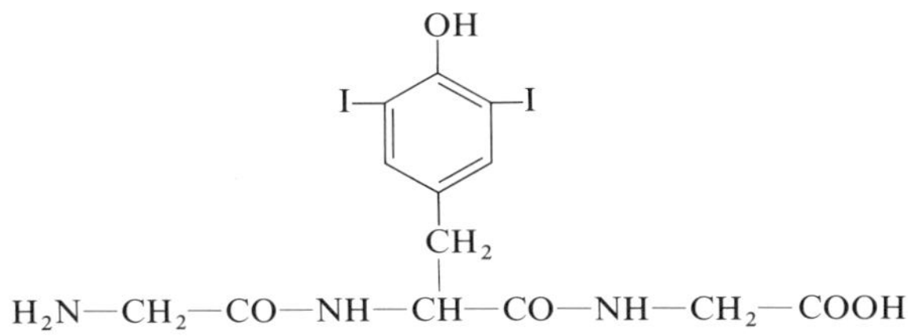

(**12**)

Felix *et al.* (1974) made a series of di- and tripeptides containing 3,4-dihydroxyphenylalanine to investigate these materials as improved drugs in therapy of Parkinson's disease. They synthesized Z_3-DOPA-OH and used this derivative to acylate amino acid esters in dicyclohexylcarbodiimide-mediated reactions. The Z_3-DOPA-OH was converted via the *N*-carboxy anhydride and treatment with benzyl alcohol to H-DOPA(Z_2)-OBzl (**13**), which was acylated by other N-protected amino acids in dicyclohexylcarbodiimide-mediated coupling reactions.

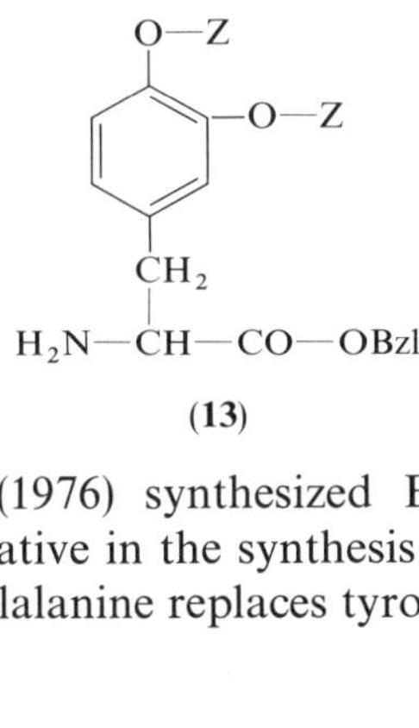

(**13**)

Banerjee and Ressler (1976) synthesized Boc-DOPA(Bzl_2)-ONp and proposed to use this derivative in the synthesis of an oxytocin analogue in which 3,4-dihydroxyphenylalanine replaces tyrosine.

XI. OPTICAL PURITY OF PEPTIDES AND AMINO ACIDS

The goal of peptide synthesis is to arrive at the end of the work with the desired peptide in a state of chemical and optical homogeneity. Side reactions leading to chemical impurities and to partially racemized products have been mentioned at various places in the preceding sections, and are particularly onerous with the hydroxyamino acids. Elimination of the resulting impurities from the peptide product requires careful attention to purification and analytical procedures as well as to proper choice of amino acid derivatives and reaction conditions. Another hazard is the difficulty of obtaining optically pure amino acid derivatives commercially; serine and threonine have been found to be among the worst offenders in this regard (see Erickson and

Merrifield, 1976, p. 411). The recent great interest in incorporation of D-amino acids into peptide hormones has exacerbated the problem, since the D isomers are often even more difficult to obtain in a state of optical purity. One well-known group of peptide chemists wasted a large amount of time on the synthesis of a group of analogues of a peptide hormone using a batch of commercial "D"-amino acid derivative which they later found to be the L isomer. This experience emphasizes the need for careful checking of all batches of commercial amino acids and derivatives for chemical and optical purity before they are used in synthesis. Unfortunately, simple measurement of optical rotation of the amino acid or derivative often cannot reveal amounts of racemized material which would be disastrous if incorporated into a peptide such as luteinizing hormone releasing hormone, where inversion of one amino acid may convert an agonist to an antagonist. Such antagonists frequently have a low amount of agonist activity, and the difficult question is whether that low activity is due to residual inherent agonist activity of the peptide or to contamination by a small amount of the peptide containing L-amino acid introduced by racemization during synthesis or use of an optically impure amino acid initially.

Determination of the optical purity of amino acids and derivatives to high precision is not easy. Measurements of optical rotation are not very precise, even when conditions (solvent, concentration, wavelength of observation) are chosen to give maximum specific rotation. Many amino acid derivatives have low specific rotations. If rotation measurements must be used, one should measure identical solutions of the two antipodes under the same conditions and show that they have equal and opposite rotations. Even in this case, the investigator may have only demonstrated the presence of similar amounts of racemization in the two optical isomers. More precise methods, such as the Manning and Moore (1968) procedure in which the amino acid is converted to a diastereomeric derivative, are required for detection of enantiomeric contamination with high sensitivity, see Kemp, Vol. 1, Chapter 7, for details.

Little is known about the ways in which the optical rotation of peptides changes with inversion of one of the optical centers. Since several asymmetric centers are involved, this relationship is not a simple one. Moreover, if the peptide is large enough to have secondary and tertiary structure, the optical rotation is influenced by conformational features (such as hydrogen bonding, α-helix content and β-structure, for example), especially if one attempts to increase the magnitude of the rotation by going to the shorter wavelengths where the conformational influences are greatest. One does not usually have available a standard peptide of firmly established optical purity for reference.

Fortunately for the peptide chemist, many diastereomeric peptides have been separated without great difficulty. In certain cases, a peptide hormone analogue has been synthesized with one DL-amino acid and the resulting enantiomeric peptides separated during purification. Partition techniques have been most useful for this purpose, either countercurrent distribution (see, e.g., Schneider and du Vigneand, 1962; Yamashiro *et al.*, 1966) or partition chromatography on Sephadex or BioGel (see, e.g., Yamashiro *et al.*, 1977). Most recently high-performance liquid chromatography has been applied to this problem with encouraging results (Rivier and Burgus, 1976; Rivier *et al.*, 1977). Perhaps the greatest difficulty is that there is no assurance that a suitable system can be found to separate any given diastereomeric pair. Investigators have reported failure to separate peptide pairs from synthesis with a DL-amino acid. Unless resolution of the pair of peptides at hand has been demonstrated in the system being used, obtaining a single symmetrical peak cannot be taken as assurance of homogeneity. This whole problem requires constant vigilance and careful attention on the part of the peptide chemist.

REFERENCES

Adams, E. (1976). *Int. J. Pept. Protein Res.* **8**, 503–516.

Adams, E. (1977). *Int. J. Pept. Protein Res.* **9**, 293–309.

Amit, B., Hazum, M., Fridkin, M., and Patchornik, A. (1977). *Int. J. Pept. Protein Res.* **9**, 91–96.

Andreeva, L. I., Lyakisheva, A. G., Antonov, V. K., and Shemyakin, M. M. (1969). *Zh. Obshch. Khim.* **39**, 2774–2778.

Banerjee, S. N., and Ressler, C. (1976). *J. Org. Chem.* **41**, 3056–3058.

Bauer, W., and Pless, J. (1975). *In* "Peptides: Chemistry, Structure and Biology" (R. Walter and J. Meienhofer, eds.), pp. 341–345. Ann Arbor Sci. Publ., Ann Arbor, Mich.

Bell, J. R., Jones, J. H., and Webb, T. C. (1975). *Int. J. Pept. Protein Res.* **7**, 235–242.

Bodanszky, M., and Ondetti, M. A. (1966). *Chem. Ind.* (*London*) pp. 26–27.

Bodanszky, M., Ondetti, M. A., Levine, S. D., and Williams, N. J. (1967). *J. Am. Chem. Soc.* **89**, 6753–6757.

Bodanszky, M., Lin, C. Y., Yiotakis, A. E., Mutt, V., and Said, S. I. (1976). *Bioorg. Chem.* **5**, 339–350.

Bodanszky, M., Fink, M. L., Klausner, Y. S., Natarajan, S., Tatemoto, K., Yiotakis, A. E., and Bodanszky, A. (1977). *J. Org. Chem.* **42**, 149–152.

Buku, A., Altmann, R., and Wieland, T. (1976). *Justus Liebigs Ann. Chem.* pp. 417–431.

Erickson, B. W., and Merrifield, R. B. (1973). *J. Am. Chem. Soc.* **95**, 3750–3756.

Erickson, B. W., and Merrifield, R. B. (1976). *In* "The Proteins" (H. Neurath and R. L. Hill, eds.), 3rd ed., Vol. 2, pp. 255–527. Academic Press, New York.

Felix, A. M. (1974). *J. Org. Chem.* **39**, 1427–1429.

Felix, A. M., Jiminez, M. H., Vergona, R., and Cohen, M. R. (1973). *Int. J. Pept. Protein Res.* **5**, 201–206.

Felix, A. M., Winter, D. P., Wang, S. S., Kulesha, I. D., Pool, W. R., Hane, D. L., and Sheppard, H. (1974). *J. Med. Chem.* **17**, 422–426.

Finn, F. M., and Hofmann, K. (1976). *In* "The Proteins" (H. Neurath and R. L. Hill, eds.), 3rd ed., Vol. 2, pp. 105–253. Academic Press, New York.

Flavin, M. (1971). *In* "Methods in Enzymology" (H. Tabor and C. W. Tabor, eds.), Vol. 17, Part B, pp. 418–422. Academic Press, New York.

Fridkin, M., Hazum, E., Tauber-Finkelstein, M., and Shaltiel, S. (1977). *Arch. Biochem. Biophys.* **178**, 517–526.

Gisin, B. F., Merrifield, R. B., and Tosteson, D. C. (1969). *J. Am. Chem. Soc.* **91**, 2691–2695.

Goodacre, J., Ponsford, F. J., and Stirling, I. (1975). *Tetrahedron Lett.* pp. 3609–3612.

Gosden, A., MacRae, R., and Young, G. T. (1977). *J. Chem. Res.* (*S*) **1**, 22–23.

Greenstein, J. P., and Winitz, M. (1961). "Chemistry of the Amino Acids," 3 vols. Wiley, New York.

Hirschmann, R. (1971). *In* "Peptides 1969" (E. Scoffone, ed.), pp. 138–144. North-Holland Publ., Amsterdam.

Hirschmann, R., Schwam, H., Strachan, R. G., Schoenewaldt, E. F., Barkemeyer, H., Miller, S. M., Conn, J. B., Garsky, V., Veber, D. F., and Denkewalter, R. G. (1971). *J. Am. Chem. Soc.* **93**, 2746–2754.

Hope, D. B., and Wälti, M. (1971). *Biochem. J.* **125**, 909–911.

Hruby, V. J., and Ehler, K. W. (1970). *J. Org. Chem.* **35**, 1690.

Inouye, K., Sakakibara, S., and Prockop, D. J. (1976). *Biochem. Biophys. Acta* **420**, 133–141.

Iselin, B., and Schwyzer, R. (1956). *Helv. Chim. Acta* **39**, 57–64.

Ito, S., and Prota, G. (1977). *J. Chem. Soc., Chem. Commun.* pp. 251–252.

Lenard, J., and Hess, G. P. (1964). *J. Biol. Chem.* **239**, 3275–3281.

Levy, D., and Carpenter, F. H. (1970). *Biochemistry* **9**, 3215–3222.

Lowbridge, J., Manning, M., Haldar, J., and Sawyer, W. (1977). *J. Med. Chem.* **20**, 1173–1176.

McOmie, J. F. W., ed. (1973). "Protecting Groups in Organic Chemistry." Plenum, New York.

Manning, J. M., and Moore, S. (1968). *J. Biol. Chem.* **243**, 5591–5597.

Matsuura, S., Niu, C.-H., and Cohen, J. S. (1976). *J. Chem. Soc., Chem. Commun.* pp. 451–452.

Mizoguchi, T., Levin, G., Woolley, D. W., and Stewart, J. M. (1968). *J. Org. Chem.* **33**, 903.

Morley, J. S., and Smith, J. M. (1968). *J. Chem. Soc. C* pp. 726–733.

Okumura, K., Sugano, H., and Miyoshi, M. (1973). Japanese Patent 73 18,229; *Chem. Abstr.* **78**, 160111 (1973).

Pawelczak, K., Rzeszotarska, B., and Nadolska, B. (1976). *Justus Liebigs Ann. Chem.* pp. 1521–1530.

Peña, C., Stewart, J. M., Paladini, A. C., Dellacha, J. M., and Santomé, J. A. (1975). *In* "Peptides: Chemistry, Structure and Biology" (R. Walter and J. Meienhofer, eds.), pp. 523–528. Ann Arbor Sci. Publ., Ann Arbor, Michigan.

Pless, J., and Bauer, W. (1973). *Angew. Chem., Int. Ed. Engl.* **12**, 147–148.

Ramachandran, J., and Li, C. H. (1963). *J. Org. Chem.* **28**, 173–177.

Rebek, J., and Feitler, D. (1974). *J. Am. Chem. Soc.* **96**, 1606–1607.

Rivier, J., and Burgus, R. (1976). *In* "Peptides 1976" (A. Loffet, ed.), pp. 235–240. Editions de l'Université de Bruxelles, Belgium.

Rivier, J., Wolbers, R., and Burgus, R. (1977). *In* "Peptides, Proceedings of the Fifth American Peptide Symposium" (M. Goodman and J. Meienhofer, eds.), pp. 52–55. Wiley, New York.

Robinson, A. B. (1974). *Proc. Natl. Acad. Sci. U.S.A.* **71**, 885–888.

Sakakibara, S. (1977). *In* "Peptides, Proceedings of the Fifth American Peptide Symposium" (M. Goodman and J. Meienhofer, eds.), pp. 436–447. Wiley, New York.

Sakakibara, S., Inouye, K., Shudo, K., Kishida, Y., Kobayashi, Y., and Prockop, D. J. (1973). *Biochim. Biophys. Acta* **303**, 198–202.

Schnabel, E. (1963). *Justus Liebigs Ann. Chem.* **667**, 179–182.

Schneider, C. H., and du Vigneand, V. (1962). *J. Am. Chem. Soc.* **84**, 3005–3008.

Schnyder, J., and Rottenberg, M. (1975). *Helv. Chim. Acta* **58**, 518–521.

Schröder, E. (1963). *Justus Liebigs Ann. Chem.* **670**, 127–136.

Schröder, E., and Lübke, K. (1965). "The Peptides," Vol. 1. Academic Press, New York.

Semkin, E. P., Smirnova, A. P., and Shchukina, L. A. (1968). *Zh. Obshch. Khim.* **38**, 2358–2363.
Shchukina, L. A., Semkin, E. P., and Smirnova, A. P. (1967). *Khim. Prir. Soedin.* **3**, 358–367.
Sieber, P., Kamber, B., Hartmann, A., Jöhl, A., Riniker, B., and Rittel, W. (1974). *Helv. Chim. Acta* **57**, 2617–2621.
Sivanandaiah, K. M., Smeby, R. R., and Bumpus, F. M. (1966). *Biochemistry* **5**, 1224–1229.
Smithwick, E. L. (1973). German Patent 2,253,924; *Chem. Abstr.* **79**, 32301 (1973).
Stewart, F. H. C. (1969). *Aust. J. Chem.* **22**, 1291.
Stewart, J. M., and Young, J. D. (1969). "Solid Phase Peptide Synthesis." Freeman, San Francisco, California.
Stewart, J. M., Ryan, J. W., and Brady, A. H. (1974). *J. Med. Chem.* **17**, 537–539.
Stewart, J. M., Peña, C., Matsueda, G. R., and Harris, K. (1976). *In* "Peptides 1976" (A. Loffet, ed.), pp. 285–290. Editions de l'Université de Bruxelles, Belgium.
Sugano, H., and Miyoshi, M. (1976). *J. Org. Chem.* **41**, 2352–2353.
Trudelle, Y., and Spach, G. (1972). *Tetrahedron Lett.* pp. 3475–3478.
Turan, A., Manning, M., Haldar, J., and Sawyer, H. (1977). *J. Med. Chem.* **20**, 1169–1172.
Vorbrüggen, H., and Krolikiewicz, K. (1975). *Angew. Chem., Int. Ed. Engl.* **14**, 818.
Wälti, M., and Hope, D. B. (1975). *J. Chem. Soc., Trans. Perkin 1* pp. 1691–1694.
Weygand, F., Steglich, W., Fraunberger, F., Pietta, P., and Schmid, J. (1968). *Chem. Ber.* **101**, 923–934.
Wieland, T. (1972). *Naturwissenschaflen* **59**, 225–231.
Wünsch, E., ed. (1974). "Houben-Weyl Methoden der organischen Chemie," Vol. 15, Parts 1 and 2. Thieme, Stuttgart.
Wünsch, E., and Wendlberger, G. (1968). *Chem. Ber.* **101**, 3659–3663.
Wünsch, E., and Wendlberger, G. (1972). *Chem. Ber.* **105**, 2508–2514.
Yajima, H., Kitagawa, K., Segawa, T., Nakano, M., and Kataoka, K. (1975). *Chem. Pharm. Bull.* **23**, 3299–3301.
Yamashiro, D. (1977). *J. Org. Chem.* **42**, 523–525.
Yamashiro, D., and Li, C. H. (1973a). *J. Org. Chem.* **38**, 591–592.
Yamashiro, D., and Li, C. H. (1973b). *J. Am. Chem. Soc.* **95**, 1310–1315.
Yamashiro, D., Gillessen, D., and du Vigneaud, V. (1966). *J. Am. Chem. Soc.* **88**, 1310–1313.
Yamashiro, D., Tseng, L. F., Doneen, B. A., Loh, H. H., and Li, C. H. (1977). *Int. J. Pept. Protein Res.* **10**, 159–166.
Yang, C. C., and Merrifield, R. B. (1976). *J. Org. Chem.* **41**, 1032–1041.
Young, J. D., Benjamini, E., Stewart, J. M., and Leung, C. Y. (1968). *Biochemistry* **6**, 1455–1460.

Chapter **5**

Differential Protection and Selective Deprotection in Peptide Synthesis

JEAN-LUC FAUCHÈRE and ROBERT SCHWYZER

I. INTRODUCTION

Contemporary chemical synthesis of pure peptides with unique sequences from their constituent amino acids requires two types of functional group protection: a *temporary protection* of α-amino and α-carboxyl groups, and

THE PEPTIDES, VOLUME 3

ISBN 0-12-304203-8

a *semipermanent protection* of side-chain functions (and, usually, of one or the other of the α-functions). The temporary protection must be relieved at certain points and at certain times during the synthesis in order to allow the condensation reactions to produce unique and correct peptide bonds. Relief of temporary protection must be *selective* and should, ideally, not impair semipermanent protection, which is usually removed only at the end of the synthesis.

These two levels of protecting group stability are the rule, and *differential protection* must provide the growing chain with temporary as well as with semipermanent protecting groups (Sections V and VI). The degree of semipermanent protection depends largely on the general strategy employed: either *minimal* or *maximal* side-chain protection (usually dictated by solubility exigencies).

Additional stability levels are required for the following special purposes: (i) for the case that during the synthesis a transition from maximal to minimal (or to an intermediate stage of) side-chain protection should become necessary; (ii) when the solid-phase synthesis of protected peptides (Section VI,B,3) relies on an intermediate stability level for the link to the resin with respect to the temporary and semipermanent protection (the latter should remain intact on cleavage of the protected peptide from its support); (iii) when the attachment of polypeptides is through a unique functional group to a polymer matrix for affinity chromatography (e.g., Fauchère and Pelican, 1975) or to proteins for other purposes (e.g., Eberle *et al.*, 1977); (iv) when there is the introduction of labeling groups at one specific side-chain function (e.g., Schwyzer and Schiller, 1971); and (v) in the preparation of asymmetrical disulfides (e.g., Kamber, 1973). Purposes iii, iv, and v may require the selective deprotection of one single functional side-chain group at any stage of the synthesis.

The diversity of available deprotection methods plays an important role in the selection and in the combination of protecting groups. The main procedures employed, as well as their limitations in terms of observed side reactions, are analyzed briefly in Section II.

The most popular method for the cleavage of protecting groups is acidolysis. The design of groups labile to acids has reached a very sophisticated level and many of the modern syntheses rely totally on them. Acidolysis is the preferred method in conventional synthesis in solution, and in solid-phase synthesis it is practically the only one used. The methodology of solid-phase synthesis is strongly dependent on differential acidolysis, which allows the successive cleavage of bonds with different stabilities toward acid. In Section III, we therefore try to compare semiquantitatively the rates of acidolytic fission of a number of the usual protecting groups and to define the levels of their stability. For two groups to occupy two levels of stability sufficiently different for practical purposes, their rate constants of cleavage must be separated by several orders of magnitude. In this favorable case, conditions can be found which allow selective acidolysis of the more labile group, although both groups are labile to acids. A table of the relative stabilities of several groups in different acidic media, and a classification of the usual groups according to the mildest conditions available for their cleavage are also presented in Section II.

A number of protecting groups that require only mild alkaline conditions for their removal have recently been introduced in peptide synthesis. These groups represent at least one additional stability level in deprotection by variation of the pH. Most of the alkali-labile groups described so far are considered in Section IV. The interesting new possibilities offered by these groups for differential protection are also analyzed in this section. Reference is made to successful syntheses.

The relative stability of two groups strongly depends on the peptide to be deprotected and on the detailed conditions used for cleavage. Therefore, the possibility of the selective removal of the more labile group has to be tested in each particular case and cannot be predicted with certainty in spite of the large body of experience accumulated in peptide synthesis. Nevertheless, with these limitations in mind, a table of selectivity which tentatively states the particular conditions under which a given temporary group could be removed in the presence of another, semipermanent group is presented in Section V.

Tables I (Section III) and III (Section V) contain most of the information necessary for the elaboration of protection schemes. Section VI describes the main classes of schemes used so far in solution and solid-phase synthesis. Special mention is made of the intermediate stability of the link between peptide and resin in solid-phase synthesis. A number of protecting group combinations described in the literature are compared and some new protecting schemes are suggested.

II. GENERAL METHODS FOR THE CLEAVAGE OF PROTECTING GROUPS

In order to increase the possibilities of selective cleavage, it is advantageous to have a large arsenal of deblocking procedures at one's disposal. The ideal situation, where a specific cleavage method would be available for each individual protecting group, is far from being realized. On the other hand, it may be desirable for a given protecting group to be cleavable by more than one reagent because of the resulting increased flexibility in the choice of splitting conditions. However, this results in a reduced selectivity for the removal of other groups. The main deblocking procedures are listed in this section, although detailed description of experimental conditions is omitted. Their combined use in the synthesis of complex polypeptides is analyzed in Sections V and VI. Two of them, acidolysis and alkaline hydrolysis, are examined in a more detailed manner in Sections III and IV.

The popularity of *acidolysis* is due to the particular ease with which a large number of protecting groups is removed by acids. The most commonly used acidic media include trifluoroacetic acid, liquid HF, hydrochloric and hydrobromic acids in glacial acetic acid, in formic acid, or in other organic solvents. The successive application of these reagents very often allows a differential deprotection (Section III). Anhydrous HF removes the majority of the usual protecting groups and is very often brought into play for final deprotection. However, because of its toxic and corrosive properties, it requires special equipment (Sakakibara *et al.*, 1967). It can often be replaced by the less hazardous boron tris(trifluoroacetate) (Pless and Bauer, 1973) or by trifluoromethanesulfonic acid (Fujii and Yajima, 1975) or fluorosulfonic acid (Yajima *et al.*, 1975). Pyridinium polyhydrogen fluoride can also replace liquid HF except for the deprotection of N^g-nitroarginine and of N^{im}-benzylhistidine (Matsura *et al.*, 1976). Methanesulfonic acid (Yajima *et al.*, 1975) is being used increasingly for global deprotection (Ogawa *et al.*, 1978; Kobuta *et al.*, 1979). In contrast to fluorosulfonic acid, it cleaves the benzyloxycarbonyl group without concomitant N^ε-alkylation of lysine (Fujii *et al.*, 1977). However, N to O acyl shift still takes place as it does in liquid HF and has to be reversed by subsequent treatment of the peptide with 2 *N* ammonia (Fujino *et al.*, 1978). Many other reagents, each with its advantages, have been used for acidolysis. Some of these, such as trimethylsilyl perchlorate (Vorbrüggen and Krolikiewicz, 1975) or *p*-toluenesulfonic acid (Goodacre *et al.*, 1975), show in certain cases an unusually high selectivity.

Although side reactions, including racemization, are always to be feared when using *alkaline hydrolysis* for deprotection, it still remains a very important deblocking procedure. Deprotection is not always mediated by the action of *hydroxyl ions*, as in β-elimination (Section IV), but can also involve *ammonia*, *hydrazine*, or other nucleophiles. Some examples are the removal of the N^{im}-2,4-dinitrophenyl group by *thiolysis* with 2-mercaptoethanol at pH 8 (Schaltiel and Fridkin, 1970), of *S*-acetamidomethyl by *rhodanolysis* (Veber *et al.*, 1972), and of resin-bound phenacyl esters by *thiophenoxide* in dimethylformamide (Mizoguchi *et al.*, 1970). Unusually stable protection by β-halogenalkyl esters or β-halogenalkyloxycarbonyl groups has been removed by the *supernucleophile* cobalt(I) phthalocyanine (Eckert and Ugi, 1976).

Another very popular type of deblocking procedures relies on reduction. It includes treatment by sodium in liquid ammonia, catalytic hydrogenation as well as electrolytic reduction. Reductive cleavage with *sodium in liquid ammonia*, introduced by Sifferd and du Vigneaud (1935) as early as 1935, has been extensively used for residues such as *S*-benzylcysteine, N^{im}-benzylhistidine, N^{ε}-tosyllysine, and N^{g}-tosylarginine (du Vigneaud *et al.*, 1953; Katsoyannis *et al.*, 1957; Schwyzer and Li, 1958). The discovery of side reactions leading in some cases to the cleavage of the Lys-Pro (Hofmann and Yajima, 1961) or Thr-Pro peptide bonds (Marglin and Merrifield, 1966) or to a demethylation or desulfurization of methionine and cysteine residues (Lübke and Klostermeyer, 1970), greatly decreased the importance of the method. In a "rapid synthesis" of oxytocin (Kisfaludy and Schön, 1975), undefined side reactions prevented the use of sodium in liquid ammonia for the simultaneous deprotection of *S*-benzylcysteine and ethoxycarbonyltyrosine. Nevertheless, the method can still be used provided some precautions are taken such as the addition of sodium amide (Katsoyannis *et al.*, 1971) or of urea (Mühlemann *et al.*, 1972) to the reaction mixture.

Catalytic hydrogenolysis over palladium black (Bergmann and Zervas, 1932) is also very common. The main uses include the removal of a number of groups of the benzyloxycarbonyl or of the benzyl ester family and elimination of the nitro group in the side chain of arginine. The principal drawback of catalytic hydrogenation is that it cannot be applied to sulfur-containing residues, which have a poisoning effect on the catalyst (see Schröder and Lübke, 1965). However, Merrifield (1964) and Medzihradszky and Medzihradszky–Schweiger (1965) have reported successful catalytic hydrogenolyses of methionine-containing peptides in the presence of freshly prepared palladium black and/or of a strong base. The extension of the method to *S*-benzylcysteine-containing peptides seems to be possible if the reduction is performed in liquid ammonia (Kuromizu and Meienhofer, 1974). A stepwise synthesis of oxytocin was achieved in this way, leading at each step to a quantitative removal of the benzyloxycarbonyl group used for N^{α}-protection. The only exception was the COOH-terminal tetrapeptide Z-Cys(Bzl)-Pro-Leu-Gly-NH_2, where not more than 60% cleavage could be obtained. The general applicability of catalytic hydrogenolysis is of great interest as it allows the ideal combination of N^{α}-benzyloxycarbonyl protecting groups with groups of the *tert*-butyl type for side-chain protection, the latter being stable during hydrogenation in liquid ammonia.

Electrolytic reduction has occasionally been employed for the deprotection of N^{g}-nitroarginine (Walshaw and Young, 1965), of tosyl peptides (Iwasaki *et al.*, 1973) of 4-picolyl esters (Camble *et al.*, 1969), and of the newly introduced protecting group, diphenyl-4-pyridylmethyl (Coyle and Young, 1976).

None of the above reduction methods can be used efficiently during the progress of solid-phase synthesis or for the cleavage of the peptide from the solid support. Even in solution, none of them is completely satisfying, although one research group (Schlatter *et al.*, 1977; Jones, 1977) has reported on the successful removal of polypeptides from the resin by hydrogenation using palladium acetate as catalyst. They tend to be replaced by the rather universal acidolytic procedure using liquid HF. As HF-cleavable protecting groups for cysteine, histidine, lysine, and arginine are now available, this is always possible when deprotection is performed at the end of the synthesis. On the other hand, a reductive splitting agent cannot be replaced by liquid HF if acid-labile groups are used as semipermanent protection during the synthesis.

Cleavage by heavy *metal ions* is also possible in some isolated cases. Mercuric ions or silver ions remove several *S*-protecting groups of cysteine: the important *S*-acetamidomethyl group (Veber *et al.*, 1972) is generally removed by mercuric ions at pH 4. It is also known that 8-quinolyloxy-carbonyl (Corey and Dawson, 1962) can be cleaved by copper(II) ions at neutral pH.

Solvolysis, another general method for removing protecting groups, has been applied in several instances. Alcoholysis with methanol in the presence of a base induces transesterification and can be used for the detachment of synthetic polypeptides from resin supports as methyl esters (Beyerman, 1972; Halpern *et al.*, 1968). Alcoholysis by 50% 2-(dimethylamino)ethanol in dimethylformamide was also shown to be very effective: the resulting dimethylaminoethyl ester can be hydrolyzed under mild conditions (Barton *et al.*, 1973). Solvolysis in water at pH 7 is one of the two steps involved in the removal of the 5-benzisoxazolylmethyleneoxycarbonyl group (Kemp and Hoyng, 1975). Heating in aqueous butanol or acetonitrile brings about the cleavage of the tosylaminocarbonyl protecting group (Künzi and Studer, 1975).

Further possibilities are offered by protecting groups that can be cleaved by *photolysis*. The use of photolabile *o*-nitrobenzyl derivatives for the protection of amine and carboxyl groups (Amit *et al.*, 1974), and for the side chains of histidine (Kalbag and Roeske, 1975) and tyrosine (Amit *et al.*, 1977) has been reported. One of the methods for the removal of 2-(3,5-dimethyl-oxyphenyl)propyl(2)oxycarbonyl (Ddz) (Birr *et al.*, 1972) is photolysis at 280 nm in tetrahydrofuran. Furthermore, Rich and Gurwara (1975) devised a photolabile anchor for the peptide chain during solid-phase synthesis, thereby permitting the preparation of peptides fully protected by acid-labile groups.

The *enzymatic* removal of suitably masked amino acids has also been described in the special case where the trypsin-labile benzyloxycarbonyl-arginyl group was chosen for N^{α}-protection (Meyers and Glass, 1975). Trypsin digestion at each cycle could be performed under mild aqueous conditions, with the particular advantage of being stereospecific. The possibility of cleaving a peptide from the usual solid supports by an enzymatic reaction (Blecher and Pfaender, 1973) is still very limited as the enzyme requires a largely hydrophilic environment.

The ingenious *safety-catch* principle involves an otherwise stable group which is made labile by chemical modification at the appropriate stage of the synthesis. So far, the main application has been to the cleavage of the polypeptide from the resin at the end of solid-phase synthesis. Examples are breaking of: sulfonamide links by sequential N-methylation and alkaline hydrolysis (Kenner *et al.*, 1971), glycol ester links by hydrolysis after dehydration of the stable glycol ester to a reactive enol ester (Wieland *et al.*, 1972), and 4-alkylthiophenyl ester links by oxidation to a 4-alkylsulfonylphenyl ester followed by β-elimination under mild alkaline conditions (Flanigan and Marshall, 1970; see also this Volume, Chapter 7). When one of the above methods is applied to the deprotection of a complex polypeptide, a number of problems of diverse origins can be encountered.

One difficulty arises from the possible side reactions of certain amino acid residues. As a rule, oxidation steps should be avoided when the peptide contains sensitive residues such as tryptophan, cysteine, methionine, tyrosine, threonine and serine. In the case of tryptophan, the sensitivity to oxidants becomes especially high in acidic media and can only partly be reduced by addition of thiols or by working in a nitrogen atmosphere. Another general rule is that exposure of the protected polypeptide to strongly alkaline conditions should be avoided because of possible racemization, transpeptidation (glutamic and aspartic acids), or deamidation (glutamine and asparagine) as side reactions (Section IV).

Although the restrictions are less severe when using strongly acidic media for deprotection, a number of unwanted alterations of the peptide may also occur when drastic conditions such as exposure to liquid HF are involved (Sakakibara, 1971). Tryptophan is known to be modified irreversibly by alkylation. Thus removal of the 2-nitrophenylsulfenyl (Nps) group by HCl in acetic acid can lead to nitrophenylthiolation (Scoffone *et al.*, 1968), removal of the *tert*-butyloxycarbonyl group by HCl in formic acid to (reversible) formylation (Ohno *et al.*, 1972), or removal of xanthenyl by HCl in tetrahydrofuran to xanthenylation of the indole ring (Southard, 1971). *tert*-Butyltrifluoroacetate, which is formed during the removal of *tert*-butyl groups by trifluoroacetic acid, can alkylate methionine and tryptophan (Lundt *et al.*, 1978; Löw *et al.*, 1978). The presence of serine, threonine, and tyrosine also imposes limitations. For example, it is well known that HBr in acetic acid can cause acetylation of the alcoholic or phenolic side chains.

Furthermore, it can not be excluded that some ring alkylation of tyrosine occurs during acidolytic deprotection even in the presence of anisole (Iselin, 1962). Treatment by liquid HF is often accompanied by N → O acyl migration at serine (Sakakibara, 1971). Cysteine and methionine further restrict the choice of deprotecting reagents. As already stated, reduction can only be applied under special conditions when any of these residues is present, and the use of scavengers is essential during acidolysis of alkyloxycarbonyl groups in the presence of methionine in order to prevent alkylation of the thioether function (e.g., Gutte, 1975). Even the use of the most popular scavenger, anisole, can lead to a side reaction: during deprotection of peptides with HF it can be acylated by the carboxyl group of glutamic acid (Rudinger, 1963; Feinberg and Merrifield, 1975; Sano and Kawanishi, 1975). Drastic deprotection methods can further lead to a destruction of the native conformation of a biologically active peptide. Thus, partial irreversible denaturation of native lysozyme was observed in liquid HF (Sharp *et al.*, 1973). This is a great handicap for the final deprotection in protein synthesis. Difficulties of this kind were also encountered during the treatment by HF of native neurotoxin I from Naja naja oxiana and of fully protected α-bungarotoxin and apamin (Ivanov *et al.*, 1979). Extensive HF induced side reactions involving the 10 cysteine residues of α-bungarotoxin were observed even in the presence of different scavengers.

Another series of restrictions is due to the presence of certain protecting groups introduced at an earlier stage of the synthesis. One danger is, for example, the conversion of N^g-nitroarginine residues into ornithine during ammonolysis (Künzi *et al.*, 1974) or HF treatment (Yamashiro *et al.*, 1972a). Another is the use of alkaline conditions for cleavage when the phthaloyl group has been chosen for semipermanent protection: opening of the phthalimido ring may lead to irreversible protection (Schröder and Lübke, 1965, p. 12). Furthermore, acidolysis with HF in anisole is accompanied by partial O → C rearrangement of *O*-benzyltyrosine) to form 3-benzyltyrosine residue (Erickson and Merrifield, 1973a).

Finally, a third cluster of limitations arises from the fact that the properties of the protecting groups have to be compatible with the conditions used for coupling. Thus, protection by the phthaloyl group is incompatible with coupling by the azide procedure involving hydrazine for the synthesis of the intermediate hydrazide. Hydrazine treatment may also attack the benzyl ester protection of the side-chain carboxyl groups of aspartic and glutamic acid (Hemmasi and Li, 1974). Furthermore, because of the acidic conditions required for the formation of the azide, no highly acid-labile groups such as 2-(4-biphenylyl)propyl(2)oxycarbonyl (Bpoc), 2-phenyl-2-propyloxycarbonyl (Poc), 2-(3,5-dimethyloxyphenyl)propyl(2)oxycarbonyl (Ddz), or 2-nitrophenylsulfenyl (Nps) are admissible in the NH_2-terminal fragment. The additive, 1-hydroxybenzotriazol, used in dicyclohexylcarbodiimide-activated coupling has been found to cleave the 2-nitrophenylsulfenyl group (König and Geiger, 1972) as well as the tosyl protection of histidine (Fujii and Sakakibara, 1974).

In spite of the number of methods available, selective deprotection remains a considerable problem, which will be examined in more detail in the next four sections.

III. DEPROTECTION BY ACIDOLYSIS

Among the methods available for the cleavage of protecting groups, acidolysis is undoubtedly the most commonly used. Reasons for this popularity are the high variability and the easy control of the splitting conditions. Even the more cumbersome and hazardous procedure involving liquid HF is used routinely nowadays. As a matter of fact, in solid-phase synthesis, variation of the pH (acidolysis or alkaline hydrolysis) is almost the only procedure employed for deprotection as long as the polypeptide is bound to the resin. It is therefore of the greatest importance to establish for each group the conditions under which it can be cleaved at a reasonable rate and with a minimum of side reactions as well as to determine the degree of selectivity of its acidolytic fission with respect to the other groups. This is best achieved with comparative kinetic studies.

It must be mentioned that a small number of protecting groups is not cleavable even under the drastic conditions of liquid HF at room temperature for up to 60 minutes: ethyloxycarbonyl, benzoyl, formyl, phthaloyl, methyl ester, ethyl ester, *N*-benzyl, *N*-*p*-methoxybenzyl, *S*-methyl, *S*-ethyl, *S*-isopropyl, *S*-acetamidomethyl (Sakakibara, 1971), 2-(*p*-tolysulfonyl)ethyl ester, and 2-(*p*-phenylazobenzylsulfonyl)ethyl ester (J. -L. Fauchère, unpublished results). The great majority of the remaining groups is cleaved by liquid HF. For this reason the HF procedure has become the method of choice for releasing unprotected peptides at the end of solid-phase synthesis.

A. Comparative Kinetic Studies

Comparative kinetic studies usually give insight into the mildest conditions available for the cleavage of a given protective group, into the possibilities of selective deprotection, and occasionally into the particular mechanism of cleavage. In the case of acidolysis, a number of such studies exist. A brief chronological review shall show how slowly the concept of acidolytic selectivity emerged and how its first qualitative evaluation developed toward more quantitative terms (see Table I).

Table I Relative Rates of Acidolytic Fission of Some Usual Protecting Groups

Protecting group	Abbrev	0.1 *N* HBr–AcOH, 20° Prot-Gly-NH-Bzl or Hippuric acid ester[a]	0.05 *N* HBr–AcOH, 20° Prot-Gly-NH-Bzl[a]	2 *N* HCl–THF–H_2O (9:1) 20°, Prot-Gly-NH-Bzl or Hippuric acid ester[a]	5.5 *N* HCl–THF–H_2O (9:1) 20°, Prot-Gly-NH-Bzl[a]
2-(4-Biphenylyl)propyl(2)oxycarbonyl	Bpoc[a]				
2-(3,5-Dimethyloxyphenyl)propyl(2)oxycarbonyl	Ddz				
2-Phenyl-2-propyloxycarbonyl	Poc				
α-2,4,5-tetramethylbenzyloxycarbonyl	Tmz				
$(cC_3H_5)CH(CH_3)OCO$ Cylcopropylmethyloxycarbonyl					
2-Nitrophenylsulfenyl	Nps	3.63	4.64	68.2	
4-Methyloxybenzyloxycarbonyl	Moz	1.38	1.60	1.18	1.60
$(1,CH_3\text{-}cC_6H_{10})OCO$ 1-Methylcyclohexyloxycarbonyl					
Diphenylmethyloxycarbonyl	Doc				
Furfuryloxycarbonyl	Foc	1.38	1.45	1.18	1.60
tert-Butyloxycarbonyl	Boc	1	1	1	1
Trityl (triphenylmethyl)	Trt	0.21	0.13	0.018	
1-Methylcyclobutyloxycarbonyl	Mcb				
Isobornyloxycarbonyl	Ibc				
$(cC_5H_9)OCO$ Cyclopentyloxycarbonyl					
1-Cyclopropylmethyloxycarbonyl	cPoc				
tert-Butyl ester	O*t*Bu	0.062		0.15	0.40
Benzhydryl ester	ODpm	0.062		0.15	1.20
Benzyloxycarbonyl	Z	$5\text{–}6 \times 10^{-4}$	3×10^{-4}		4.5×10^{-4}
4-Phenylazobenzyloxycarbonyl	Paz				
4-Chlorobenzyloxycarbonyl	4 Cz				
2-Chlorobenzyloxycarbonyl	2 Cz				
4-Methyloxyphenylazobenzyloxycarbonyl	Mpaz				
3-Nitrobenzyloxycarbonyl	3 Nz				
4-Cyanobenzyloxycarbonyl	CNz				
2,4-Dichlorobenzyloxycarbonyl	2,4 Dcz				
3,4-Dichlorobenzyloxycarbonyl	3,4 Dcz				
3-Chlorobenzyloxycarbonyl	3 Cz				
2,6-Dichlorobenzyloxycarbonyl	2,6 Dcz				
Benzyl ester (Asp(OBzl)-Glu(OBzl)-Ser(Bzl)-Thr(Bzl))	OBzl				0.022
O-Benzyltyrosine	Tyr(Bzl)				

80% aq AcOH, 22°–25° Prot-Gly-OEt[b,c]	3% TFA–$CHCl_3$, 22°–25° Prot-Gly-OEt[d]	0.5% TFA–$CHCl_3$, 22°–25° Prot-Gly-OEt[d]	90% aq TFE, 22°–25° "pH" 4, Prot-Gly-Val-OMe, Prot-Val-Val-OMe[e]	TFA, 25°, Prot-Phe-OH, Prot-Phe-Ala-OMe, Prot-Gly-OH; Prot-Leu-OH[f,g]	HCOOH, 25° Prot-Phe-OH, Prot-Phe-Ala-OMe[g]	50% TFA–CH_2Cl_2, 20° Lys (Prot)[h]	50% TFA–CH_2Cl_2, 20° side chain of Ser, Thr, Tyr, Asp, Glu, Cys[i]	1.65 *N* HBr–AcOH, 25° Prot-Gly-OH[j]	2 *N* HBr–AcOH, 25° Prot-Gly-OEt[k]	5.5 *N* HBr–AcOH, 20° Prot-Gly-NH-Bzl or Hippuric acid ester[a]	50% TFA–CH_2Cl_2, 20° linkage to resin[l,m]
000		1	1								
400		0.18									
700		0.11	0.1–0.2								
	4200	0.033									
					7200						
	250										
					5400						
1											
1	1				2700						
,000			40–800								
				150	150						
				175							
								18.4			
				7.5							
				1	(1)	1	1	1	1	1	
									0.65		
						0.35			0.60		
						0.016			0.40		
									0.15		
									0.07		
									0.05		
						0.012					
						0.006					
						0.0012					
						0.001					
										3.3×10^{-4}	
							0.018–0.026				
							1.25				

(continued)

Table 1 *(Continued)*

Protecting group	Abbrev	0.1 *N* HBr–AcOH, 20° Prot-Gly-NH-Bzl or Hippuric acid ester[a]	0.05 *N* HBr–AcOH, 20° Prot-Gly-NH-Bzl[a]	2 *N* HCl–THF–H_2O (9:1) 20°, Prot-Gly-NH-Bzl or Hippuric acid ester[a]	5.5 *N* HCl–THF–H_2O (9:1) 20°, Prot-Gly-NH-Bzl[a]
S-4-methyloxybenzyl	Cys(Mob)				
$OCH_2C_6H_4$—Ⓟ					
$OCH_2C_6H_4OCH_2C_6H_4$—Ⓟ					
$OCH_2C_6H_4CH_2CONHCH_2$—Ⓟ					
$N_2H_2COOC(CH_3)_2CH_2CH_2C_6H_4OCH_2C_6H_4$—Ⓟ					

[a] Losse *et al.* (1968).
[b] Sieber and Iselin (1968).
[c] Birr *et al.* (1972).
[d] Matsueda and Stewart (1975).
[e] Riniker *et al.* (1975).
[f] Calculated from Fujino *et al.* (1972).
[g] Brady *et al.* (1977).
[h] Erickson and Merrifield (1973b).
[i] Erickson and Merrifield (1973a).

Boissonnas and Preitner (1953) evaluated the stability of *N*-benzyloxycarbonyl, *N*-phenyloxycarbonyl, *N*-tosyl, *N*-formyl, *N*-phthaloyl, *N*-*p*-chlorobenzyloxycarbonyl, *N*-tolyloxycarbonyl, *N*-ethyloxycarbonyl, *N*-isobutyloxycarbonyl, and *N*-allyloxycarbonyl groups. They stated conditions that should allow a successive cleavage of the first five groups. Bláha and Rudinger (1965a) compared the rate of deprotection of a number of *p*- and *m*-substituted benzyloxycarbonylglycines and benzyloxycarbonylpeptides in HBr—acetic acid. In a similar study Bláha and Rudinger (1965b) measured the cleavage rates of several cycloalkyloxycarbonylglycines showing that the reactivity of the protecting group can be varied within wide limits by suitable structural modifications. Kinetic measurements of Homer *et al.* (1965) as well as those of Meienhofer (1966) with benzyloxycarbonyl- or substituted benzyloxycarbonylamino acids or small peptides showed,

80% aq AcOH, 22°–25° Prot-Gly-OEt[b,c]	3% TFA–$CHCl_3$, 22°–25° Prot-Gly-OEt[d]	0.5% TFA–$CHCl_3$, 22°–25° Prot-Gly-OEt[d]	90% aq TFE, 22°–25° "pH" 4, Prot-Gly-Val-OMe, Prot-Val-Val-OMe[e]	TFA, 25°, Prot-Phe-OH, Prot-Phe-Ala-OMe, Prot-Gly-OH; Prot-Leu-OH[f,g]	HCOOH, 25° Prot-Phe-OH, Prot-Phe-Ala-OMe[g]	50% TFA–CH_2Cl_2, 20° Lys (Prot)[h]	50% TFA–CH_2Cl_2, 20° side chain of Ser, Thr, Tyr, Asp, Glu, Cys[i]	1.65 *N* HBr–AcOH, 25° Prot-Gly-OH[j]	2 *N* HBr–AcOH, 25° Prot-Gly-OEt[k]	5.5 *N* HBr–AcOH, 20° Prot-Gly-NH-Bzl or Hippuric acid ester[a]	50% TFA–CH_2Cl_2, 20° linkage to resin[l,m]
								0.20			
											1
											5.05
											0.01
											4.55

Bláha and Rudinger (1965b).
Meienhofer (1965).
Mitchell *et al.* (1976).
Calculated from Wang (1975).
Abbreviations: Poc, 2-phenylpropyl(2)oxycarbonyl; Tmz, α-2,4,5-tetramethylbenzyloxycarbonyl; Foc, furfuryloxy-rbonyl; Mcb, 1-methylcyclobutyloxycarbonyl; Ibc, isobornyloxycarbonyl; cPoc, 1-cyclopropylmethyloxycarbonyl; oc, diphenylmethyloxycarbonyl; Ⓟ, copoly(styrene–divinylbenzene); TFE, 2,2,2-trifluorethanol.

on the one hand that the rate of splitting was rather insensitive to the nature of the amino acid or peptide, on the other that the reactivity was increased when the ring substituent of benzyloxycarbonyl was electron-donating and decreased when it was electron-withdrawing.

When it became evident that the introduction of protecting groups of the *tert*-butyl type into peptide synthesis (Schwyzer, 1966) would be most successful, the lability of these new groups in acidic media had to be compared quantitatively with that of groups of the benzyloxycarbonyl family. In a detailed investigation, Losse *et al.* (1968) determined relative rate constants, k_{rel}, for the splitting of a number of groups in HBr–acetic acid, in HCl–tetrahydrofuran, in glacial acetic acid, and in trifluoroacetic acid. Among the many k_{rel} values reported, the fundamentally important value for *tert*-butyloxycarbonyl: benzyloxycarbonyl was found to be 1700 in 0.1 *N* HBr–acetic acid at 20°C.

Sieber and Iselin (1968) measured the first-order rate constants for the cleavage of a series of *N*-aralkoxycarbonylglycine ester derivatives under acidic conditions. The rate constants in 80% acetic acid at room temperature were found to vary greatly depending on the stability of the corresponding aralkylcarbonium ion. Some of the groups investigated were cleaved up to 60,000 times faster than *N*-*tert*-butyloxycarbonyl, at much the same rate as *N*-trityl. The relative rates were shown to depend strongly on the reaction conditions. As a result, 2-(4-biphenylyl)propyl(2)oxycarbonyl (Bpoc) was proposed as the most suitable *N*-protecting group of the series as it was cleaved 3,000 times faster than *tert*-butyloxycarbonyl and still gave rise to sufficiently stable amino acid derivatives.

The situation is more complicated in solid-phase synthesis. The physical state of the resin (swelling etc.) plays a decisive role during deprotection. Karlsson *et al.* (1970) compared the three following methods for the cleavage of *tert*-butyloxycarbonyl: (1) 1 *N* HCl in acetic acid, (2) 4 *N* HCl in dioxane, and (3) 50% trifluoroacetic acid in methylene chloride at 25°C for 30 min. They found that the latter was the fastest, probably because of the better swelling of the polymer in this mixture. The reaction time had to be reduced to 10 min in order to minimize the acidolytic cleavage of the 4-alkylbenzyl ester link to the resin. However, sterically hindered *tert*-butyloxycarbonyl derivatives such as Boc-Ile-Ala-Ⓟ (-Ⓟ = resin) required 30 min, leading to losses from the resin. Additional experiments by the same authors (Ragnarsson *et al.*, 1970) led to the statement that benzyloxycarbonyl as an ε-amino protecting group for lysine is insufficiently stable if 50% trifluoroacetic acid in methylene chloride is used for N^{a}-deprotection. On the other hand, complete stability of the 4-alkylbenzyl ester link to the resin was observed under the conditions for splitting the Bpoc group. Schnabel *et al.* (1971) found *N*-*p*-chlorobenzyloxycarbonyl and *N*-*m*-chlorobenzyloxycarbonyl to be more suitable for N^{ε}-protection during N^{α}-Boc cleavage, and *N*-*p*-nitrobenzyloxycarbonyl to be too stable toward HF acidolysis. They reported a good selectivity of Boc splitting in the presence of benzyloxycarbonyl during conventional syntheses if either 70% aqueous trifluoroacetic acid or boron trifluoride etherate in glacial acetic acid were used. The drawback of the 70% aqueous trifluoroacetic acid procedure was that γ-benzyl ester groups on glutamic acid side chains were still partially cleaved.

In a further fundamental study series, Erickson and Merrifield (1974) evaluated the stability of benzyl-type side-chain protecting groups in the solid-phase procedure when 50% trifluoroacetic acid in methylene chloride was used for N^{α}-deprotection. *O*-Benzyltyrosine, *O*-benzyloxycarbonyltyrosine, N^{ε}-benzyloxycarbonyllysine, N^{ε}-4-chlorobenzyloxycarbonyllysine, *S*-*p*-methyloxybenzylcysteine and *O*-benzylserine were found to be inadequate, giving rise to either intramolecular rearrangements or premature cleavage

(Erickson and Merrifield, 1973a). The investigation of the suitability of six chlorinated benzyloxycarbonyl groups for the N^{ε}-protection of lysine during a solid-phase synthesis of decalysyl-valine showed that chain branching could be effectively prevented by employing either the 2,4-dichlorobenzyloxycarbonyl or the 2-chlorobenzyloxycarbonyl groups (Erickson and Merrifield, 1973b). The apparent first-order rate constants of cleavage of these two groups was found to be $k = 4.9 \times 10^{-8}$ and 6.3×10^{-8} sec^{-1}, respectively (the unsuitable benzyloxycarbonyl group had $k = 3.9 \times 10^{-6}$ sec^{-1}). The two chlorinated groups are still completely cleaved by liquid HF within 1 hr at 0°C. Four other side-chain protecting groups, namely tosyl and 2,4-dinitrophenyl for histidine, 2-chlorobenzyl for serine, and 4-methylbenzyl for cysteine, were found to allow the selective cleavage of the N^{α}-*tert*-butyloxycarbonyl group as they were completely stable in 50% trifluoroacetic acid in dichloromethane for over 200 hr at 20°C.

Matsueda and Stewart (1975) compared the cleavage rates of a series of new 1-arylalkoxycarbonyl groups with that of *tert*-butyloxycarbonyl. They selected 2,4,5-trimethylbenzyloxycarbonyl (Tmz) as a suitably labile N^{α}-protecting group. The deprotection rate of Tmz-Gly-OEt in 0.5 to 3% trifluoroacetic acid in chloroform was 4,200 times higher than for the *tert*-butyloxycarbonyl and in the same order of magnitude as for the 2-(3,5-dimethyloxyphenyl)propyl(2)oxycarbonyl (Ddz) group.

Another important contribution was the discovery by Riniker *et al.* (1975) that the use of 90% aqueous trifluorethanol as a solvent permits the selective cleavage of N^{α}-trityl (Trt) in the presence of Bpoc or 2-phenyl-2-propyloxycarbonyl (Poc) under controlled acidic conditions (apparent pH 4). The ratio of the rate constants for Trt:Bpoc was at least 300 except in the cases of N^{α}-Trt-glycyl peptides where Bpoc had to be replaced by the 5 to 10 times more stable Poc in order to obtain good selectivity. The value of 300 can be considered as satisfactory when the differential deprotection step is used only once during a synthesis, as for example in that of insulin by the same authors (Sieber *et al.*, 1977). Brady *et al.* (1977) compared the stability of some new acid-labile protecting groups: 1-methylcyclobutyloxycarbonyl (Mcb) was 18 times more stable than *tert*-butyloxycarbonyl in formic acid whereas cyclobutyloxycarbonyl was about 2,700 times more stable than Boc in trifluoroacetic acid.

Finally, the labilities of some polypeptide linkages to the copoly (styrene–divinylbenzene) support in solid-phase synthesis were determined. Wang (1975) assessed the cleavage rate of the peptide ester bond to the conventional hydroxymethyl resin as well as that of the peptide hydrazide bond to the 3-(*p*-benzyloxyphenyl)-1,1-dimethylpropyloxycarbonylhydrazide resin. The cleavage from both supports was rapid in 50% trifluoroacetic acid in methylene chloride. It was negligible under the conditions of Bpoc splitting (0.5%

TFA–CH_2Cl_2). Mitchell *et al.* (1976) found the peptidyl-4-oxymethylphenyl-acetamidomethyl resin to be about 100 times more stable than the conventional peptidyl-oxymethyl resin to cleavage under the same conditions.

Studies such as those briefly reviewed above give a quantitative impression of the stability of protecting groups toward acidolysis. An aim of such studies would be an absolute classification of the protecting groups according to their lability in acidic media. To attain such a goal would require a great number of very systematic studies because the relative rates of cleavage differ strongly when solvent, apparent pH, nature of the protected compound, state of swelling of the resin, or other factors are varied. An example was given by Birr *et al.* (1972) in the case of the 2-(3,5-dimethyloxyphenyl)propyl-(2)oxycarbonyl (Ddz) group. The cleavage rate ratio of Ddz: Boc at 20°C was 1,400 in 80% aqueous acetic acid and 2,800 in 50% aqueous acetic acid, whereas at 50°C it was 2,800 in glacial acetic acid and 10,500 in 50% aqueous acetic acid. Another example is the discrepancy of the k_{rel} values of 2-nitrophenylsulfenyl (Nps) group cleavage with respect to *tert*-butyloxy-carbonyl when determined in HBr in glacial acetic acid on the one hand ($k_{rel} = 3.6$ to 4.6) or in 2 *N* HCl in TFA–H_2O (9;1) on the other ($k_{rel} = 68.2$) (Losse *et al.*, 1968). A third typical example of the strong solvent dependence of acidolytic cleavage is the fact that trityl in the model compound Trt-Gly-benzylamide is cleaved 6 to 8 times more slowly than *tert*-butyloxycarbonyl in HBr in acetic acid whereas it can be split selectively in the presence of Boc in glacial acetic acid (Loose *et al.*, 1968). A number of other examples exist which have been compiled in Table I.

Nevertheless, a strict comparison of the acid lability of protecting groups is possible provided rigorously identical cleavage conditions are employed. Conventional synthesis and especially solid-phase synthesis usually make use of standardized methods of cleavage. As far as the kinetic studies are carried out under these standard conditions, they will be relevant to the synthetic procedures.

It will therefore be very useful when a new group is being introduced into peptide synthesis to compare its lability toward acids with that of some well-known groups in terms of relative rates of cleavage under the standardized conditions. This practice, which has been followed for some new protecting groups (Fujino *et al.*, 1972; Birr *et al.*, 1972; Brady *et al.*, 1977), is possible at least in the numerous cases where cleavage progresses as a first-order or a pseudo first-order reaction. Another practicable way is to report relative half-times ($t_{1/2}$) for removal of the protection. Insofar as one deals with first-order kinetics, $t_{1/2}$ contains the same information as the rate constant k and has the advantage of having more intuitive value. Such data are to be preferred to the more qualitative statements as, for example, the time required for complete cleavage. Even if the complete time dependence of the cleavage is reported, a comparative study should be performed under identical conditions with at least one well-known protecting group.

The results of the available kinetic studies are listed in Table I. Values are expressed in terms of relative deprotection rates k_{rel} under defined conditions. Three ranges of stability have been distinguished so that k_{rel} values are calculated with respect to benzyloxycarbonyl (Z) for very stable groups, to 2-(4-biphenylyl)propyl(2)oxycarbonyl (Bpoc) for very labile groups, or to *tert*-butyloxycarbonyl (Boc) for groups of intermediate stability. The groups listed include some of the acid-labile groups usually chosen for NH_2-terminal, side-chain or COOH-terminal protection, as well as some of the acid-labile linkages between polypeptide and resin as used in solid-phase synthesis. The relative cleavage rate from the resin is based on the classical 4-alkylbenzyl ester linkage.

Comparison of the stability of two groups is only valid under identical cleavage conditions, i.e., only within a single column of Table I. Discrepancies of k_{rel} in different columns are explained by the dependence of k_{rel} on the particular conditions of cleavage and on the nature of the peptide to be deprotected.

B. Stability Levels and Selectivity

When looking at Table 1, the immediate question arises as to which k_{rel} values will allow a *selective* cleavage of the more labile protecting group. In the frequent case of two first-order reactions, a $k_{rel} = 460{:}1$ will, theoretically, allow the removal of 99% of the more labile group and leave 99% of the more stable group intact. On this simple basis, at least four stability levels can be defined between the most stable and the most labile N-protection of Table I, i.e., 2,6-dichlorobenzyloxycarbonyl (2,6 Dcz) and 2-(4-biphenylyl)-propyl(2)oxycarbonyl (Bpoc). O and S protection poses similar, but not inherently different problems. Usually, only three levels are employed in practical work. The first level contains benzyloxycarbonyl (Z) and groups more stable than Z, the second *tert*-butyloxycarbonyl and groups of similar stability, and the third the very labile groups such as Ddz or Bpoc. However, examples are known in which a fourth level is utilized, e.g., the selective removal of the trityl group in the presence of the Bpoc group (Riniker *et al.*, 1975).

The repeated removal of N^{α}-protection in the presence of acid-labile side-chain protection requires a ratio higher than 460:1. For solid-phase synthesis, a $k_{rel} \geqslant 1.5 \times 10^4$ has been suggested as a criterion for practicability (Erickson and Merrifield, 1973b). In this case, and with prevailing first-order kinetics, less than 0.05% of the more stable group is lost during the removal of 99.95% of the more labile group. Such a high k_{rel} imposes severe restrictions on the possibilities of acid-labile protecting group combination and allows essentially only two stability levels. The lower level is generally used for N^{α}-protection, the higher one for side-chain protection *and* resin attachment. The final, more drastic acid treatment is thus usually designed to cleave simultaneously the acid-labile protecting groups from the peptide *and* the peptide from the resin.

Kinetic studies are still lacking for a considerable number of groups. Nevertheless, an approximate stability level can be assigned to every group on the basis of more qualitative laboratory experiments. Usually, the groups are tested with more or less standardized acidic media used for deprotection, and the mildest medium is determined that will, within a reasonable time, bring about a "complete" cleavage of the protecting group in question. In the following, we shall consider four commonly used acidic media and attempt a classification of protecting groups on this basis. It should be kept in mind that, for many of the groups specified below, acidolysis is not the only or even the best cleavage method, but just the one considered in this context of cleavage specificity.

1. Dilute Trifluoroacetic Acid

Dilute, aqueous trifluoroacetic acid and dilute, aqueous acetic acid were first used to remove N^{α}-trityl groups in the presence of semipermanent *N*-tosyl (Schwyzer and Sieber, 1956) or *N*-*tert*-butyloxycarbonyl and *tert*-butyl ester groups (Schwyzer *et al.*, 1960). Dilute trifluoroacetic acid in organic solvents, e.g., 0.5–15% TFA in CH_2Cl_2 or in $CHCl_3$, (as well as dilute HCl in organic solvents), was introduced later into solid-phase synthesis or classical solution synthesis for similar purposes. Examples of groups removed by such dilute acid are as follows:

1. (*N*) trityl, Trt (Schwyzer and Sieber, 1956; Schwyzer *et al.*, 1960; Losse *et al.*, 1968).
2. (*N*) 2-nitrophenylsulfenyl, Nps (2 equiv of HCl in organic solvents; Goerdeler and Holst, 1959; Zervas *et al.*, 1963; Najjar and Merrifield, 1966).
3. (*N*) 2-(4-biphenylyl)propyl(2)oxycarbonyl, Bpoc (Sieber and Iselin, 1968).
4. (*N*) 2-benzoyl-1-methylvinyl, Bmv (Southard *et al.*, 1969).
5. (*N*) 2-(3,5-dimethyloxyphenyl)propyl(2)oxycarbonyl, Ddz (Birr *et al.*, 1972).
6. (*N*) 2-phenylpropyl(2)oxycarbonyl, Poc (Sandberg and Ragnarsson, 1974).
7. (*N*) α,2,4,5-tetramethylbenzyloxycarbonyl, Tmz (Matsueda and Stewart, 1975).
8. (*N*) diphenylphosphinyl, Dpp (Kenner *et al.*, 1976).

9. (*N*) 2-(4-phenylazophenyl)propyl(2)oxycarbonyl, Azc (Tun-Kyi and Schwyzer, 1976).

10. (*O*, ester) α-(4-Ⓟ-phenyl)-benzyl ester (-Ⓟ = resin; Southard *et al.*, 1971).

Although the conditions for cleaving these groups appear to be very mild, it cannot generally be avoided that some of the remaining groups of the next category (concentrated trifluoroacetic acid) will be attacked to a certain extent, especially in repetitive deblocking procedures. Groups of the highest stability levels (HBr–acetic acid and liquid HF) will be left essentially intact.

2. *Concentrated Trifluoroacetic Acid*

The class of protecting groups that can be removed by pure or only slightly diluted trifluoroacetic acid (e.g., 90% TFA) includes as main representative those of the *tert*-butyl type (Schwyzer *et al.*, 1960) which occur as either urethane, ester, ether, or thioether derivatives. Because of their relative stability in dilute trifluoroacetic acid (as well as in alkali and against hydrogenolysis), these groups have been extensively used for semipermanent protection (Schwyzer, 1966). Later, groups with similar stability characteristics were devised and prepared. The following are the ones most commonly used:

1. (*N*) *tert*-butyloxycarbonyl, Boc (Schwyzer *et al.*, 1960; Schwyzer, 1966; Karlsson *et al.*, 1970; Schnabel *et al.*, 1971).
2. (*N*) 2-furfuryloxycarbonyl, Foc (Jeschkeit *et al.*, 1966).
3. (*N*) adamantyloxycarbonyl, Adc (Haas *et al.*, 1966).
4. (*N*) diphenylmethyloxycarbonyl, Doc (Sieber and Iselin, 1968).
5. (*N*) *tert*-amyloxycarbonyl, Aoc (Sakakibara *et al.*, 1969).
6. (*N*) isobornyloxycarbonyl, Ibc (Fujino *et al.*, 1972).
7. (*N*) 1-methylcyclobutyloxycarbonyl, Mcb (Brady *et al.*, 1977).
8. (N^{im}) *tert*-butyloxycarbonyl, Boc (Yamashiro *et al.*, 1972a).
9. (*O*, ester) *tert*-butyl ester, O*t*Bu (Schwyzer *et al.*, 1960; Schwyzer, 1966).
10. (*O*, ester) diphenylmethyl ester, ODpm (Stelakatos *et al.*, 1966).
11. (*O*, ester) 4-methoxybenzyl ester, OMob (Stewart, 1968b).
12. (*O*, ester) 2,4,6-trimethylbenzyl ester, OTmb (Stewart, 1968c).
13. (*O*, ester) pentamethylbenzyl ester, OPmb (Stewart, 1968a).
14. (*O*, ester) piperonyl ester, OPon (Stewart, 1971).
15. (*O*, ether) *tert*-butyl ether, *t*Bu (Beyerman and Bontekoe, 1962).
16. (*O*, ester) 4-(Ⓟ-alkoxy)benzyl ester (-Ⓟ = resin; Wang, 1975).
17. Ⓟ-*tert*-alkyloxycarbonylhydrazide (-Ⓟ = resin; Wang, 1972). (*N*)-5-Dibenzosuberyl, Sub (Pless, 1976) is also a member of this group although it behaves as an exception, being stable in strongly acidic media such as 4 *N* HBr in acetic acid or 5 *N* HCl in dioxane.

3. HBr in Glacial Acetic Acid

Typical of this class of protecting groups are benzyloxycarbonyl and derivatives thereof with electron-attracting substituents (the reaction becomes increasingly second-order and also slower; Homer *et al.*, 1965). The acidic reagents and their use for removing the benzyloxycarbonyl group were first described by Ben-Ishai and Berger (1952), Ben-Ishai (1954), and Boissonnas and Preitner (1953), although other organic solvents for HBr and liquid HBr may also be used (see Wünsch, 1974). Typical examples are as follows:

1. (*N*) benzyloxycarbonyl, Z (Ben-Ishai and Berger, 1952; Schwyzer, 1954; Losse *et al.*, 1968).
2. (*N*) 2-chlorobenzyloxycarbonyl, 2 Cz (Meienhofer, 1966).
3. (*N*) 3-chlorobenzyloxycarbonyl, 3 Cz (Bláha and Rudinger, 1965a).
4. (*N*) 4-chlorobenzyloxycarbonyl, 4 Cz (Meienhofer, 1966).
5. (*N*) 4-cyanobenzyloxycarbonyl, 4 CNz (Noda *et al.*, 1970).
6. (*N*) 2-bromobenzyloxycarbonyl, 2 Bz (Yamashiro *et al.*, 1972b).
7. (*N*) 2,4-dichlorobenzyloxycarbonyl, 2,4 Dcz (Erickson and Merrifield, 1973b).
8. (*N*) 4-dihydroxyborylbenzyloxycarbonyl, Dobz (Kemp and Roberts, 1975).
9. (N^{im}) benzyloxycarbonyl, Z (Ontjes and Anfinsen, 1970).
10. (*N*) diisopropylmethyloxycarbonyl, Dmc (Sakakibara *et al.*, 1970).
11. (*N*) 5-benzisoxazolylmethyleneoxycarbonyl, Bic (Kemp and Hoyng, 1975).
12. (*N*) cyclobutyloxycarbonyl, cBoc (Brady *et al.*, 1977).
13. (N^{im}) 2,2,2-trifluoro-1-benzyloxycarbonylaminoethyl, Ztf (Weygand *et al.*, 1967).
14. (N^{im}) toluenesulfonyl, Tos (Coy *et al.*, 1973).
15. (*O*, ester) benzyl, OBzl (Ben-Ishai and Berger, 1952; MacLaren *et al.*, 1958; Bláha and Rudinger, 1965a; Marshall and Merrifield, 1965; Stewart, 1966).
16. (*O*, ester) 4-nitrobenzyl, ONb (increased stability compared to OBzl; Schwarz and Arakawa, 1959; Schwyzer and Sieber, 1959).
17. (*O*, ester) 4-bromobenzyl, OBb (Yamashiro *et al.*, 1973).
18. (*O*, ester) 4-chlorobenzyl, OCb (Kisfaludy and Löw, 1965).
19. (*O*, ether) benzyl, Bzl (Rivier, 1974; see Wünsch, 1974).
20. (*O*, ether) 3-bromobenzyl, 3Bb (Yamashiro *et al.*, 1973).
21. (*O*, ether) 2,6-dichlorobenzyl, Dcb (Erickson and Merrifield, 1973a).
22. (*O*, ether) 2,2,2-trifluoro-1-benzyloxycarbonylaminoethyl, Ztf (Weygand *et al.*, 1968).
23. (*S*) 4-methoxybenzyl, Mob (Hammerström *et al.*, 1970).

24. (*S*) 4-methylbenzyl, Meb (Erickson and Merrifield, 1974).
25. (*S*) 3,4-dimethylbenzyl, 3,4 Dmb (Yamashiro *et al.*, 1972b).
26. (*O*, ester) 4-Ⓟ-benzyl ester (-Ⓟ = resin; Merrifield, 1963).

4. *Liquid Hydrogen Fluoride*

The groups of this class require liquid HF for acidolysis (method, see Sakakibara *et al.*, 1967). A number of groups, such as benzyl and 4-nitrobenzyl, that are quite difficult to cleave by HBr in acetic acid, are more easily removed by liquid HF (unless one resorts to reductive methods). The ones mentioned below are not attacked by the milder procedures:

1. (N^g) nitro, -NO_2 (Sakakibara, 1971).
2. (N^g) *p*-toluenesulfonyl, Tos (Mazur and Plume, 1968).
3. (*N*) 2,6-dichlorobenzyloxycarbonyl, 2,6 Dcz (Erickson and Merrifield, 1973b).
4. (*S*) *tert*-butyl, *t*Bu (Sakakibara, 1971).
5. (*S*) benzyl, Bzl (Sakakibara, 1971).
6. [2-(4-hydroxymethylphenyl)-acetamido]methyl-Ⓟ, OCH_2Pam (Mitchell *et al.*, 1976).

This classification by stability levels shall be employed in Section V for the elaboration of protection schemes.

IV. ALKALI-LABILE PROTECTING GROUPS

Alkaline hydrolysis has been much less extensively used as a deblocking procedure than acidolysis. The reason for this probably lies in the fact that alkaline conditions are known to promote some unwanted side reactions, including base-catalyzed racemization (Young, 1967), deamidation of asparagine (Robinson and Tedro, 1973), and $\alpha \rightarrow \beta$ transpeptidation of aspartyl peptides (Iselin and Schwyzer, 1962). Another possible pitfall is β-elimination of *O*-acyl derivatives of serine and threonine (Patchornik and Sokolovsky, 1964). For these reasons, a very limited number of alkali-labile protecting groups has been proposed up to now.

Nevertheless, the advantage of the introduction of an additional selectivity level with respect to pH variation is being more generally realized, and the number of base-sensitive groups is bound to increase in the future. In order to minimize the side reactions mentioned above, groups will have to be developed that are cleavable under "mild" alkaline conditions. The following short review of the presently available groups shows that some progress towards this goal has recently been achieved.

One of the first alkali-labile protecting groups that was introduced into peptide synthesis was the trifluoracetyl, Tfa (Weygand and Csendes, 1952). Although it is readily removed by treatment with alkali, it has found only limited use for N^{α}-protection mainly because of the concern about possible racemization during the coupling steps with Tfa-amino acids.

Phthaloyl (Pht <), another group known since long can also lead to racemization if not introduced with the aid of *N*-ethoxycarbonylphthalimide (Nefkens *et al.*, 1960). Since dilute alkali leads generally not to cleavage but to the formation of a stable phthalimidic acid derivative (Rudinger *et al.*, 1960), removal is restricted to hydrazinolysis (see Wünsch, 1974).

A thoroughly new concept appeared with the application of β-elimination. The success of this method led to the development of a series of new groups. β-Elimination is readily achieved under weakly alkaline conditions provided the leaving group is situated in a position β to a sulfonyl or other strongly electron-attracting group. It proceeds according to the following mechanism (Kader and Stirling, 1964):

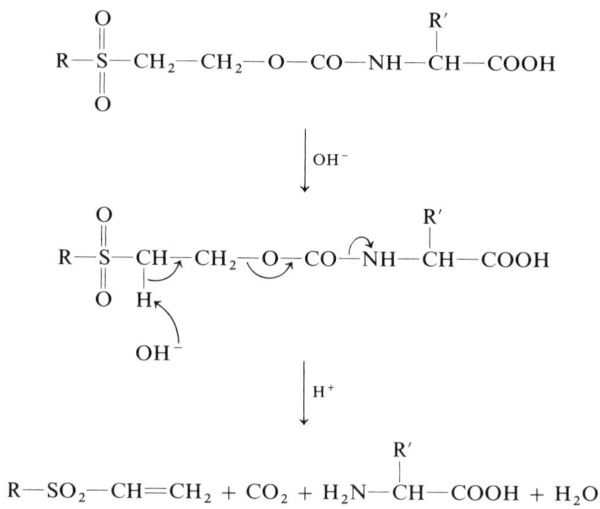

R = alkyl, R′ = amino acid side chain

Since the first report of Kader and Stirling (1964), who described 2-(4-toluenesulfonyl)ethyloxycarbonyl (Tsc) as an amine protecting group, a number of alkylsulfonylethyl-containing groups have been proposed for carboxyl as well as for amino protection (Table II). With the aim of cleaving the polypeptide from the solid support by treatment with alkali, the substituted resin 2-hydroxyethylsulfonylmethyl–copoly(styrene–divinylbenzene) was also synthesized (Tesser *et al.*, 1976) for use in solid-phase synthesis.

Furthermore, the interesting property of alkylthioethyloxycarbonyl amine protecting groups and of alkylthioethyl esters of being stable toward acids and alkali during the synthesis and convertible to the corresponding alkali-labile alkylsulfonylethyl derivatives by catalytic oxidation (Schultz *et al.*, 1963) has promoted the use of these parent groups. They are also included in Table II, although the fact that an oxidation step is involved in their activation seriously limits their scope of application.

Conditions for the elimination of alkylsulfonylethyl derivatives are usually mild, and depend somewhat on the nature of the peptide (J. L. Fauchère, unpublished observations). As an example, the COOH-terminal 2-(4-toluenesulfonyl)ethyl ester (OTse) was readily saponified at pH 11 for 10 min at room temperature (Eberle *et al.*, 1975).

β-Elimination is not restricted to sulfonyl-containing groups. Wünsch and Spangenberg (1971) found that 2-cyano-*tert*-butyloxycarbonyl (Cyc) can be removed via this mechanism in a 0.3 N K_2CO_3 solution within six hours at room temperature. 9-Fluorenylmethoxycarbonyl (Fmoc) (Carpino and Han, 1972) was also found to be an amine protecting group that can be cleaved by ammonia, piperidine, or ethanolamine via the same mechanism. Kunz (1976a) proposed 2-triphenylphosphonoethyloxycarbonyl (Pec) as a new protecting group susceptible to β-elimination under mildly alkaline conditions.

All the groups assembled in Table II are completely stable under acidic conditions, indeed a property of paramount importance as far as selectivity is concerned. The other essential quality of these groups is, as already stated, their easy elimination at pH values which generally do not need to exceed 11. These conditions are usually not sufficient for the splitting of the conventional methyl, ethyl, and benzyl esters, for which an undesirable exposure of the protected polypeptide to strongly alkaline conditions is often indispensable (especially for proline esters; Schwyzer and Sieber, 1956). In contrast, the phenyl ester (OPh) (Kenner and Seely, 1972) used for COOH-terminal protection was removed in 15 min at 20°C at pH 10.5. Similarly, phenyl esters used for the attachment of the polypeptide to the resin in solid-phase synthesis were shown to be very susceptible to ammonolysis (Blake and Li, 1971). Furthermore, the ester linkages between peptides and polymers of the type Ⓟ—$CH_2N(COCH_3)(CH_2)_nOH$ [-Ⓟ = copoly-(styrene–divinylbenzene)] are readily hydrolyzed with less than a twofold excess of 0.05 N aqueous NaOH in dioxane–ethanol (1:1) (Tilak and Hollinden, 1968b). Finally, phenacyl ester links (in solid-phase synthesis) could be cleaved with 0.5 N aqueous NaOH in dioxane, by ammonia, or by hydrazine in methanol (Mizoguchi *et al.*, 1970).

The introduction of alkali-labile groups has, by enlarging the pH range for differential deprotection, allowed the elaboration of new protection

Table II Protecting Groups Cleaved *via* β-Elimination

Name	Structure	Abbrev	Reference
4-Toluenesulfonylethyloxycarbonyl	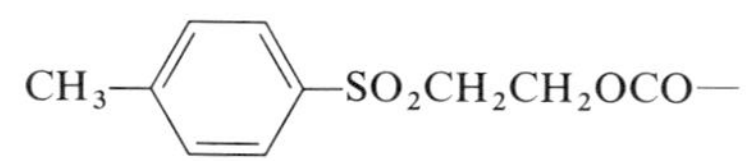	Tsc	Kader and Stirling, 1964
4-Toluenethioethyloxycarbonyl	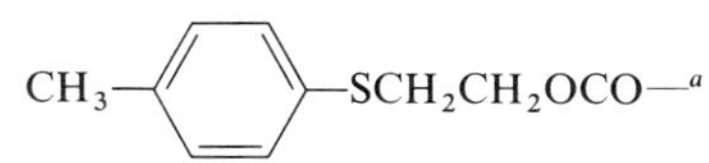 [a]	—	Kader and Stirling, 1964
Methylsulfonylethyloxycarbonyl	$CH_3SO_2CH_2CH_2OCO$—	Msc	Tesser, 1975
Methylthioethyloxycarbonyl	$CH_3SCH_2CH_2OCO$— [a]	—	Kunz, 1976b
2-(Triphenylphosphono)ethyloxycarbonyl	$(C_6H_5)_3P^+CH_2CH_2OCO$—	Pec	Kunz, 1976a
2-Cyano-*tert*-butyloxycarbonyl	$NCCH_2C(CH_3)_2OCO$—	Cyc	Wünsch,and Spangenberg, 1971
9-Fluorenylmethyloxycarbonyl	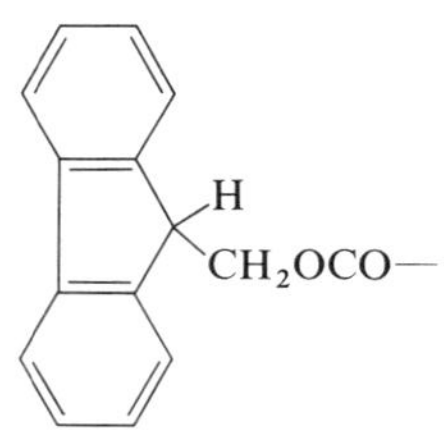	Fmoc	Carpino and Han, 1972

4-Toluenesulfonylethyl ester	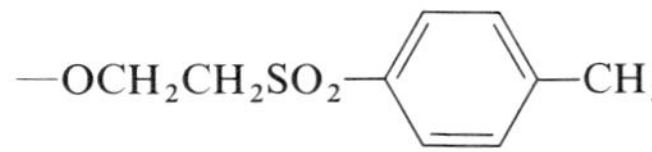	OTse	Miller and Stirling, 1968
Methylsulfonylethyl ester	$—OCH_2CH_2SO_2CH_3$	OMse	Hardy *et al.*, 1968
4-Phenylazobenzylsulfonylethyl ester	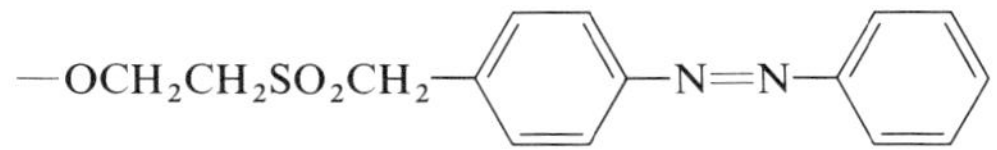	OPse	Vogel, 1975; Diaz *et al.*, 1979
Resin-bound benzenesulfonyl ester	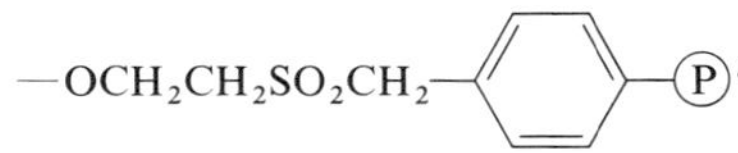	—	Tesser *et al.*, 1976
4-Nitrophenylthioethyl ester	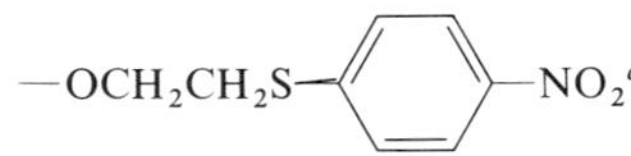	—	Amaral, 1969

[a] Cleavable after conversion to the corresponding sulfonyl derivative.
[b] —(P), Copoly(styrene–divinylbenzene) resin

schemes (see Section VI). The combination of these groups with those cleaved by acidolysis has been utilized with success in several instances.

Protection of the lysine side chain with alkali-labile groups in conjunction with acid-labile groups for N^{α}-protection has been one of the main applications. N^{ε}-Trifluoroacetyllysine [Lys(Tfa)] was used during the synthesis of fragments of staphyloccocal nuclease (Anfinsen *et al.*, 1967), of insulin (Tilak and Hollinden, 1968a), and of fragments of cytochrome c (Borin *et al.*, 1976). N^{ε}-Phthaloyllysine [Lys(Pht)] (Costopanagiotis *et al.*, 1963) and N^{ε}-methylsulfonylethyloxycarbonyl [Lys(Msc)] (Eberle *et al.*, 1975) were found to be suitable for the synthesis of α-melanotropin and related peptides. In the latter case, the following new combination of protecting groups was used: Boc or Z for the N^{α}-amino groups, Msc for the N^{ε}-amino group of lysine-(11) and OTse for the carboxylic acid groups of COOH-terminal glycine-(10) in α-MSH-(11–13)-tripeptide amide and α-MSH-(1–10)-decapeptide, respectively. This provided a facile and mild selective deprotection of either the α-amino groups by acidolysis or of the other groups by β-elimination in alkali (Eberle *et al.*, 1975), (Eberle *et al.*, 1977). Base-sensitive groups were further employed for the protection of the COOH-terminal carboxyl function in the synthesis of a gramicidin S analogue (Ludescher and Schwyzer, 1972) as well as in that of a corticotropin derivatives (J. L. Fauchère, unpublished results). A heptapeptide from the A-chain of insulin was synthesized using the Tse-ester as carboxyl protecting group. The ester could be saponified selectively in the presence of the *S*-ethylthiocysteinyl residue (Seidel *et al.*, 1976). The applicability of β-elimination for appropriately derivatized copoly(styrene–divinylbenzene) resins was also investigated and found to proceed satisfactorily in the synthesis of a protected corticotropin fragment esterified to a hydroxyethylsulfonylmethyl resin (Buis *et al.*, 1976).

A solid-phase synthesis of dihydrosomatostatin (Chang and Meienhofer, 1978) has proven convincingly that fluorenylmethyloxycarbonyl (Fmoc) can be used for the protection of the α-amino function of the growing chain in combination with *tert*-butyl type side-chain blocking. Two other solid-phase syntheses of acyl carrier protein (65–74) (Atherton *et al.*, 1978a) and of β-endorphin (Atherton *et al.* 1978b) afforded higher yields when using N-$^{\alpha}$Fmoc-amino acids in place of N^{α}-*tert*-butyloxycarbonylamino acids.

A recent stepwise synthesis of somatostatin with isolation of intermediates in solution (Diaz *et al.*, 1979) has given another proof of the usefulness of alkali-labile protecting groups. The procedure made use of (*p*-phenylazobenzylsulfonyl)ethyloxy (OPse) for the protection of the COOH-terminal carboxyl group of cysteine and of 2-methylsulfonylethyloxycarbonyl (Msc) for the side chain of lysine. Only the cysteine side chain was protected with a non alkali-labile group, namely, acetamidomethyl (Acm). The *tert*-butyloxycarbonyl group was used for N^{α}-protection of the growing chain, except for the N^{α}-terminal alanine, where again Msc was preferred. Yields were high at every coupling and deprotecting step. The fully protected peptide: Msc-Ala-Gly-Cys(Acm)-Lys(Msc)-Asn-Phe-Phe-Trp-Lys(Msc)-Thr-Phe-Thr-Ser-Cys(Acm)-OPse was obtained in chemically pure form. At the end of the synthesis, this tetradecapeptide was partially deprotected in one step with a strong base (Eberle *et al.*, 1975) during 5 min at room temperature to give [Cys(Acm)^{3}, Cys(Acm)^{14}]-somatostatin. This synthesis is an illustration of the ideal combination of the *tert*-butyloxycarbonyl group for N^{α}-protection with the Msc and OPse groups, which appear to be completely stable under the conditions of Boc cleavage.

V. COMPATIBILITY OF PROTECTING GROUPS

By definition, protection in peptide synthesis must be reversible: protecting groups are only an artificial means of orienting the synthesis, and naturally occurring peptides are free of them. Hence, the blocking groups have to be removed either during the extension of the chain (*temporary protecting groups*) or after completion of the synthesis (*semipermanent protecting groups*). The central problem is, of course, to achieve selectivity in deprotection. Conditions have to be established for each temporary group such that it can be removed completely without affecting the semipermanent groups.

Such a selective cleavage is not possible for groups which have similar stabilities towards a given deprotection method. Thus, none of the three groups 2-(4-biphenylyl)propyl(2)oxycarbonyl (Bpoc), 2-nitrophenylsulfenyl (Nps), or 2-(3,5-dimethyloxyphenyl)propyl(2)oxycarbonyl (Ddz) can be removed selectively by acidolysis. Similarly, the benzyl ester and the benzyloxycarbonyl groups are not distinguishable by catalytic hydrogenolysis (see Wünsch, 1974). Use of different mechanisms may be an alternative e.g., thiolysis for Nps, photolysis for Ddz, or alkaline hydrolysis for benzyl ester cleavage. In other cases, the order in which two groups can be removed selectively is predetermined. Thus, tosyl cannot be removed before the benzyl ester, nor Boc before Bpoc, nor the methyl ester before the methylsulfonylethyloxycarbonyl (Msc) group. This state of affairs clearly delineates which of the two groups can be used as a semipermanent and which as a temporary group in the same segment. Finally, the situation may arise where two or more methods exist for differential cleavage of two groups. Examples are the benzyloxycarbonyl group and its 4-methyloxy derivative (Moz) which can

Table III Selective Deprotection Procedures

Semipermanent protecting group	Amine				
	Bpoc	Ddz	Azc	Trt	Nps
NH_2					
Boc	dTFA[b]	dTFA[b]		H^+, TFE pH 4[n]	CH_3CSNH_2[x] NTBA[d]
Z	dTFA[b]	dTFA[b]	dTFA[o]	TFA[a]	HCl[y]
2,4 Dcz	dTFA[b]	dTFA[b]	dTFA[o]	TFA[b]	HCl[b]
Tfa	dTFA[a]	dTFA[cc]	dTFA[c]	TFA[a]	HCl[r]
Msc, Tsc	dTFA[c]	dTFA[c]	dTFA[o]	TFA[c]	HCl[c]
Pht	dTFA[aa]	dTFA[cc]	dTFA[c]	TFA[a]	HCl[c]
Tos	dTFA[aa]	dTFA[cc]	dTFA[c]	TFA[a]	HCl[c]
COOH					
OMe. OEt	aq AcOH[n]	dTFA[cc]	dTFA[o]	aq AcOH[n]	HCl[y]
O*t*Bu	dTFA[b]	dTFA[e]		H^+, TFE pH 4[n]	CH_3CSNH_2[r] NTBA[d]
OBzl	HCl[i]	dTFA[e]	dTFA[c]	TFA[a]	HCl[y] CH_3CSNH_2[u]
OMse, OTse, OPse	dTFA[k]	dTFA[c]	dTFA[o]	TFA[c]	HCl[c]
OPh	HCl[s]	dTFA[c]	dTFA[c]	TFA[c]	HCl[c]
OH(Ser, Thr)					
Bzl	HCl[i]	dTFA[l]	dTFA[c]	TFA[a]	HCl–AcOH[z]
*t*Bu	aq AcOH[n]	dTFA[e]		aq AcOH[n]	HBr–MeOH[m]
2 Cb	dTFA[c]	dTFA[c]	dTFA[c]	TFA[c]	HCl[c]
OH (Tyr)					
Bzl	HCl[i]	dTFA[c]	dTFA[c]	TFA[a]	HCl[c]
*t*Bu	aq AcOH[n]	dTFA[e]		H^+, TFE pH 4[n]	HBr–MeOH[m]
Dcb	dTFA[c]	dTFA[c]	dTFA[c]	TFA[c]	HCl[c]
N^g					
NO_2	dTFA[i]	dTFA[cc]	dTFA[c]	TFA[c]	HCl–AcOH[z] CH_3CSNH_2[u]
Tos	HCl[i]	dTFA[cc]	dTFA[c]	TFA[c]	HCl[c]
N^{im}					
Bzl	dTFA[i]	TFA[gg]	TFA[gg]	TFA[gg]	TFA[gg]
Tos	dTFA[c]	dTFA[cc]	TFA[gg]	TFA[gg]	HCl[a] NTBA[d]
Dnp	dTFA[c]	dTFA[e]	dTFA[o]	TFA[c]	HCl[c]
SH					
Bzl	dTFA[c]	dTFA[c]	dTFA[c]	TFA[a]	HCl[c] NTBA[d]

nporary protecting group

			Carboxyl			
Boc	Moz	Z	O*t*Bu	OMe, OEt	OBzl	OTse, OPse[pp]
	H_2(Pd) in liq NH_3[v]	H_2(Pd)[a]		OH^- [a]	H_2(Pd) in liq NH_3[v]	*m*OH^- [j,k]
$_3$ etherate[g]	TFA[t]		TFA[dd]	OH^- [hh]	OH^- [a]	*m*OH^- [j]
A[l]	TFA[b]		TFA[b]	OH^- [c]	OH^- [c]	*m*OH^- [c]
A[a]		H_2(Pd)[r]				
OOH[w]	TFA[a]	HBr–AcOH[bb]	TFA[a]		H_2(Pd)[a]	
A[j]						
l–HCOOH[j]	TFA[c]	H_2(Pd)[j]	TFA[c]		H_2(Pd)[c]	
l[oo]	TFA[a]	HBr–AcOH[a]				
A[a]	TFA[a]	H_2(Pd)[a]	TFA[a]		H_2(Pd)[a]	
		HBr–TFA[ff]	TFA[a]	OH^- [ll]	H_2(Pd)[a]	*m*OH^- [c]
A[a]	TFA[a]	H_2(Pd)[a]	TFA[a]		H_2(Pd)[a]	
	H_2(Pd) in liq NH_3[v]	HBr–AcOH[a]	TFA[a]	OH^- [a]	H_2(Pd) in liq NH_3[v]	*m*OH^- [j,k]
		H_2(Pd)[a]				
A[l]	TFA[t]		TFA[l]			
l–HCOOH[j,k]	TFA[c]	HBr–AcOH[j]	TFA[j,k]			
l[s]	TFA[c]	H_2(Pd)[s]	TFA[c]		H_2(Pd)[c]	
A[l]	TFA[a]		TFA[l]	OH^- [jj]	OH^- [a]	*m*OH^- [c]
	H_2(Pd) in liq NH_3[v]	H_2(Pd)[a]		OH^- [n]	H_2(Pd) in liq NH_3[v]	*m*OH^- [c]
A[l]	TFA[c]		TFA[l]	OH^- [c]	OH^- [c]	*m*OH^- [c]
l–AcOH[ee]	TFA[a]		TFA[a]		OH^- [a]	*m*OH^- [c]
	H_2(Pd) in liq NH_3[v]	H_2(Pd)[a]		OH^- [n]	H_2(Pd) in liq NH_3[v]	*m*OH^- [c]
A[l]	TFA[c]		TFA[l]		OH^- [c]	*m*OH^- [c]
A[f]	TFA[t]	HBr–AcOH[dd]	TFA[a]	OH^- [hh]	OH^-	*m*OH^- [c]
A[f]	TFA[t]	H_2(Pd)[t]	TFA[a]	OH^- [ll]	H_2(Pd) in liq NH_3[v]	*m*OH^- [c]
A[gg]	TFA[gg]		TFA[a]	OH^- [a]	OH^- [a]	*m*OH^- [c]
A[l]	TFA[c]		TFA[l]	OH^- [a]	OH^- [a]	*m*OH^- [c]
A[l]	TFA[c]	HBr–TFA[mm] H_2(Pd)[nn]	TFA[l]		H_2(Pd)[nn]	
FA[g]	TFA[kk] H_2(Pd) in liq NH_3[v]	H_2(Pd) in liq NH_3[v]	TFA[a]	OH^- [ll]	H_2(Pd) in liq NH_3[v]	*m*OH^- [c]

(continued)

Table III (*Continued*)

Semipermanent protecting group	Amine				
	Bpoc	Ddz	Azc	Trt	Nps
Mob	dTFA[c]	dTFA[c]	dTFA[c]	TFA[c]	HCl[c] CH_3CSNH_2[u]
Meb	dTFA[c]	dTFA[c]	dTFA[c]	TFA[c]	HCl[c]
Trt	aq AcOH[n]	TFA[cc]		aq AcOH[n]	
Acm	aq AcOH[n]	dTFA[e]	dTFA[o]	aq AcOH[n]	CH_3CSNH_2[r] HCl[c]
Amide					
NH_2	aq AcOH[n]	dTFA[c]	dTFA[c]	TFA[a]	HCl[c]
Ddm	dTFA[c]	dTFA[e]	dTFA[c]		

[a] Schröder and Lübke (1965, pp. 72–75).
[b] See Table I, Section III,A.
[c] Not yet tested.
[d] Juillerat and Bargetzi, 1976.
[e] Birr, 1973; Trudelle and Birr, 1975.
[f] Yamashiro *et al.*, 1972a.
[g] Gutte and Merrifield, 1971.
[h] König and Geiger, 1970.
[i] Blake and Li, 1973; Wang and Merrifield, 1971.
[j] Eberle *et al.*, 1975; Tesser, 1975; J. L. Fauchère, unpublished results.
[k] Diaz *et al.*, 1979.
[l] Erickson and Merrifield, 1973b, 1974.
[m] Wünsch, 1967.
[n] Riniker *et al.*, 1969; Kamber and Rittel, 1969; Sieber *et al.*, 1977.
[o] Tun-Kyi and Schwyzer, 1976.
[p] Veber *et al.*, 1972.
[q] Schnabel *et al.*, 1971.
[r] Borin *et al.*, 1976; Moroder *et al.*, 1973.
[s] Galpin *et al.*, 1976; Kenner and Seely, 1972.
[t] Koyama *et al.*, 1976b; Ogawa *et al.*, 1976.
[u] Tomatis *et al.*, 1973.
[v] Meienhofer and Kuromizu, 1974.
[w] Tomatis *et al.*, 1977; Guggi *et al.*, 1976.

be removed either by catalytic reduction or by acidolysis in the presence of the *N*-toluenesulfonylethyloxycarbonyl (Tsc), *N*-methylsulfonylethyloxycarbonyl (Msc), and *S*-acetamidomethyl (Acm) groups.

One of the most versatile ways of achieving selectivity in deprotection is differential solvolysis. By using both acidic and basic conditions, a broad range of stabilities can be covered. It was shown in Sections III and IV that

Temporary protecting groups

			Carboxyl			
Boc	Moz	Z	O*t*Bu	OMe, OEt	OBzl	OTse, OPse[pp]
FA[l]	TFA[c]		TFA[l]	OH^-[c]	OH^-[c]	*m*OH^-[c]
Cl–EtOH[ee]						
FA[l]	TFA[c]		TFA[l]	OH^-[c]	OH^-[c]	*m*OH^-[c]
F_3 etherate[q]				OH^-[n]	OH^-[a]	*m*OH^-[c]
FA[p]	TFA[c] H_2(Pd) in liq NH_3[v]	H_2(Pd) in liq NH_3[v] HF[p]	TFA[p]	OH^-[p]	H_2(Pd) in liq NH_3[v]	*m*OH^-[k]
FA[a]	TFA[t]	H_2(Pd)[t]	TFA[a]	OH^-[m]	H_2(Pd)[a]	*m*OH^-[j]
	H_2(Pd)[c]	H_2(Pd)[r]		OH^-[h]	H_2(Pd)[w]	*m*OH^-[c]

[x] Kessler and Iselin, 1966.
[y] Vičar *et al.*, 1976.
[z] Najjar and Merrifield, 1966.
[aa] W. Rittel, personal communication.
[bb] Anfinsen *et al.*, 1967.
[cc] (1% in CH_2Cl_2.) C. Birr, personal communication.
[dd] Inouye *et al.*, 1976.
[ee] Noda *et al.*, 1971.
[ff] Shimizu *et al.*, 1976.
[gg] Fujii and Sakakibara, 1974.
[hh] Yajima *et al.*, 1976a.
[jj] Ivanov *et al.*, 1976.
[kk] Yajima *et al.*, 1974.
[ll] Siemion *et al.*, 1973.
[mm] Chillemi and Merrifield, 1969.
[nn] Stewart *et al.*, 1972.
[oo] Schwyzer *et al.*, 1963.
[pp] Abbreviations: Azc, 2-(*p*-phenylazophenyl)isopropyloxycarbonyl; OTse 2-(4-tolylsulfonyl)ethyl ester; OPse, 2-(*p*-phenylazobenzylsulfonyl)ethyl ester; OMse, 2-methylsulfonylethyl ester; Tsc, 2-(*p*-tolylsulfonyl)ethyloxycarbonyl; Msc, 2-methylsulfonylethyloxycarbonyl; Ddm, 4,4′-dimethoxybenzhydryl; dTFA, dilute TFA; TFE, 2,2,2-trifluorethanol; NTBA, 3-nitro-4-mercaptobenzoic acid; *m*OH^-, mild alkaline conditions, pH < 11.

many stability levels can be utilized in this range, allowing a sequential removal of blocking groups.

Two stability levels also exist with respect to reduction methods. Groups of the lower level such as *N*-benzyloxycarbonyl and substituted derivatives thereof, benzyl ester and its substituted derivatives, *S*-benzyl, *S*-4-nitrobenzyl (Schröder and Lübke, 1965, pp. 72–75), N^g-nitro (Schröder and Lübke, 1965,

p. 169), *N*-allyloxycarbonyl (Schröder and Lübke, 1965, p. 35), *N*-1,1-dimethyl-2-propynyloxycarbonyl (Southard *et al.*, 1971), and *N*-piperidinoxycarbonyl (Stevenson and Young, 1969) are cleavable by catalytic hydrogenolysis. Groups of the higher level require the more drastic reduction by sodium in liquid ammonia. These include the N^{ω}-tosyl groups of arginine and lysine and the *S*-benzyl group of cysteine. While for Lys(Tos) reduction by Na in liquid NH_3 is the obligatory procedure (Mazur and Plume, 1968), treatment by HF is preferred for tosylarginine and benzylcysteine.

An alternative to the use of stability levels toward a single cleavage method is the successive application of two different mechanisms for splitting. Some possible combinations are acidolysis followed by reduction, and vice versa, acidolysis followed by alkaline hydrolysis or elimination, or acidolysis followed by a special procedure such as photolytic, enzymatic, or ionolytic cleavage (see Section II).

Temporary protecting groups generally block the functions at which the chain is to be extended, i.e., the NH_2-terminal amine and the COOH-terminal carboxyl groups, whereas semipermanent groups protect the side chain of trifunctional residues. Following this classification, a number of protecting groups are considered and their labilities in defined media compared in Table III. Conditions under which a temporary group can be selectivity split off in the presence of a given semipermanent group are given. Reference is made in most cases, either to Schröder and Lübke (1965, Table I, pp. 72–75; some of the data that are no longer considered valid have been corrected) or to the more recent literature. Blanks in the table mean that no method is known (to the authors!) where selective cleavage was achieved.

The value of such summarizing tables has occasionally been questioned. It should be stressed that selectivity has to be tested in each individual case as it depends strongly on the nature of the peptide to be deprotected and on the exact conditions used for splitting. The statement in Table III that a given group can be removed in the presence of another one is strictly valid only in the particular example quoted.

Only a limited number of groups could be incorporated into Table III. Data on the selectivity during cleavage of the huge number of groups not mentioned here can be found in the chapters describing the individual protecting groups.

VI. DIFFERENTIAL PROTECTION

The development of a protection scheme for the synthesis of a given peptide is governed by a number of factors that have been partially treated in the preceding sections. Whenever a protecting group is to be utilized, it is very helpful to have a summary (e.g., Table III, Section V) giving conditions under which the selective removal of widely used groups is possible. The other factors that must be taken into account range from the nature of the methods available for deprotection (Section II) and the possible side reactions of sensitive amino acid residues (Section II) to the compatibility of the chosen protecting groups with the coupling procedures involved. Although the number of the available groups is rather high, and although very often many different methods can be applied for the splitting of a single group, these factors rapidly restrict the possible combinations and the elaboration of a flexible scheme can be a considerable problem.

Depending on the functional groups to be blocked or left unprotected, synthesis schemes will vary between the two extreme tactics of maximal or minimal protection. Minimal protection, i.e., blocking of N^{α}- and N^{ε}-amine groups and of the thiol function of cysteine only, has proven possible in several instances (Hirschmann and Veber, 1973; Hofmann *et al.*, 1969). Maximal protection, i.e., protection of all functional groups that could potentially undergo a side reaction, has also been used in many syntheses in solution (Wünsch *et al.*, 1968; Ivanov *et al.*, 1976). It is the accepted procedure in solid-phase synthesis, where even the side chains of asparagine and glutamine are often protected (Stewart *et al.*, 1976) and sometimes those of methionine (Gutte and Merrifield, 1971) and tryptophan (Stewart *et al.*, 1976) are blocked.

The synthesis of a polypeptide is frequently achieved in two different stages. First, protected oligopeptide intermediates are synthesized independently in a stepwise manner; i.e., starting at the carboxylic end, one amino acid is added at a time. Stepwise elongation of the peptide from the NH_2-terminal end is also possible (e.g., Matsueda *et al.*, 1975) but is generally impaired by greater disadvantages. As stepwise synthesis is only applicable for fragments of moderate length, a different technique is brought into play in the second stage. Use is then made of a coupling method for the condensation of the segments. Final deprotection yields the desired peptide. Exceptions are known where use was made exclusively of either the stepwise approach, as in the synthesis of the 27-peptide secretin by Bodanszky *et al.* (1967), or of the condensation of dipeptidic segments, as in the synthesis of angiotensin (Schwyzer *et al.*, 1958) and of the decapeptide LHRH (Geiger *et al.*, 1971).

For the general strategy of stepwise synthesis followed by segment condensation, the selection of the temporary and semipermanent protecting groups is determined by both the method chosen for final deprotection and by the procedures involved for coupling the segments. Some strategies of selection are analyzed in the following Section, first for synthesis in solution, then for solid-phase synthesis, with reference to syntheses already achieved.

A. Protection Schemes for Synthesis in Solution

The final deprotection step in the synthesis of a polypeptide is a very complex process simultaneously involving many different reactions at many different sites. Furthermore, deprotection is very often accompanied by the formation of reactive intermediates, such as carbonium ions in the acidolysis of urethane-type protecting groups (Sieber and Iselin, 1968) or radicals in photolysis (e.g., Amit *et al.*, 1977), or by the formation of active species such as acylating or thioacylating agents, e.g., during removal of the 2-nitrophenylsulfenyl group (Yamashiro and Li, 1973). The use of scavengers such as anisole, phenol, indole, tryptophan, dimethyl sulfide, mercaptoethanol, or ethanedithiol only partially protects the peptide from side reactions. For these reasons, great attention must be devoted to this last step and the synthesis has to be planned in such a way as to render final deprotection easy and safe.

In the early days of peptide synthesis, Schwyzer (1966) proposed a simplifying principle, according to which the semipermanent groups should all be of the *tert*-butyl type, thereby allowing a simultaneous and mild deprotection by pure or slightly diluted trifluoroacetic acid as the final step. This strategy is often still used, with some variations in the choice of the N^{α}-protection and of the coupling method for the fragments.

Coupling methods are always accompanied by a risk of racemization when the COOH-terminal amino acid to be activated is not glycine or proline. The azide coupling method which is generally safe, requires blocking of the COOH-terminal carboxyl group by the methyl ester and eventual hydrazinolysis. However, in order to avoid exposure of the protected peptide to hydrazine before coupling, the methyl ester has sometimes been replaced by *tert*-butyloxycarbonylhydrazide (e.g., Camble and Cotton, 1975). This kind of protection is, of course, not compatible with semipermanent *tert*-butyl derivatives in the side chains, whereas benzyloxycarbonylhydrazide would be (Guttman *et al.*, 1975). When another coupling method is selected, the COOH-terminal carboxyl protector must generally be removed before condensation, a fact which has to be considered in conjunction with the nature of the side-chain blocking groups. In this context, the trimethylsilyl ethyl ester (OTms) can be useful as terminal carboxyl protector as it is cleavable by fluoride ions and, in contrast to the methyl ester, does not require alkaline conditions for its hydrolysis (Riniker *et al.*, 1979).

Another important factor in the elaboration of a protecting scheme is the presence of disulfide bridges in the desired peptide. Synthesis is planned to allow formation of the bridges either in each segment before condensation, in the protected peptide immediately before final deprotection, or even as the last step of the synthesis after removal of the semipermanent groups.

Syntheses of the biologically active compounds listed in Table IV were achieved using only groups of the *tert*-butyl-type for semipermanent protection. An exception is the adamantyloxycarbonyl group (Adc) in protected glucagon and proinsulin connecting peptide: this group, although not of the *tert*-butyl-type, is still removable under the same conditions. The side chains of arginine and histidine are generally left unprotected; so are those of serine, threonine and tyrosine when not protected by the *tert*-butyl ether group.

Variations in these schemes come mainly in the choice of different groups for N^{α}-protection. The *tert*-butyloxycarbonyl group can be used in segments as long as no other groups of the same type are introduced in the side chains or in the COOH-terminus (e.g., Schwyzer and Karlaganis, 1973). When such groups are present, a different protecting agent, removable either acidolytically under milder conditions such as the 2-(4-biphenylyl)propyl(2)oxycarbonyl (Bpoc) group or another mechanism (e.g., hydrogenolysis of benzyloxycarbonyl) must be employed. Extensive use has been made of the benzyloxycarbonyl group for N^{α}-protection because its removal by catalytic reduction generally gives rise to fragments in a state of high purity (Morley, 1967; Sieber *et al.*, 1972). Many syntheses have involved very acid-labile groups such as Bpoc and Trt (Kamber and Rittel, 1969; Andreatta *et al.*, 1973; Sieber *et al.*, 1977) or Nps (Wünsch, 1967). Nevertheless, it must be kept in mind that, for example, Bpoc is only about 3000 times less stable than Boc, a value which is not sufficient for complete selectivity of Bpoc-splitting (Table I, Section III) in more than one step.

Another widely accepted principle is the utilization of very stable groups for side-chain protection. This allows more flexibility in the choice of temporary groups and eliminates branching at the side chains during solid-phase synthesis, where the intermediates cannot be purified. The advantage of this strategy lies in the possibility of the ideal combination of the acid-stable semipermanent groups with acid-labile groups in the stability range of the *tert*-butyloxycarbonyl group for N^{α}-protection. The main disadvantage is that final deprotection requires exposure of the peptide to very strong reagents such as anhydrous HF or sodium in liquid ammonia, which greatly enhances the risk of side reactions (Section II).

Following this scheme, Koyama *et al.* (1976b), synthesized human corticotropin (ACTH) of the revised sequence, using Lys(Z), Arg(NO_2), Arg(Tos), Glu(OBzl), Asp(OBzl) as well as COOH-terminal benzyl ester and NH_2-terminal 4-methyloxybenzyloxycarbonyl (Moz) protection. Serine, tyrosine, histidine, methionine and asparagine were left unprotected. The Moz group

Table IV Differential Protection Providing *tert*-Butyl Derivatives of the Final Protected Polypeptide[a]

Peptide synthesized	Amino acid residues	Intermediate segments	α-Amine protection		α-Carboxyl protection		Thiol protection (Cys)	References
			Terminal	Growing chain	Terminal	Growing chain		
Porcine α-thyrocalcitonin	32	5	Boc	Z, Bpoc, Trt	NH_2	OMe, NHNHBoc	Trt	Kamber and Rittel, 1969; Riniker *et al.*, 1969
Human ACTH	39	4	Boc	Z	O*t*Bu	OMe		Sieber *et al.*, 1972
Human parathyroid hormone	34	6	Boc	Bpoc, Z, Trt	O*t*Bu			Andreatta *et al.*, 1973
Human insulin	51	3	Boc	Bpoc, Trt	O*t*Bu	OMe	Trt, Scm[c], Acm	Sieber *et al.*, 1977
Human gastrin	17	3	pGlu	Z, Boc	NH_2	OMe		Morley, 1967
Glucagon	29	5	Adc	Z, Nps	O*t*Bu	OMe		Wünsch, 1967
Proinsulin connecting peptide[b]	33	6	Adc	Z, Boc	O*t*Bu	OMe		Geiger *et al.*, 1969
S-Peptide	20	3	Boc	Boc	OH	NHNHZ		Hofmann *et al.*, 1969
Motilin	22	6	Boc	Z, Boc	O*t*Bu	OMe		Wünsch *et al.*, 1976
Scotophobin analogue[b]	15	5	Boc	Z, Boc	NH_2	O*t*Bu, NHNHZ		Guttmann *et al.*, 1975

[a] Exceptions: Adc in glucagon and proinsulin connecting peptide.
[b] Not biologically active.
[c] Abbreviations: Scm, S—CO—OCH_3.

was removed by trifluoroacetic acid with anisole as scavenger, and deblocking of all protected functions of the 39-peptide ester was performed by HF in the presence of ethanedithiol. The same scheme was applied with success to the synthesis of porcine (Yajima *et al.*, 1976a) and bovine ACTH (Koyama *et al.*, 1976a) as well as to the synthesis of the 33-peptide amide corresponding to the sequence of desulfated porcine cholecystokinin–pancreozymin (Yajima *et al.*, 1976b). Porcine motilin was obtained (Shimizu *et al.*, 1976) in a similar way with the N^{ω} groups of arginine and lysine protected by the tosyl group. The benzyl ester protection of glutamic acid was only employed during coupling and removed at the next step together with the N^{α}-benzyloxycarbonyl group. Final deblocking was carried out in one step by treatment with sodium in liquid ammonia. Use of the 2,6-dichlorobenzyl group for tyrosine, of the benzyl group for serine and of the 2-chlorobenzyloxycarbonyl group for lysine in the synthesis of the COOH-terminal 11-peptide of vasointestinal peptide (VIP) (Bodanszky *et al.*, 1976) allowed the final deprotection to be performed under milder conditions, namely by catalytic hydrogenation. The synthesis of human fibrinopeptide A (Channabasavaiah and Sivanandaiah, 1976) with the help of Glu(OBzl), Asp(OBzl), Ser(Bzl) and COOH-terminal benzyl ester is another such example. α-Bungarotoxin, with 74 amino acid residues, one of the longest peptide yet whose synthesis was attempted in solution, was also obtained by the method of maximal (hydrophobic) protection (Ivanov *et al.*, 1976). The protecting groups used were: benzyl ether for serine and threonine, *p*-methoxylbenzyl for cysteine, 2,4-dichlorobenzyl for tyrosine, N^{g}-tosyl (for arginine, benzhydryl for asparagine and glutamine, and the benzyl ester for glutamic and aspartic acid. These hydrophobic groups led to sufficient retention of solubility of the four segments, which were built up in a stepwise manner using the *tert*-butyloxycarbonyl group as the temporary protection for α-amine groups. However considerable difficulties were encountered during the removal of the side-chain protecting groups by HF leading to extensive polymerisation and affording practically no toxic material (Ivanov *et al.*, 1979). The success of the synthesis will depend on the ease of the removal of these side-chain protecting groups by HF and of the formation of the five disulfide bridges, which have not yet been described.

In addition to the two main methods that have been mentioned, there are of course other strategies that can be useful for special purposes. An interesting example is the procedure involving derivatives of the 4-picolyl group for the protection of amine, hydrazide, hydroxylic, and carboxylic functions, as in the synthesis of fragments of the porcine gastric inhibitory peptide (Macrae and Young, 1975). These groups can be used in conjunction with *tert*-butyloxycarbonyl for N^{α}-protection and can be removed by catalytic hydrogenolysis or electrolytic reduction. The main advantage is that they provide a weakly basic handle by which the synthetic polypeptide can be purified at each step on an acidic ion-exchanger (Young, 1973).

Another important route is the assembly of segments protected in the side chains and at the COOH-terminus by groups removable via β-elimination in weak alkaline media. Some applications have already been reported in Section IV and their number is bound to increase in the future. The use of groups removable by β-elimination as temporary protection for the N^{α}-amine function should also prove feasible and has probably only been retarded by the fact that the appropriate N^{α}-protected amino acids are not commercially available. Successful use of the alkali-labile phenyl ester for COOH-terminal protection combined with adamantyloxycarbonyl (Adc) and *tert*-butyl derivatives for side-chain protection has been reported for the synthesis of a 75-residue lysozyme fragment (Galpin *et al.*, 1976) and of human big gastrin I (Choudhury *et al.*, 1976).

B. Protection Schemes for Solid-Phase Peptide Synthesis

1. The Bond between Peptide Chain and Polymer Support

From conventional methods in solution to solid-phase synthesis, it is evident that the main new element is the nature of the attachment of the polypeptide to the resin. In the synthesis of deprotected polypeptides, this bond can generally be in the same stability range as the semipermanent groups and be cleaved with them in the same operation. If a protected peptide has to be synthesized for segment condensation, the bond to the resin requires additional selectivity level for deprotection. On one hand, the stability of this bond has to be assured during the growing of the polypeptide chain; on the other hand, the protected peptide must be cleavable from the resin under conditions which do not attack the semipermanent groups. Another major restriction in the development of protecting schemes in solid-phase synthesis is the fact that catalytic hydrogenolysis cannot be employed as long as the peptide is attached to the resin. For this reason, the heterogeneous part of solid-phase synthesis, mainly relies on differential acidolysis for deprotection. Depending on the bond involved, the cleavage from the resin is made either by *strong acidolysis* [HBr in TFA, liquid HF, boron tris(trifluoroacetate) in TFA] for the conventional benzyl ester or by *mild acidolysis* (50% TFA in CH_2Cl_2) for the *tert*-alkyl ester (Wang and Merrifield, 1969), the *tert*-alkyloxybenzyl ester (Wang and Merrifield, 1971), and the 4-alkoxybenzyl ester (Wang, 1972). As already stated in Section IV, *alkaline hydrolysis*, *aminolysis*, and *hydrazinolysis* have also been employed. It is of strategic importance to notice that the two last methods liberate peptides that are derivatized at the carboxyl terminus, either with an amide function (many natural peptides occur as amides) or with a hydrazide function that can be converted to the azide for further coupling. The synthetic polypeptide is also obtained in a derivatized form, as an ester, when base-catalyzed *transesterification* is used for cleavage (Beyerman, 1972; Barton *et al.*, 1973).

2. Synthesis of Unprotected Peptides

The only requirement for solid-phase synthesis of an unprotected peptide is that the removal of the N^{α}-protecting group can occur under conditions that leave both the link to the resin and all side-chain protecting groups intact. The most common protection scheme employs the *tert*-butyloxycarbonyl group to block the α-amine function, together with the 4-alkylbenzyl ester link to the resin [mostly copoly(styrene–divinylbenzene)]. Final treatment of the substituted resin by liquid HF in the presence of scavengers liberates a fully deprotected polypeptide. Erickson and Merrifield (1974) have proposed a combination of protecting groups that is compatible with the conditions used for N^{α}-deprotection, i.e., 50% TFA in CH_2Cl_2. These authors have shown that the tyrosine benzyl ether [Tyr(Bzl)] and *O*-benzyloxycarbonyl [Tyr(Z)] protection, the lysine N^{ε}-benzyloxycarbonyl [Lys(Z)] and 4-chlorobenzyloxycarbonyl [Lys(4Cz)], the cysteine *S*-4-methyloxybenzyl [Cys(Mob)], and the serine benzyl ether [Ser(Bzl)] protection, which are too labile should be replaced by the sufficiently stable Tyr(Dcb) [2,6-dichlorobenzyl], Lys(3Cz) [N^{ε}-3-chlorobenzyloxycarbonyl], Cys(Meb) [*S*-4-methylbenzyl], and Ser(2Cb) [2-chlorobenzyl ether]. Although N^{im}-tosyl- and N^{im}-2,4-dinitrophenylhistidine are both stable enough in 50% TFA, the tosyl group is preferable (Fujii and Sakakibara, 1974) because it can be removed by liquid HF, the 2,4-dinitrophenyl group requires a separate thiolysis step for splitting (Lin *et al.*, 1972). This combination of protecting groups is generally completed with Arg(Tos) or Arg(NO_2), Asp(OBzl), Glu(OBzl), Thr(Bzl), Gln(Mbh), Asn(Mbh), and sometimes with Trp(For) and Met(O). Substituted benzyl esters and benzyl ethers such as the 4-bromobenzyl ester (OBb) for aspartic acid, the 4-bromobenzyl ether (4Bb) for serine and threonine (Yamashiro, 1977), or the 4-chlorobenzyl ester (OCb) and 4-nitrobenzyl ester (ONb) for glutamic and aspartic acid (Prestidge *et al.*, 1976) are recommended as being more stable than the unsubstituted groups. Maximal protection is a general feature of solid-phase synthesis because of both the excess of carboxylic component used and the lack of purity control during the progress of the synthesis.

Representative syntheses carried out according to this scheme (with slight modifications in some cases) are ribonuclease A (Gutte and Merrifield, 1971), lysozyme (with unprotected His) (Sharp *et al.*, 1973), cobratoxin [with Cys(Mob)] and His(Z) (Aoyagi *et al.*, 1972), β-endorphin (with His(Boc), Lys(2Bz) and Tyr(2Bz) (Li *et al.*, 1976), somatostatin [with Cys(Mob)] (Rivier, 1974), β-melanotropin, and analogues [with Trp(For) and His(Boc)] (Lemaire *et al.*, 1977).

If very acid-labile groups, such as Bpoc, Ppoc, Ddz, Bmv, or Nps, are used for N^α-protection, more labile groups can also be selected for the side chains, including groups of the *tert*-butyl type. The advantage of this scheme lies in the mildness of the conditions required for N^α- and final deprotection. Unsubstituted benzyloxycarbonyl, benzyl esters, and benzyl ethers as side-chain protectors are then completely stable, as shown in the synthesis of a fragment of growth hormone with 42 residues (Blake and Li, 1973). Furthermore, the easily removable 4-alkoxybenzyl ester link to the resin has been shown to be sufficiently resistant during the growing of the peptide chain (Wang *et al.*, 1974). Although the combination of very labile groups for the α-amine-functions (e.g., Ddz; Trudelle and Birr, 1975) with *tert*-butyl derivatives for side-chain protection has proven possible, selectivity may not be completely achieved, especially in repetitive processes (see Table I, Section III). Use of some other mechanism than acidolysis for N^α-deprotection, such as thiolysis for Nps or photolysis for Ddz, may add to the advantages of this scheme.

3. *Synthesis of Protected Peptides*

The synthesis of large peptides very often requires the preparation of protected segments in a stepwise manner, followed by the assembly of the segment, generally in solution. In another approach, the segments are coupled sequentially to a growing chain attached to a solid support. Examples are the synthesis of bovine pancreatic trypsin inhibitor (Yajima *et al.*, 1974) and of the protected sequence 1 to 47 of urogastrone (Camble and Petter, 1976).

Three stability levels are needed as N^α-deprotection; cleavage from the resin and final side-chain deprotection must be performed sequentially in a selective manner. One possible scheme involves labile groups such as Bpoc, Ddz, Ppoc, Bmv, or Nps for N^α-protection (splitting with dilute trifluoroacetic acid [0.5 to 2%) in CH_2Cl_2], a 4-alkoxybenzyl ester link to the resin (50% TFA in CH_2Cl_2 for cleavage from the resin), and side-chain protecting groups that are not removed by this reagent (see Section VI,B,2). After coupling of the fragments, the remaining groups are removed either by catalytic hydrogenation, by sodium in liquid ammonia or, more often, by liquid HF. Thus, the two fragments Z-Asp(Bzl)-Arg(NO_2)-Val-Tyr(Bzl)-

$NHNH_2$ and Val-His(Bzl)-Pro-Phe were prepared on a *tert*-alkyloxycarbonylhydrazide and a 4-alkoxybenzyl ester resin, respectively, to give angiotensin II after coupling in solution and removal of the protecting groups by HF (Wang, 1972).

Solid-phase synthesis of protected segment is still a new field. After its introduction by Merrifield (1962), the general opinion was that, with sufficient technical improvement, it would furnish peptides of any length using only the stepwise strategy. A number of achievements still tend to support this idea, and the advantages of a mixed strategy (stepwise solid-phase synthesis followed by coupling of the segments) are only slowly being realized. For this reason, only a limited number of theoretically possible schemes have been tested. Thus, the following four schemes, which should be practicable, have either never or only rarely been used. *Scheme 1:* side-chain groups which are not removed by HBr in TFA, cleavage by HF or sodium in liquid ammonia; N^{α}-Boc or more labile groups, cleavage by acidolysis; cleavage from the resin by HBr in TFA. *Scheme 2:* link to the resin is cleaved via a mechanism other than acidolysis, e.g., via alkaline hydrolysis (phenacyl ester or methylsulfonylethyl ester links (Section IV)); N^{α}-Boc; side-chain groups more stable than Z toward acidolysis; final deprotection by HF. *Scheme 3:* N^{α}-Msc or Tsc, cleavage via β-elimination under mild alkaline conditions; conventional benzyl ester link to the resin, cleavage by 50% TFA in CH_2Cl_2; side-chain stable groups cleaved by HF. *Scheme 4:* N^{α}-groups in the stability range of Bpoc or Ddz; side-chain groups in the stability range of Boc in acidic media; alkali-labile link to the resin.

For obvious reasons, mild schemes such as 4, which avoid completely the use of strong acids, are to be preferred insofar as their use is not connected with a lack of selectivity during deprotection.

The general impression emerges that, in spite of the numerous refinements in the design of new groups and in the use of deprotection methods, many potentially useful combinations of groups have still not been tested and some deprotection conditions, such as mild alkaline hydrolysis, still not optimally applied.

C. Orthogonal Protection

Making use of a mathematical term, Barany and Merrifield (1977) have introduced the notion of an orthogonal system of protecting groups. In such a system, each class of protecting groups is independant during deprotection, i.e., any protecting group of one class can be removed selectively (in any order) in the presence of any member of another class. If P_{μ} and P_{ν} are two optional blocking groups of the classes μ and ν respectively, the co-deprotection ϕ is zero for any method of cleavage applied to the orthogonal system.

$$\phi(P_\mu, P_\nu) = 0$$

Hence, if there are three classes, there must be three different cleavage procedures, and each class will be sensitive to one and resistent to the others.

Orthogonal combinations of groups may be found in some instances. One example has been given by the same authors (Barany and Merrifield, 1977) and makes use of the thiol-sensitive N^α-protector dithiasuccinoyl (Dts), of acid-labile *tert*-butyl-type protectors for the side chains and of the photolabile 2-nitrophenylester derivative (Rich and Gurwara, 1975) as link to the resin. A second example tested in solution involves the alkali-labile Msc- and OPse for side chains *and* COOH-terminal carboxyl, *tert*-butyloxycarbonyl for N^α-amine, and acetamidomethyl for cysteine side chains (Diaz *et al.*, 1979).

The question arises as to whether the ultimate goal of strategy in peptide synthesis would be to design protection schemes relying strictly on orthogonal systems. The concept of orthogonality of protecting groups is very attractive from a theoretical point of view, and it can stimulate the search for protectors that can be removed by one mechanism only. In addition, it has the merit of stressing the risk involved in methods relying on sensitivity differences towards acidolysis for deprotection. On the other hand, orthogonality is likely to represent an excessive requirement for the practice of synthetic strategy, as it does not exploit the possibilities offered by the predetermined order of deprotection. Hence, benzyloxycarbonyl (Z) and *tert*-butyloxycarbonyl do not belong to two orthogonal classes because Z is also sensitive to acids, while practice has amply proven usefulness and safety of this classical combination provided Z is the first of the two groups to be removed (by catalytic hydrogenation). Although strategy in peptide synthesis may tend to idealized schemes, such as one orthogonal system of blocking groups for one synthesis, it will have to put up for a long time to come with the restricted number of available groups and of deprotection methods. Use of different levels of stability towards acidolysis in a defined order may not be ideal but still represent in many instances a practicable route.

VII. CONCLUDING REMARKS

The assembling of amino acid residues into defined polypeptide chains without protecting groups is possible in biosynthetic processes. In this case, apart from the correct template sequence of nucleotides, it requires an impressive biosynthetic machinery. In the laboratory of the peptide chemist, a minimal recourse to protecting groups is necessary. However, this cumbersome detour permits the correct synthesis of defined sequences of natural or even unnatural amino acids.

The ideal choice of a strictly orthogonal combination of protecting groups for a given synthesis is not yet practicable in the general case. It is still advisable to rely on different levels of stability for selective deprotection. A general tendency is the choice of groups that are cleaved by mild procedures. If a hierarchy of the amino-protecting groups should be established, the next group in importance after benzyloxycarbonyl (Z) and *tert*-butyloxycarbonyl (Boc) would be one which is acidolytically very labile, probably 2-nitrophenylsulfenyl (Nps). The most widely employed COOH-terminal protection in solution are the methyl (OMe) and the *tert*-butyl (O*t*Bu) esters. The general admissibility of alkali-labile blocking groups is an important discovery of the last few years. Their involvement in protection schemes increases substantially the number of possible combinations.

There is still no ideal blocking group that is introduced easily and without racemization and that is stable during synthesis but readily removed at any desired step. There are no completely satisfactory protection means for the side chains of arginine and histidine, so that intense research is still needed in these fields. Other limitations come from the side reactions very often observed during final deprotection, a process where a number of reactions occur simultaneously and a number of active species are produced. A shift toward milder methods of cleavage may provide better results.

Finally, a serious problem is associated with the change of solubility of the lengthening chain, partly due to the presence of protecting groups. Attempts to use systematically lipophilic blocking can lead to difficulties during final deprotection, whereas the search for water-soluble protecting groups has not been very fruitful to date. Although no general solution is available, synthesis on resins substantially reduces this problem.

Differential protection and selective deprotection—now also largely employed for the chemical synthesis of polynucleotides—still remain the only efficient tools for the production of unequivocally defined polypeptides.

REFERENCES

Amaral, J. S. (1969). *J. Chem. Soc. C* pp. 2495–2497.

Amit, B., Zehavi, U., and Patchornik, A. (1974). *Is. J. Chem.* **12**, 103–113.

Amit, B., Hazum, E., Fridkin, M., and Patchornik, A. (1977). *Int. J. Pept. Protein Res.* **9**, 91–96.

Andreatta, R. H., Hartmann, A., Jöhl, A., Kamber, B., Maier, R., Riniker, B., Rittel, W., and Sieber, P. (1973). *Helv. Chim. Acta* **56**, 470–473.

Anfinsen, C. B., Ontjes, D., Ohno, M., Corley, L., and Eastlake, A. (1967). *Proc. Natl. Acad. Sci. U.S.A.* **58**, 1806–1811.
Aoyagi, H., Yonezawa, H., Takahashi, N., Kato, T., Izumiya, N., and Yang, C. C. (1972). *Biochim. Biophys. Acta* **263**, 823–826.
Atherton, E., Fox, H., Harkiss, D., Logan, C. J., Sheppard, R. C., and Williams, B. J. (1978a). *J. Chem. Soc. Chem. Comm.* p. 537–539.
Atherton, E., Fox, H., Harkiss, D., Sheppard, R. C. (1978b). *J. Chem. Soc. Chem. Comm.* p. 539–540.
Barany, G., and Merrifield, B. B. (1977). *J. Am. Chem. Soc.* **99**, 7363–7365.
Barton, M. A., Lemieux, R. U., and Savoie, J. Y. (1973). *J. Am. Chem. Soc.* **95**, 4501–4506.
Ben-Ishai, D. (1954). *J. Org. Chem.* **19**, 62–66.
Ben-Ishai, D., and Berger, A. (1952). *J. Org. Chem.* **17**, 1564–1570.
Bergmann, M., and Zervas, L. (1932), *Ber. Dtsch. Chem. Ges.* **65**, 1192–1201.
Beyerman, H. C. (1972). *In* "Progress in Peptide Research" (S. Lande, ed.), pp. 25–32 Gordon & Breach, New York.
Beyerman, H. C., and Bontekoe, J. S. (1962). *Recl. Trav. Chim. Pay-Bas* **81**, 691–698.
Bijl, W. A., van Nispen, J. W., and Greven, H. M. (1979). *In* "Peptides 1978" (I. Z. Siemion and G. Kupryszewski, eds.) pp. 569–572. Wroclaw Univ. Press, Wroclaw, Poland.
Birr, C. (1973). *Justus Liebigs Ann. Chem.* pp. 1652–1662.
Birr, C., Lochinger, W., Stahnke, G., and Lang, P. (1972). *Justus Liebigs Ann. Chem.* **763**, 162–172.
Blàha, K., and Rudinger, J. (1965a). *Collect. Czech. Chem. Commun.* **30**, 585–598.
Blàha, K., and Rudinger, J. (1965b). *Collect. Czech. Chem. Commun.* **30**, 599–604.
Blake, J., and Li, C. H. (1971). *Int. J. Pept. Protein Res.* **3**, 185–189.
Blake, J., and Li, C. H. (1973). *Int. J. Pept. Protein Res.* **5**, 123–125.
Blecher, H., and Pfaender, P. (1973). *Justus Liebigs Ann. Chem.* pp. 1263–1268.
Bodanszky, M., Ondetti, M. A. Levine, S. D., and Williams, N. J. (1967). *J. Am. Chem. Soc.* **89**, 6753–6757.
Bodanszky, M., Lin, C. Y., Yiotakis, A. E., Mutt, V., and Said, S. I. (1976). *Bioorg. Chem.* **5**, 339–350.
Boissonnas, R. A., and Preitner, G. (1953). *Helv. Chim. Acta* **36**, 875–886.
Borin, G., Filippi, B., Moroder, L., and Marchiori, F. (1976). *In* "Peptides 1976" (A. Loffet, ed.), pp. 233–238 Editions de l'Université de Bruxelles, Belgium.
Brady, S., Hirschmann, R., and Veber, D. (1977). *J. Org. Chem.* **42**, 143–146.
Buis, J. T., Tesser, G. I., and Nivard, R. (1976). *Tetrahedron* **32**, 2321–2335.
Camble, R., and Cotton, R. (1975). *In* "Peptides 1974" (Y. Wolman, ed.), pp. 17–25. Wiley, New York.
Camble, R., and Petter, N. N. (1976). *In* "Peptides 1976" (A. Loffet, ed.) pp. 299–307. Editions de l'Université de Bruxelles, Belgium.
Camble, R., Garner, R., and Young, G. T. (1969). *J. Chem. Soc. C* pp. 1911–1916.
Carpino, L., and Han, G. (1972). *J. Org. Chem.* **37**, 3404–3409.
Chang, C. D., and Meienhofer, J. (1978). *Int. J. Pept. Protein Res.* **11**, 246–249.
Channabasavaiah, K., and Sivanandaiah, K. M. (1976). *Int. J. Pept. Protein Res.* **8**, 323–329.
Chillemi, F., and Merrifield, R. B. (1969). *Biochemistry* **8**, 4344–4346.
Choudhury, A. M., Kenner, G. W., Moore, S., Ramage, R., Richards, P. M., Thorpe, W. D., Moroder, L., Wendlberger, G., and Wünsch, E. (1976). *In* "Peptides 1976" (A. Loffet, ed.), pp. 257–261. Editions de l'Université de Bruxelles, Belgium.
Corey, E. J., and Dawson, R. L. (1962). *J. Am. Chem. Soc.* **84**, 4899–4904.
Coy, D. H., Coy, E. J., and Schally, A. V. (1973). *J. Med. Chem.* **16**, 1140–1143.
Coyle, S., and Young, G. T. (1976). *In* "Peptides 1976" (A. Loffet, ed.), p. 205. Editions de l'Université de Bruxelles, Belgium.

Diaz, J., Guégan, R., Beaumont, M., Benoit, J., Clément, J., Fauchard, C., Galtier, D., Millan, J. Muneaux, C., Muneaux, Y., Vedel, M., and Schwyzer, R. (1979). *Bioorg. Chem.*, **8**, 429–442.

du Vigneaud, V., Ressler, C., Swan, J. M., Roberts, C. W., Katsoyannis, P. G., and Gordon, S. (1953). *J. Am. Chem. Soc.* **75**, 4879–4880.

Eberle, A., Fauchère, J. L., Tesser, G. I., and Schwyzer, R. (1975). *Helv. Chim. Acta* **58**, 2106–2129.

Eberle, A., Kriwaczek, V. M., and Schwyzer, R. (1977). *FEBS Lett.* **80**, 246–250.

Eckert, H., and Ugi, I. (1976). *Angew. Chem.* **88**, 717–718.

Erickson, B. W., and Merrifield, R. B. (1973a). *J. Am. Chem. Soc.* **95**, 3750–3756.

Erickson, B. W., and Merrifield, R. B. (1973b). *J. Am. Chem. Soc.* **95**, 3757–3763.

Erickson, B. W., and Merrifield, R. B. (1974). *Isr. J. Chem.* **12**, 79–85.

Fauchère, J. L., and Pelican, G. M. (1975). *Helv. Chim. Acta* **58**, 1984.

Felix, A., Jimenez, M., Mowles, T., and Meienhofer, J. (1978). *Int. J. Pept. Protein Res.* **11**, 329–339; **12**, 184.

Feinberg, R. S., and Merrifield, R. B. (1975). *J. Am. Chem. Soc.* **97**, 3485–3496.

Flanigan, E., and Marshall, G. R. (1970). *Tetrahedron Lett.* p. 2403–2406.

Fujii, N., and Yajima, H. (1975). *Chem. Pharm. Bull.* **23**, 1596–1603.

Fujii, N., Funakoshi, S., Sasaki, T., and Yajima, H. (1977). *Chem. Pharm. Bull.* **25**, 3096–3098.

Fujii, T., and Sakakibara, S. (1974). *Bull. Chem. Soc. Jpn.* **47**, 3146–3151.

Fujino, M., Shinagawa, S., Nishimura, O., and Fukuda, T. (1972). *Chem. Pharm. Bull.* **20**, 1017–1020.

Fujino, M., Wakimasu, M., Shinagawy, S., Kitada, C., and Yajima, H. (1978). *Chem. Pharm. Bull.* **26**, 539–548.

Galpin, I. J., Handa, B. K., Hudson, D., Jackson, A. G., Kenner, G. W., Ohlsen, S. R., Ramage, R., Singh, B., and Tyson, R. G. (1976). *In* "Peptides 1976" (A. Loffet, ed.), pp. 247–256. Editions de l'Université de Bruxelles, Belgium.

Geiger, R., Jäger, G., König, W., and Volk, A. (1969). *Z. Naturforsch., Teil B* **24**, 999–1004.

Geiger, R., König, W., Wissmann, H., Geisen, K., and Enzmann, F. (1971). *Biochem. Biophys. Res. Comm.* **45**, 767–773.

Goerdeler, J., and Holst, A. (1959). *Angew. Chem.* **71**, 775.

Goodacre, J., Ponsford, R. J., and Stirling, I. (1975). *Tetrahedron Lett.* pp. 3609–3612.

Gregory, R. A., Tracy, H. J., Harris, J. I., Runswick, M. J., Moore, S., Kenner, G. W., and Ramage, R. (1979). *Hoppe-Seyler's Z. Physiol. Chem.* **360**, 73–80.

Guggi, A., Tomatis, R., Salvadori, S., and Rocchi, R. (1976). *Int. J. Pept. Protein Res.* **8**, 115–124.

Gutte, B. (1975). *J. Biol. Chem.* **250**, 889–904.

Gutte, B., and Merrifield, R. B. (1971). *J. Biol. Chem.* **246**, 1922–1941.

Guttman, H., Weinstein, B., Bartschot, R., and Tam, P. S. (1975). *Experientia* **31**, 285–288.

Haas, W. L., Krumkalns, E. V., and Gerzon, K. (1966). *J. Am. Chem. Soc.* **88**, 1988–1992.

Halpern, B., Chew, L., Close, V., and Patton, W. (1968). *Tetrahedron Lett.* p. 5163–5164.

Hammerström, K., Lunkenheimer, W., and Zahn, H. (1970). *Makromol. Chem.* **133**, 41–51.

Hardy, P. M., Rydon, H. N., and Thompson, R. C. (1968). *Tetrahedron Lett.* pp. 2525–2526.

Hemmasi, B., and Li, C. H. (1974). *Int. J. Pept. Protein Res.* **6**, 303–307.

Hirschmann, R., and Veber, D. F. (1973). *In* "The Chemistry of Polypeptides" (P. G. Katsoyannis, ed.), pp. 125–142. Plenum, New York.

Hofmann, K., and Yajima, H. (1961). *J. Am. Chem. Soc.* **83**, 2289–2293.

Hofmann, K., Visser, J. P., and Finn, F. M. (1969). *J. Am. Chem. Soc.* **91**, 4883–4887.

Homer, R. B., Moodie, R. B., and Rydon, H. N. (1965). *J. Chem. Soc.* pp. 4403–4409.

Inouye, K., Sumitomo, Y., and Shin, M. (1976). *Bull. Chem. Soc. Jpn.* **49**, 3620–3628.

Iselin, B. (1962). *Helv. Chim. Acta* **45**, 1510–1515.

Iselin, B., and Schwyzer, R. (1962). *Helv. Chim. Acta* **45**, 1499–1509.
Ivanov, V. T., Mikhaleva, I. I., Volpina, O. M., Myagkova, M. A., and Deigin, V. I. (1976). *In* "Peptides 1976" (A. Loffet, ed.), p. 219. Editions de l'Université de Bruxelles, Belgium.
Ivanov, V. T., Tsetlin, V. I., Mikhaleva, I. I., Volpina, O. M., Nurridinova, A. R., Utkin, N. Yu., Arseniev, A. S., Pashkov, V. S., Karlsson, A. R., Surin, A. M., and Bystrov, V. F. (1979). *In* "Peptides 1978" (I. Z. Sièmion and G. Kupryszewski, eds.) pp. 41–62. Wroclaw University Press, Wroclaw, Poland.
Iwasaki, T., Matsumoto, K., Matsuoka, M., Takahashi, T., and Okumuta, K. (1973). *Bull. Chem. Soc. Jpn.* **46**, 852–855.
Jeschkeit, H., Losse, G., and Neubert, K. (1966). *Chem. Ber.* **99**, 2803–2812.
Jones, D. A., Jr., (1977). *Tetrahedron Lett.* **33**, 2853.
Juillerat, M., and Bargetzi, J. P. (1976). *Helv. Chim. Acta* **59**, 855–866.
Kader, A. T., and Stirling, C. J. (1964). *J. Chem. Soc.* p. 258–266.
Kalbag, S. M., and Roeske, R. W. (1975). *J. Am. Chem. Soc.* **97**, 440–441.
Kamber, B. (1973). *Helv. Chim. Acta* **56**, 1370–1381.
Kamber, B., and Rittel, W. (1969). *Helv. Chim. Acta* **52**, 1074–1081.
Karlsson, S., Lindeberg, G., Porath, J., and Ragnarsson, U. (1970). *Acta Chem. Scand.* **24**, 1010–1014.
Katsoyannis, P. G., Gish, D. T., and du Vigneaud, V. (1957). *J. Am. Chem. Soc.* **79**, 4516–4520.
Katsoyannis, P. G., Zalut, C., Tometsko, A., Tilak, M., Johnson, S., and Trakatellis, A. C. (1971). *J. Am. Chem. Soc.* **93**, 5871–5877.
Kemp, D. S., and Hoyng, C. F. (1975). *Tetrahedron Lett.* pp. 4625–4628.
Kemp, D. S., and Roberts, D. C. (1975). *Tetrahedron Lett.* pp. 4629–4632.
Kenner, G. W., and Seely, J. H. (1972). *J. Am. Chem. Soc.* **94**, 3259.
Kenner, G. W., McDermott, J. R., and Sheppard, R. C. (1971). *J. Chem. Soc., Chem. Commun.* pp. 636–637.
Kenner, G. W., Moore, G. A., and Ramage, R. (1976). *Tetrahedron Lett.* pp. 3623–3626.
Kessler, W., and Iselin, B. (1966). *Helv. Chim. Acta* **49**, 1330–1344.
Kisfaludy, L., and Löw, M. (1965). *Acta Chim. Acad. Sci. Hung.* **44**, 33–35.
Kisfaludy, L., and Schön, I. (1975). *Acta Chim. Acad. Sci. Hung.* **84**, 227–228.
König, W., and Geiger, R. (1970). *Chem. Ber.* **103**, 2041–2051.
König, W., and Geiger, R. (1972). *In* "Chemistry and Biology of Peptides" (J. Meienhofer, ed.), pp. 343–350. Ann Arbor Sci. Publ., Ann Arbor, Michigan.
König, W., Teetz, V. Tripier, D., and Volk, A. (1979). *In* "Peptides, 1978" (I. Z. Sièmion and G. Kupryszewski, eds.) pp. 619–623. Wroclaw Univ. Press, Wroclaw, Poland.
Koyama, K., Kawatani, H., Yajima, H., Fujino, M., and Nishimura, O. (1976a). *Chem. Pharm. Bull.* **24**, 2106–2111.
Koyama, K., Watanabe, H., Kawatani, H., Iwai, J., and Yajima, H., (1976b). *Chem. Pharm. Bull.* **24**, 2558–2563.
Kubota, M., Hirayama, T., Nagase, O., and Yajima, H. (1979). *Chem. Pharm. Bull.* **27**, 1050.
Kunz, H. (1976a). *Chem. Ber.* **109**, 2670–2683.
Kunz, H. (1976b). *Chem. Ber.* **109**, 3693–3706.
Künzi, H., and Studer, R. O. (1975). *Helv. Chim. Acta* **58**, 139–146.
Künzi, H., Manneberg, M., and Studer, R. O. (1974). *Helv. Chim. Acta* **57**, 566–572.
Kuromizu, K., and Meienhofer, J. (1974). *J. Am. Chem. Soc.* **96**, 4978–4981.
Lemaire, S., Yamashiro, D., Rao, A. J., and Li, C. H. (1977). *J. Med. Chem.* **20**, 155–158.
Li, C. H., Lemaire, S., Yamashiro, D., and Doneen, B. A. (1976). *Biochem. Biophys. Res. Commun.* **71**, 19–25.
Lin, M. C., Gutte, B., Caldi, D., Moore, S., and Merrifield, R. B. (1972). *J. Biol. Chem.* **247**, 4768–4774.

Löw, M., Kisfaludy, L., and Sohar, P. (1978). *Hoppe-Seyler's Z. Physiol. Chem.* **359**, 1643–1651.
Losse, G., Zeidler, D., and Grieshaber, T. (1968). *Justus Liebigs Ann. Chem.* **715**, 196–203.
Ludescher, U., and Schwyzer, R. (1972). *Helv. Chim. Acta* **55**, 2052–2060.
Lübke, K., and Klostermeyer, H. (1970). *Adv. Enzymol.* **33**, 445–525.
Lundt, B. F., Johansen, N. L., Vorlund, A., and Murkussen, J. (1978). *Int. J. Pept. Protein Res.* **12**, 258–268.
Maclaren, J. A., Savige, W. E., and Swan, J. M. (1958). *Aust. J. Chem.* **11**, 345–359.
Macrae, R., and Young, G. T. (1975). *J. Chem. Soc. Perkin I* pp. 1185–1191.
Marglin, A., and Merrifield, R. B. (1966). *J. Am. Chem. Soc.* **88**, 5051–5052.
Marshall, G. R., and Merrifield, R. B. (1965). *Biochemistry* **4**, 2394–2401.
Matsueda, G. R., and Stewart, J. M. (1975). *In* "Peptides: Chemistry, Structure and Biology" (R. Walter and J. Meienhofer, eds.), p. 333–339. Ann. Arbor Sci. Publ., Ann. Arbor, Michigan.
Matsueda, R., Maruyama, H., Kitazawa, E., Takahagi, H., and Mukaiyama, T. (1975). *J. Am. Chem. Soc.* **97**, 2573–2575.
Matsuura, S., Niu, C. H., and Cohen, J. S. (1976). *J. Chem. Soc. Chem. Comn.* p. 451–452.
Mazur, R. H., and Plume, G. (1968). *Experientia* **24**, 661.
Medzihradszky, K., and Medzihradszky-Schweiger, H. (1965). *Acta Chim. Acad. Sci. Hung.* **44**, 15–18.
Meienhofer, J. (1966). *In* "Peptides 1963" (L. Zervas, ed.), pp. 55–59. Pergamon, Oxford.
Meienhofer, J., and Kuromizu, K. (1974). *Tetrahedron Lett.* p. 3259–3262.
Merrifield, R. B. (1962). *Fed. Proc., Fed. Am. Soc. Exp. Biol.* **21**, 412.
Merrifield, R. B. (1963). *J. Chem. Soc.* **85**, 2149–2154.
Merrifield, R. B. (1964). *J. Org. Chem.* **29**, 3100–3102.
Meyers, C., and Glass, J. D. (1975). *Proc. Natl. Acad. Sci. U.S.A.* **72**, 2193–2196.
Miller, A. W., and Stirling, C. J. (1968). *J. Chem. Soc.* pp. 2612–2617.
Mitchell, A. R., Erickson, B., Ryabtsev, M. N., Hodges, R. S., and Merrifield, R. B. (1976). *J. Am. Chem. Soc.* **98**, 7357–7362.
Mizoguchi, T., Shigezane, K., and Takamura, N. (1970). *Chem. Pharm. Bull.* **18**, 1465.
Morley, J. S. (1967). *In* "Peptides 1966" (H. C. Beyerman, A. van de Linde, and Maassen van den Brink, eds.), pp. 226–234. North-Holland Publ., Amsterdam.
Moroder, L., Borin, G., Marchiori, F., and Scoffone, E. (1973). *Biopolymers* **12**, 477–492.
Mühlemann, M., Titov, M. I., Schwyzer, R., and Rudinger, J. (1972). *Helv. Chim. Acta* **55**, 2854–2860.
Najjar, V. A., and Merrifield, R. B. (1966). *Biochemistry* **5**, 3765–3770.
Nishimura, O., Kitada, C., and Fujino, M. (1978). *Chem. Pharm. Bull.* **26**, 1576–1585.
Nishimura, O., and Fujino, M. (1976). *Chem. Pharm. Bull.* **24**, 1568–1575.
Nefkens, G. H., Tesser, G. I., and Nivard, F. (1960). *Recl. Trav. Chim. Pays-Bas* **79**, 688–698.
Noda, K., Terada, S., and Izumiya, N. (1970). *Bull. Chem. Soc. Jpn.* **43**, 1883–1885.
Noda, K., Terada, S., Mitsuyasu, N., Waki, M., Kato, T., and Izumiya, N. (1971). *Naturwissenschaften* **58**, 147–148.
Ogawa, H., Kubota, M., and Yajima, H. (1976). *Chem. Pharm. Bull.* **24**, 2428–2434.
Ogawa, H., Sugiura, M., Yajima, H., Sakurai, H., and Tsuda, K. (1978). *Chem. Pharm. Bull.* **26**, 1549–1557.
Ohno, M., Tsukamoto, S., and Izumiya, N. (1972). *J. Chem. Soc. Chem. Commun. D* p. 663–664.
Ontjes, D. A., and Anfinsen, C. B. (1970). *In* "Peptides: Chemistry and Biochemistry" (B. Weinstein and S. Lande, eds.), pp. 79–98. Dekker, New York.
Patchornik, A., and Sokolovsky, M. (1964). *J. Am. Chem. Soc.* **86**, 1206–1212.
Pless, J., and Bauer, W. (1973). *Angew. Chem., Int. Ed. Engl.* **12**, 147–148.
Pless, J. (1976). *Helv. Chim. Acta* **59**, 499–512.

Prestidge, R. L., Harding, D. R., and Hancok, W. S. (1976). *J. Org. Chem.* **41**, 2579–2883.
Ragnarsson, U., Karlsson, S., and Lindeberg, G. (1970). *Acta Chem. Scand.* **24**, 2821–2825.
Rich, D. H., and Gurwara, S. K. (1975). *J. Am. Chem. Soc.* **97**, 1575–1579.
Riniker, B., Brugger, M., Kamber, B., Sieber, P., and Rittel, W. (1969). *Helv. Chim. Acta* **52**, 1058–1073.
Riniker, B., Kamber, B., and Sieber, P. (1975). *Helv. Chim. Acta* **58**, 1086–1094.
Riniker, B., Eisler, K., Kamber, B., Müller, H., Rittel, W., and Sièber, P., (1979). *In* "Peptides 1978" (I. Z. Sièmion and G. Kupryszewski, eds.), pp. 631–634. Wroclaw Univ. Press, Wroclaw, Poland.
Rivier, J. Kaiser, R., and Galyean (1978). *Biopolymers* **17**, 1927–1938.
Robinson, A. B., and Tedro, S. (1973). *Int. J. Pept. Protein Res.* **5**, 275–278.
Rudinger, J. (1963). *In* "Peptides 1962" (G. T. Young, ed.), pp. 133–164. Pergamon, Oxford.
Rudinger, J., Krupička, J., Zaoral, M., and Cernik, V. (1960). *Collect. Czech. Chem. Commun.* **25**, 3338–3343.
Sakakibara, S. (1971). *In* "Chemistry and Biochemistry of Amino Acids, Peptides and Proteins" (B. Weinstein, ed.), Vol. 1, pp. 51–85. Dekker, New York.
Sakakibara, S., Shimonishi, Y., Kishida, Y., Okada, M., and Sugihara, H. (1967). *Bull. Chem. Soc. Jpn.* **40**, 2164–2167.
Sakakibara, S., Honda, I., Takada, K., Miyoshi, M., Ohnishi, T., and Okumura, K. (1969). *Bull. Chem. Soc. Jpn.* **42**, 809–811.
Sakakibara, S., Fukuda, T., Kishida, Y., and Honda, I. (1970). *Bull. Chem. Soc. Jpn.* **43**, 3322.
Sandberg, B., and Ragnarsson, U. (1974). *Int. J. Pept. Protein Res.* **6**, 111–119.
Sandberg, B., and Ragnarsson, U. (1975). *Int. J. Pept. Protein Res.* **7**, 503–504.
Sano, S., and Kawanishi, S. (1975). *J. Am. Chem. Soc.* **97**, 3480.
Schaltiel, S., and Fridkin, M. (1970). Biochemistry, 9, 5122.
Schlatter, J. M. and Mazur, R. H. (1977). *Tetrahedron Lett.* **33**, 2851–2852.
Schnabel, E., Klostermeyer, H., and Berndt, H. (1971). *Justus Liebigs Ann. Chem.* **749**, 90–108.
Schröder, E., and Lübke, K. (1965). *In* "The Peptides," Vol. 1, p. 27. Academic Press, New York.
Schultz, H. S., Freyermuth, H. B., and Buch, S. R. (1963). *J. Org. Chem.* **28**, 1140–1142.
Schwarz, H., and Arakawa, K. (1959). *J. Am. Chem. Soc.* **81**, 5691–5695.
Schwyzer, R. (1954). *Helv. Chim. Acta* **37**, 647–649.
Schwyzer, R. (1966). *Naturwissenschaften* **53**, 189–197.
Schwyzer, R., and Karlaganis, G. (1973). *Justus Liebigs Ann. Chem.* pp. 1298–1309.
Schwyzer, R., and Li, C. H. (1958). *Nature* (*London*) **182**, 1669–1670.
Schwyzer, R., and Schiller, P. W. (1971). *Helv. Chim. Acta* **54**, 897–904.
Schwyzer, R., and Sieber, P. (1956). *Angew. Chem.* **68**, 518.
Schwyzer, R., and Sieber, P. (1959). *Helv. Chim. Acta* **42**, 972–977.
Schwyzer, R., Iselin, B., Kappeler, H., Riniker, B., Rittel, W., and Zuber, H. (1958). *Helv. Chim. Acta* **41**, 1287–1295.
Schwyzer, R., Rittel, W., Kappeler, H., and Iselin, B. (1960). *Angew. Chem.* **72**, 915–917.
Schwyzer, R., Costopanagiotis, A., and Sieber, P. (1963). *Helv. Chim. Acta* **46**, 870–889.
Scoffone, E., Fontana, A., and Rocchi, R. (1968). *Biochemistry* **7**, 971–979.
Seidel, C., Klauschenz, E., Kaufmann, K. D., and Kunzek, H. (1976). *J. Prakt. Chem.* **318**, 229–247.
Sharp, J. J. Robinson, A. B., and Kamen, M. D. (1973). *J. Am. Chem. Soc.* **95**, 6097–6108.
Shimizu, F., Imagawa, K., Mihara, S., and Yanaihara, N. (1976). *Bull. Chem. Soc. Jpn.* **49**, 3594–3596.
Sieber, P., and Iselin, B. (1968). *Helv. Chim. Acta* **51**, 614–622.
Sieber, P., Rittel, W., and Riniker, B. (1972). *Helv. Chim. Acta* **55**, 1243–1248.
Sieber, P., Kamber, B., Hartmann, A., Jöhl, A., Riniker, B., and Rittel, W. (1977). *Helv. Chim. Acta* **60**, 27–37.

Siemion, I. Z., Konopinska, D., Wiejak, S., Rzeszotarska, B., and Najbar, Z. (1973). *In* "Peptides 1972" (H. Hanson and H.-D. Jakubke, eds.), pp. 210–213. North-Holland Publ., Amsterdam.
Sifferd, R. H., and du Vigneaud, V. (1935). *J. Biol. Chem.* **108**, 753–761.
Southard, G. L. (1971). *In* "Peptides 1969" (E. Scoffone, ed.), pp. 168–170. North-Holland Publ., Amsterdam.
Southard, G. L., Brooke, G. S., and Pettee, J. M. (1969). *Tetrahedron Lett.* p. 3505–3508.
Southard, G. L., Zaborowski, B. R., and Pettee, J. M. (1971). *J. Am. Chem. Soc.* **93**, 3302–3303.
Stelakatos, G. C., Paganou, A., and Zervas, L. (1966). *J. Chem. Soc. C* p. 1191–1199.
Stevenson, D., and Young, G. T. (1969). *J. Chem. Soc. C* pp. 2389–2398.
Stewart, F. H. C. (1966). *Aust. J. Chem.* **19**, 1067–1083.
Stewart, F. H. C. (1968a). *Aust. J. Chem.* **21**, 1327–1335.
Stewart, F. H. C. (1968b). *Aust. J. Chem.* **21**, 2543–2550.
Stewart, F. H. C. (1968c). *Aust. J. Chem.* **21**, 2831–2834.
Stewart, F. H. C. (1971). *Aust. J. Chem.* **24**, 2193–2197.
Stewart, J. M., Knight, M., Paiva, A. C., and Paiva, T. (1972). *In* "Progress in Peptide Research" (S. Lande, ed.) pp. 59–64. Gordon & Breach, New York.
Stewart, J. M., Pena, C., Matsueda, G. P., and Harris, K. (1976). *In* "Peptides 1976" (A. Loffet, ed.), pp. 285–290. Editions de l'Université de Bruxelles, Belgium.
Tesser, G. I. (1975). *In* "Peptides, 1974" (Y. Wolman, ed.), pp. 53–56. Wiley, New York. Israel University Press, Jerusalem.
Tesser, G. I., Buis, J. T., Wolters, E. T., and Bothé-Helmes, E. G. (1976). *Tetrahedron* **32**, 1069–1072.
Tilak, M. A., and Hollinden, C. S. (1968a). *In* "Peptides 1968" (E. Bricas, ed.) pp. 173–178. North-Holland Publ., Amsterdam.
Tilak, M. A., and Hollinden, C. S. (1968b). *Tetrahedron Lett.* pp. 1297–1300.
Tomatis, R., Guggi, A., Menegatti, E., Guarneri, M., Rocchi, R., and Benassi, C. A. (1973). *Gazz. Chim. Ital.* **103**, 1129–1141.
Tomatis, R., Guggi, A., Salvadori, S., and Rocchi, R. (1977). *Int. J. Pept. Protein Res.* **9**, 137–147.
Trudelle, Y., and Birr, C. (1975). *Int. J. Pept. Protein Res.* **7**, 403–410.
Tzougraki, C., Makofske, R., Gabriel, T., Wang, S., Kutny, R., Meienhofer, J., and Li, C. H. (1978). *J. Am. Chem. Soc.* **100**, 6248–6249.
Tun-Kyi, A., and Schwyzer, R. (1976). *Helv. Chim. Acta* **59**, 1642–1648.
Tun-Kyi, A. (1978). *Helv. Chim. Acta* **61**, 1086–1090.
Veber, D. F., Milkowski, J. D., Varga, S., Denkewalter, R. G., and Hirschmann, R. (1972). *J. Am. Chem. Soc.* **94**, 5456–5461.
Vičar, J., Gut, V., Fric, I., and Bláha, K. (1976). *Collect. Czech. Chem. Commun.* **41**, 3467–3473.
Vogel, R. (1975). Ph.D. Thesis, University of Basel.
Vorbrüggen, H., and Krolikiewicz, K. (1975). *Angew. Chem., Int. Ed. Engl.* **14**, 818.
Walshaw, K. B., and Young, G. T. (1965). *J. Chem. Soc.* pp. 786–788.
Wang, S. S. (1972). *In* "Chemistry and Biology of Peptides" (J. Meienhofer, ed.), pp. 179–182. Ann Arbor Sci. Publ., Ann Arbor, Michigan.
Wang, S. S. (1975). *J. Org. Chem.* **40**, 1235–1239.
Wang, S. S., and Merrifield, R. B. (1969). *J. Am. Chem. Soc.* **91**, 6488–6491.
Wang, S. S., and Merrifield, R. B. (1971). *In* "Peptides 1969" (E. Scoffone, ed.), pp. 74–83. North-Holland Publ., Amsterdam.
Wang, S. S., Yang, C. C., Kulesha, I. D., Sonenberg, M., and Merrifield, R. B. (1974). *Int. J. Pept. Protein Res.* **6**, 103–109.
Weygand, F., and Csendes, E. (1952). *Angew. Chem.* **64**, 136.
Weygand, F., Steglich, W., and Pietta, P. (1967). *Chem. Ber* **100**, 3841–3849.

Weygand, F., Steglich, W., Fraunberger, F., Pietta, P., and Schmid, J. (1978). *Chem. Ber.* **101**, 923–934.

Wieland, T., Birr, C., and Fleckenstein, P. (1972). *Justus Liebigs Ann. Chem.* **756**, 14–19.

Wünsch, E. (1967). *Z. Naturforsch., Teil B* **22**, 1269–1276.

Wünsch, E., ed. (1974). "Houben-Weyl Methoden der Organischen Chemie," Vol. 15, Parts 1 and 2. Thieme, Stuttgart.

Wünsch, E., and Spangenberg, R. (1971). *Chem. Ber.* **104**, 2427–2429.

Wünsch, E., Jaeger, E., and Scharf, R. (1968). *Chem. Ber.* **101**, 3664–3670.

Wünsch, E., Wendlberger, G., and Stocker, H. (1976). *Hoppe-Seyler's Z. Physiol. Chem.* **457**, 459–465.

Yajima, H., Kiso, Y., Okada, Y., and Watanabe, H. (1974). *J. Chem. Soc., Chem. Commun.* p. 106–107.

Yajima, H., Kiso, Y., Ogawa, H., Fujii, N., and Irie, H. (1975). *Chem. Pharm. Bull.* **23**, 1164–1166.

Yajima, H., Kaname, K., Kiso, Y., Tanaka, A., and Nakamura, M. (1976a). *Chem. Pharm. Bull.* **24**, 492–499.

Yajima, H., Mori, Y., Koyama, K., Tobe, T., Setoyama, M., Adachi, H., Kanno, T., and Saito, A. (1976b). *Chem. Pharm. Bull.* **24**, 2794–2802.

Yamashiro, D. (1977). *J. Org. Chem.* **42**, 523–525.

Yamashiro, D., and Li, C. H. (1973). *J. Org. Chem.* **38**, 2594–2597.

Yamashiro, D., Blake, J., and Li, C. H. (1972a). *J. Am. Chem. Soc.* **94**, 2855–2859.

Yamashiro, D., Noble, R., and Li, C. H. (1972b). *In* "Chemistry and Biology of Peptides" (J. Meienhofer, ed.), pp. 197–202. Ann Arbor Sci. Publ., Ann Arbor, Michigan.

Yamashiro, D., Noble, R., and Li, C. H. (1973). *J. Org. Chem.* **38**, 3561–3565.

Yamashiro, D., and Li. C. H. (1978). *J. Am. Chem. Soc.* **100**, 5174–5179.

Young, G. T. (1967). *In* "Peptides 1966" (H. Beyerman, A. van de Linde, and W. Maassen van den Brink, eds.), pp. 55–66. North-Holland Publ., Amsterdam.

Young, G. T., (1973). *In* "The Chemistry of Polypeptides" (P. G. Katsoyannis, ed.) pp. 43–58. Plenum, New York.

Zervas, L., Borovas, D., and Gazis, E. (1963). *J. Am. Chem. Soc.* **85**, 3660–3666.

Chapter **6**

Peptide Synthesis with Minimal Protection of Side-Chain Functions

JOHN K. INMAN

THE PEPTIDES, VOLUME 3

ISBN 0-12-304203-8

I. INTRODUCTION

A. General Comments and Aims

A major consideration in planning the synthesis of a peptide is the potential involvement of side-chain functional groups in undesired reactions during formation of peptide bonds or removal of protecting groups. Thirteen of the 20 amino acids mainly comprising proteins and many biologically active peptides possess a third functional group. The strategies and tactics of synthesis should be chosen so that serious side reactions with these third functions can be avoided.

Successful approaches to synthesis require or make prudent the covalent masking of at least some types of side chains. Often, liberal use of protecting groups is made on a precautionary basis in order to avoid side reactions during peptide bond formation. Yet, there are some special advantages to be gained in minimizing side-chain protection. It is the aim of this chapter to put into order a selection of facts, ideas, and experiences that should aid the peptide chemist in deciding whether or not to seek these advantages and, if so, in determining the pattern of protection that is likely to be minimal for his chosen peptide; many of the risks and means of such syntheses will also be covered. Although the literature reports a great number of peptides that have been successfully prepared without extensive side-chain protection, the author is aware of only a few attempts to define explicitly minimal protection or to discuss the corresponding strategies and tactics that should be employed. For some previous discussions of this topic the reader is referred to reviews by Hirschmann and Veber (1973) and Wünsch (1971, 1974), and the monograph of Bodanszky *et al.* (1976, Chapters 4 and 8).

B. Degrees and Patterns of Side-Chain Protection

1. Definitions

In the discussions that follow, the term *tactics* will refer to the means (methods and reagents) for forming peptide bonds, introducing protecting groups and removing the same. The *strategy* of a synthesis will cover the overall plan which includes dividing the complete sequence into smaller targets or intermediate sequences termed *segments* (heretofore commonly referred to as fragments), determining a pattern of protection and a deprotection schedule, deciding on the order of joining segments, etc. Any peptide bond-forming step where the acylating component, or *carboxyl component*, is a tripeptide or larger and the nucleophilic partner, or *amine component*, is at least a dipeptide will be called a *segment condensation.* Because of common practice, it is useful to include stepwise additions of dipeptides along with single amino acids in *stepwise elongation* schemes for the building of peptides or segments from their carboxyl ends.

2. The Very Minimum of Protection

In all approaches to peptide synthesis by presently known chemical methods, protection is obligatory or indispensible for the following functional groups:

1. The α-amino group of the carboxyl component (amino acid or peptide).
2. The ε-amino group of lysine (or δ-amine of ornithine) in either carboxyl or amine component.
3. The free sulfhydryl group of cysteine in either component.

Beyond this presently irreducible minimum in the use of protecting groups, one is faced with a myriad of options. It may be helpful first to order these options in some manner regarding the degree of side-chain blocking employed, such as, very minimum, minimal, intermediate, and full. As an aid in this task the amino acids are grouped in Table I according to the frequency with which their side chains have been protected in practice. The very minimum of protection, of course, refers to using only obligatory protection (group 1,) which means leaving unblocked the ω-carboxyls of glutamic and aspartic residues. The latter amino acids fall into group 2, designated "conditionally obligatory," because there are definitive situations governing the

Table I Categories of Amino Acids According to Usage of Side-Chain Protecting Groups

Side-chain protection is	Amino acid	Side-chain functionality
1. Obligatory	Lys	ε-Amine
	Orn	δ-Amine
	Cys (with free SH)	Sulfhydryl
2. Conditionally obligatory	Asp	Carboxyl
	Glu	Carboxyl
	(Cys) only if $(Cys)_2$ is considered "unprotected"	—SH/—S—S—
3. Optional and frequently used	Arg	Guanidino N's
	Tyr	Phenolic —OH
	Ser	Primary aliphatic —OH
	Thr	Secondary aliphatic —OH
	His	Imidazole N's
4. Optional and infrequently used	Gln	Primary carboxamide
	Asn	Primary carboxamide
	Met	Aliphatic thioether
	Trp	Indole ring positions
5. Never required	Gly, Ala, Val, Leu, Ile, Phe, Pro	H (essentially nonfunctional)

requirement for protecting their side groups: the carboxyl side chains can be left unprotected in *amine* components in cases where carboxyl components are activated in a separate vessel (e.g., in the conventional azide, mixed anhydride, or active ester methods); they must be protected when activation plus coupling is a "one-pot" process, for example, in the use of carbodiimides (with or without additives) or special coupling reagents.

Where aspartic and glutamic acids occur in *carboxyl* components the circumstances are different: their side chains cannot be left unprotected when they are carboxyl-terminal residues of a non-COOH-terminal segment. Otherwise, any form of activation is apt to produce a cyclic α,ω-anhydride intermediate leading to a mixture of α and ω peptides. Non-COOH-terminal aspartic and glutamic acid residues in carboxyl components may be left unprotected when certain kinds of activation are used. The formation of azides from hydrazinolyzed esters or protected hydrazides has been employed almost exclusively in this situation. Selective COOH-terminal activation could also be accomplished by converting an inactive ester group to an active ester, for example by changing a peptide 4-(methylthio)phenyl ester to its 4-(methylsulfonyl)phenyl ester by carful oxidation with *m*-chloroperoxybenzoic acid (Johnson and Ruettinger, 1970). In such cases, the peptide could contain tryptophan but not methionine, cysteine, or cystine.

The only known way to introduce unprotected aspartic or glutamic acid directly as single amino acids in stepwise elongations is through their *N*-carboxyanhydrides (Denkewalter *et al.*, 1966). This tactic was employed by the group at Merck Sharp and Dohme Research Laboratories in the synthesis of pancreatic ribonuclease S-protein by a classical segment condensation approach (Denekewalter *et al.*, 1969; Strachan *et al.*, 1969; Jenkins *et al.*, 1969; Veber *et al.*, 1969; Hirschmann *et al.*, 1969). Their work constitutes the only extensive synthesis accomplished at the very minimum level of protection (see Section III).

The next most direct way of adding unprotected aspartic or glutamic acid residues to a peptide segment is through the use of side-chain protecting groups that are removed with the α-amine blocking function before the next residue is coupled. For instance, one can employ activated forms of N^{α}-Z, ω-*O*-Bzl- or N^{α}-Boc, ω-O*t*Bu-dicarboxylic amino acids and next remove both groups simultaneously by hydrogenolysis or acidolysis, respectively.

The very minimum degree of protection could include synthesis of cysteine-containing peptides without the use of an extrinsic masking group for the sulfhydryl function. It has long been known that symmetrical cystine peptides can be prepared directly by allowing the disulfide linkage to serve as the protecting function. The finished "H" peptide then may be reduced to a cysteine peptide or cleaved by oxidative sulfitolysis. The use of this tactic was beautifully exemplified in the synthesis of insulin B chain by Zahn and Schmidt (1970a,b) who started with bis-Z-L-cystine bis-*p*-nitrophenyl (ONp) ester at each of the two cysteine positions and finally joined the two separate cystine peptides by an azide-promoted condensation. In other respects, however, this synthesis did not employ a minimum of side-chain protection. A very similar approach was employed in the synthesis of oxytocin by Hase and Walter (1973).

3. *Minimal, Intermediate, and Full Protection*

Beyond the very minimum of protection, which has rarely been used, a rough spectrum exists in the usage of side-chain blocking groups as indicated in Table I. It would be better to say that various patterns of usage are followed. "Minimal" ("minimum" or "a minimum of") side-chain protection, acknowledged as such by Beacham *et al.* (1971), Moroder *et al.* (1972), Yanaihara *et al.* (1977), and Kasafirek *et al.* (1977), applies to "classical" or solution-phase syntheses involving segment condensation strategies where, for the most part, tyrosine, serine, threonine, and histidine are introduced unprotected; aspartic acid, glutamic acid, and arginine are protected through part of the synthesis (commonly not beyond completion of minor or major segments); and, asparagine, glutamine, or methionine (as sulfoxide) are protected infrequently as special problems arise. Examples of these syntheses are presented in Section III.

"Intermediate" use of masking would apply to blocking, additionally, tyrosine, histidine, and occasional serine side chains; aspartic and glutamic acids would be left protected until the late or terminal stages of synthesis. Truly full or complete protection of all 13 trifunctional amino acids is rarely, if ever, used. The fullest protection of third functions is practiced in solid-phase synthesis where asparagine and glutamine are usually introduced as active esters with the ω-amide groups unblocked. In the "global" protection school of classical synthesis (Wünsch, 1971; Sakakibara, 1977), the side chains of aspartic acid, glutamic acid, tyrosine, serine, and threonine are uniformly blocked (as with O*t*Bu and *t*Bu or, alternatively, with OBzl and Bzl groups). Yet, arginine and histidine residues often are introduced unblocked or else they are deprotected soon thereafter.* The positive charges of protonated histidine and arginine side chains can increase solubility in polar solvents to a considerable extent, compensating for the opposing effect of hydrophobic blocking groups on other residues (Wünsch, 1971). However,

* The guanidino group of arginine is protected somewhat by protonation because of its highly basic character; however, in this chapter the term, protection, will mean the use of covalently attached groups in all cases.

a number of peptides have been synthesized with arginine residues protected throughout without any serious solubility problems arising (Sakakibara, 1977). Other examples of global or maximal protection are found in syntheses reported by Wünsch (1967, 1972). Sieber *et al.* (1970), and Jäger *et al.* (1974).

C. Advantages and Disadvantages of Minimal Side-Chain Protection

A number of factors should be considered before deciding whether or not to use minimal side-chain protection in a synthesis. Many points that have been considered by various workers are summarized below on a pro and con basis. Particular experiences with these issues will be discussed in Sections II and III. After considering the known facts, one may yet have to assess relative advantages and disadvantages through preliminary or trial syntheses. The possible *advantages* of minimal side-chain protection are as follows:

1. Avoidance of side reactions during removal of side-chain protecting groups.
2. Avoidance of side reactions involving a protecting group during the removal of temporary blocking (of α-NH_2 or α-COOH), carboxyl activation or coupling reactions.
3. Increased solubility in polar and semiaqueous solvents, especially in the case of large, complex peptides.
4. Enhanced opportunities for purification of intermediates (discussed below).
5. Increased economy in time and materials.
6. Avoidance of side reactions during introduction of unnecessary protecting groups as might be attempted for a semisynthesis.
7. Increased likelihood that di- or tripeptide intermediates will be available (or described) from previous syntheses since there are fewer combinations of side-chain structures in the unprotected state.

The possible *disadvantages* of minimal side-chain protection are as follows:

1. Side reactions involving unprotected third functions occurring during formation of peptide bonds (activation or coupling).
2. Side reactions occurring during removal of temporary protecting groups (α-NH_2, α-COOH) or an occasional side-chain masking group.
3. Low solubility in activation and coupling solvents, especially in cases of small to intermediate peptides rich in polar side chains.
4. Loss of opportunities to increase solubility by using special "solubilizing" or "bifunctional" protecting groups (Wünsch, 1971; although they can be used also with obligatory protection; see Coyle *et al.*, 1975).

5. Restrictions in choice of strategies and coupling tactics necessary for minimizing the above problems.

The first two items in each list cover most of the potential side reactions which need consideration. Examples of these problems are dealt with in succeeding sections. In general, it should be pointed out that precautionary use of side-chain masking does not necessarily do away with side-reaction problems, but can lead to trouble in certain instances where minimal protection would work satisfactorily. Acidolytic cleavage of benzyl- or *tert*-butyl-based protecting groups can release benzyl halides, benzyl and *tert*-butyl carbonium ions and *tert*-butyl trifluoroacetate (Lundt *et al.*, 1978) which may not be completely trapped by scavengers before reacting with the same or other side chains. Benzyl group migration on the tyrosine ring during deblocking has been particularly troublesome (Engelhard and Merrifield, 1978; Bodanszky *et al.*, 1978; Lundt *et al.*, 1979). On the other hand, potential side reactions involving unprotected third functions during activation and coupling restrict the types of strategies that can be employed. As peptides grow larger these potential side reactions become increasingly important. In general, methods of synthesis that rest on repetitive operations in stepwise elongations using substantial excesses of activated carboxyl components will require fairly full protection of side chains. Emphasis in this regard would be placed on methods employing strong acylating functions or reagents such as anhydrides or carbodiimides (without additives), methods allowing infrequent purification of intermediates, and stepwise elongations exceeding 9 or 10 units. Thus, repetitive methods such as standard solid-phase synthesis (Merrifield, 1963), the excess mixed anhydride method (Tilak, 1970; Beyerman, 1972), synthesis on polymers (Bayer and Mutter, 1972), etc., have not been considered in this review. However, recently Coy and Branyas (1979) reported successful solid-phase syntheses of a number of small peptides (up to 9 residues) using carbodiimide coupling and no side-chain protection for tyrosine, serine or arginine.

Low solubility problems remain a major stumbling block in many solution-phase syntheses. Although solubilities of small, polar peptides in coupling solvents can be improved by adding side-chain protecting groups, solubilities of large segments are sometimes reduced by extensive protection. In semi-aqueous solvents commonly used in purification steps, the usual hydrophobic blocking groups can create solubility problems for all sizes of peptides. For the effective purification of intermediates, unprotected side chains provide (1) ionic groups allowing for ion-exchange chromatography and electrophoresis, (2) the native structures and conformations necessary for affinity chromatography on antibody columns, and (3) subtle differences in character which aid in adsorption chromatographic and countercurrent distribution separations.

Also, highly protected peptides have significantly higher molecular weights which allow for less efficient separations by gel permeation methods. These solubility and purification enhancing factors constitute, perhaps, the most important advantages of minimal side-chain protection.

In industrial scale peptide synthesis minimal protection can result in fewer steps, cheaper starting materials, and fewer deprotection problems (Feurer, 1977). Similar economies can be gained in research syntheses of numerous small peptide derivatives and analogues where the side-reaction problems can be solved in a general way in preliminary experiments.

Semisyntheses (Smyth, 1975), where amino acids or peptides are added to natural peptides, proteins, or their constitutive segments, (see Section III,C) place a special demand on the tactics of minimal side-chain protection since attempts to use any but indispensable protection invites a host of unexpected by-products from both introducing and removing blocking groups.

II. TACTICS AND PROBLEMS CONNECTED WITH USE OF MINIMAL SIDE-CHAIN PROTECTION

A. General Comments on Tactics

Nearly all tactics employed in solution-phase peptide synthesis have been incorporated into minimal protection approaches, at least with regard to NH_2- and COOH-terminal blocking, obligatory side-chain masking, peptide bond formation, and purifications. The reader should refer to other reviews for general discussions and details of these methods. This section will deal with special considerations for minimizing side reactions involving unprotected side chains during peptide bond-forming steps.

The commonly used methods of peptide bond formation rely upon augmentation of the driving force (free energy change) as an indirect means for increasing reaction rates to practical levels. In general such "overactivation" (Brenner, 1967) can be expected to lead to an increase in the proportional extent of side reactions in relation to the desired reaction. Of course, it is this ratio in the outcome which must be considered rather than acylating power per se. Thus, many small peptides of trifunctional amino acids have been successfully prepared with the strongly activating mixed anhydride or carbodiimide methods. In the case of larger peptides it has usually been considered wise not to try one's luck but rather to scale down the acylating power of these methods with active ester-forming additives such as *N*-hydroxysuccinimide (HOSu; Weygand *et al.*, 1966), 1-hydroxybenzotriazole (HOBt; König and Geiger, 1970a), *N*-hydroxy-5-norbornene-2,3-dicarboximide (HONb; Fujino *et al.*, 1974b), or *p*-chloro- or *p*-nitrobenzenesulfohydroxamic acid (Yajima *et al.*, 1973a).

In instances where economy dictates the use of excess carboxyl component for improving yields, prolonged reaction times should be carefully avoided: a precise reaction time should be optimal, beyond which the ratio of by-products to product will continue to increase with little practical gain in yield. For minimal protection approaches it would be better, in general, to work with an excess of the amine component. In principle, the results would be a lower number and proportion of side products leading to easier purifications and, thus, higher actual yields; also, time of reaction would not be so critical. The above considerations have received perhaps too little attention. Experimental findings on side-chain by-products (discussed below) repeatedly point to the use of excess base in the coupling mixture as a major source of trouble. The use of the required, stoichiometric amount (and in some instances less) of tertiary amine should be the rule with minimal side-chain protection even with relatively mild acylating methods. In individual cases important improvements in yield and by-product formation can be realized on an empirical basis by changing the additive or active ester leaving group, the base employed, or the solvent. Certainly, if amine components are not largely dissolved, carboxyl components may be in effective excess throughout most of the coupling reaction.

Selective reactivities of acylating reagents must be considered. Acylimidazoles are more prone to promote O-acylation (see Section II,C,2) than acyl azides. The latter have been a cornerstone in those minimal protection strategies that have depended heavily upon numerous segment condensations. Acyl azides, while being fairly reactive, are substantially amine-specific under the usual conditions of peptide bond formation. They are readily formed from hydrazinolyzed esters or blocked hydrazides. Unprotected amino acid residues have often been introduced as COOH-terminal methyl or ethyl esters in dipeptides or small segments and then condensed to an amine after hydrazinolysis and treatment with nitrous acid. The latter reaction can best be carried out in very cold, nonaqueous solutions as described by Honzl and Rudinger (1961) or by some variation of this procedure. Azides have been generated also *in situ* by such reagents as diphenyl phosphorazidate and used in syntheses with unprotected side chains (Shioiri *et al.*, 1972; Shioiri and Yamada, 1974a,b,c; Ozawa *et al.*, 1977; Hamada *et al.*, 1977).

Table II lists many intermediates that may be especially useful in the stepwise introduction of those amino acids that are most definitive regarding the usage of minimal side-chain protection (note groups 2 and 3 of Table I and Section I,B,3). The commonly available methyl and ethyl esters were not included. Emphasis was placed on active (or activated) esters whose synthesis and properties have been described. Derivatives whose side-chain protection may be promptly removed simultaneously with the α-amine blocking group

Table II References to Syntheses and Properties of Some Amino Acid Derivatives[a] Useful for Stepwise Additions in Minimal Protection Strategies

Derivative	References[b]
Aspartic acid	
Asp-NCA	Hirschmann *et al.* (1971)
Boc-Asp(O*t*Bu)-OH	Schröder and Klieger (1964b); Schnabel (1967); Voelter *et al.* (1972)
Boc-Asp(O*t*Bu)-OH, DCHA	Wünsch and Zwick (1963); Schröder and Klieger (1964b)
Boc-Asp-O*t*Bu	Ondetti *et al.* (1968a); Tomasz (1971)
Boc-Asp(ONp)-O*t*Bu	Ondetti *et al.* (1968a)
Boc-Asp(O*t*Bu)-OPfp	Kisfaludy *et al.* (1973)
Boc-Asp(O*t*Bu)-OTcp	Broadbent *et al.* (1967)
Z-Asp(OBzl)-OH	Bryant *et al.* (1959); Benoiton (1962); Losse *et al.* (1964)
Z-Asp-OBzl	Losse *et al.* (1964)
Z-Asp(OBzl)-OH · DCHA	Klieger *et al.* (1961)
Z-Asp(OBzl)-ONo	Bodanszky *et al.* (1974a)
Z-Asp(OBzl)-ONp	Guttmann (1961); Losse *et al.* (1964); Zervas and Hamalidis (1965); Itoh (1970)
Z-DAsp(OBzl)-ONp	Mazur *et al.* (1969)
Z-Asp(ONp)-OBzl	Losse *et al.* (1964)
Z-Asp(OPfp)-OBzl	Kisfaludy *et al.* (1970)
Z-Asp(OSu)-OBzl	Hardy *et al.* (1972)
Z-Asp(OBzl)-OTcp	Pless and Boissonnas (1963); Davey *et al.* (1966)
Z-DAsp(OBzl)-OTcp	Davey *et al.* (1966)
Moz-Asp(O*t*Bu)-OH · DCHA	Schröder and Klieger (1964b)
Moz-Asp(OBzl)-OH	Weygand and Nintz (1965)
Moz-Asp(OBzl)-ONp	Yajima and Kiso (1974a)
Glutamic acid	
Glu-NCA	Hirschmann *et al.* (1971)
Boc-Glu(O*t*Bu)-OH	Schröder and Klieger (1964a); Schnabel (1967)
Boc-Glu-O*t*Bu	Tomasz (1971)
Boc-Glu(O*t*Bu)-OObt[c]	Jäger (1973)
Boc-Glu(O*t*Bu)-OCH_2CN	Schröder and Klieger (1964a)
Boc-Glu(O*t*Bu)-OPcp	Visser *et al.* (1970)
Boc-Glu(O*t*Bu)-OPfp	Kisfaludy *et al.* (1973)
Boc-Glu(O*t*Bu)-OSu	Hofmann *et al.* (1971)
Boc-Glu(O*t*Bu)-OTcp	Broadbent *et al.* (1967)
Ibc-Glu(O*t*Bu)-OH[d]	Jäger and Geiger (1973)
Z-Glu(OBzl)-OH	Hanby *et al.* (1950); Weygand and Hunger (1962)
Z-Glu(OBzl)-OH · DCHA	Klieger *et al.* (1961); Klieger and Gibian (1962); Weygand and Hunger (1962)
Z-Glu-OBzl	Klieger and Gibian (1962); Weygand and Hunger (1962)
Z-Glu-OBzl · DCHA	Klieger and Gibian (1962); Weygand and Hunger (1962)
Z-Glu(OBzl)-ONo	Bodanszky *et al.* (1974a)
Z-Glu(OBzl)-ONp	Goodman and Stueben (1959)
Z-Glu(ONp)-OBzl	Wendel *et al.* (1972)
Z-Glu(OSu)-OBzl	Hardy *et al.* (1972)

(*continued*)

Table II *(Continued)*

Derivative	References[b]
Moz-Glu(O*t*Bu)-OH	Schröder and Klieger (1964a)
Moz-Glu(O*t*Bu)-OH · DCHA	Schröder and Klieger (1964a)
Moz-Glu(O*t*Bu)-N_2H_3	Schröder and Klieger (1964a)
Moz-Glu(OBzl)-OH · DCHA	Weygand and Hunger (1962)
Moz-Glu-OBzl	Weygand and Hunger (1962)
Moz-Glu(OBzl)-ONp	Yajima and Kiso (1974a)
Arginine	
Arg-NCA	Hirschmann *et al.* (1971); Weinert *et al.* (1971)
$Arg(Z)_2$-NCA	Hayakawa *et al.* (1969)
Adc-$Arg(Adc)_2$-OH[e]	Jäger and Geiger (1970)
Adc-$Arg(Adc)_2$-ONp	Jäger and Geiger (1970)
Adc-$Arg(Adc)_2$-OSu	Jäger and Geiger (1970)
Boc-Arg(Boc)-OH	Schnabel (1967)
Z-Arg-ONp · HNO_3/picrate	Glass and Pelzig (1978)
Z-$Arg(NO_2)$-OH	Hofmann *et al.* (1956); Gibian and Schröder (1961)
Z-$Arg(NO_2)$-N_2H_2-Boc	Tomatis *et al.* (1976a)
Z-$Arg(NO_2)$-OPcp	Kovacs *et al.* (1967)
Z-$Arg(NO_2)$-OPfp	Kisfaludy *et al.* (1970)
Z-$Arg(Z)_2$-OH	Zervas *et al.* (1957); Zervas *et al.* (1959); Wünsch and Wendlberger (1967); Smithwick and Shuman (1974)
Z-$Arg(Z)_2$-ONp	Wünsch and Wendlberger (1967)
Z-$Arg(Z)_2$-OSu	Anderson *et al.* (1964)
Z-$Arg(Z)_2$-OTcp	Jäger *et al.* (1974)
Moz-$Arg(Adc)_2$-OH	Jäger and Geiger (1970)
Moz-$Arg(Adc)_2$-OSu	Jäger and Geiger (1970)
Moz-$Arg(NO_2)$-OH	Klieger (1969); Nagasawa *et al.* (1973)
Moz-$Arg(NO_2)$-OPcp	Klieger (1969)
Moz-$Arg(Moz)_2$-OH	Smithwick and Shuman (1974)
Moz-$Arg(Z)_2$-OH	Weygand and Nintz (1965)
Tyrosine	
Tyr-NCA	Hirschmann *et al.* (1971)
Aoc-Tyr(*t*Bu)-OH[f]	Hase *et al.* (1972)
Aoc-Tyr(*t*Bu)-OH · DCHA	Hase *et al.* (1972)
Aoc-Tyr(*t*Bu)-OSu	Hase *et al.* (1972)
Boc-Tyr-N_2H_3	Schröder (1963)
Boc-Tyr-N_2H_2-Z	Storey *et al.* (1972)
Boc-Tyr-ONp	Schnabel *et al.* (1968); Meienhofer *et al.* (1970); Hase and Walter (1973)
Boc-Tyr-OPcp	Jones *et al.* (1973); Kamber (1973)
Boc-Tyr-OSu	Schnabel *et al.* (1968); Wolman *et al.* (1972)
Boc-Tyr-OTcp	Mehlis *et al.* (1972)
Boc-Tyr(Boc)-OH · DCHA	Schnabel *et al.* (1968)
Boc-Tyr(Boc)-ONp	Schnabel *et al.* (1968)
Boc-Tyr(Boc)-OSu	Schnabel *et al.* (1968)
Boc-Tyr(Boc)-OTcp	Mehlis *et al.* (1972)
Boc-Tyr(*t*Bu)-OH · DCHA	Jost (1971)
Boc-Tyr(*t*Bu)-N_2H_3	Schröder (1963)

Table II (*Continued*)

Derivative	References[b]
Boc-Tyr(*t*Bu)-OSu	Jost (1971)
Z-Tyr-N_2H_3	Harington and Pitt-Rivers (1944); Determann and Wieland (1961)
Z-Tyr-N_2H_2-Boc	Guttmann (1961)
Z-Tyr-OTcp	Pless and Boissonnas (1963)
Z-Tyr(Bzl)-OH	Morley (1967)
Z-Tyr(Bzl)-ONo	Bodanszky *et al.*, (1974a)
Z-Tyr(Bzl)-ONp	Itoh (1970)
Z-Tyr(Bzl)-OTcp	Morley (1967)
Z-Tyr(Z)-OH	Katchalski and Sela (1953)
Z-Tyr(Z)-N_2H_2-Boc	Guttmann (1961); Zahn and Fölsche (1968)
Z-Tyr(Z)-ONp	Mazur and Schlatter (1963)
Z-Tyr(Z)-ON < Pht[g]	Kaurov *et al.* (1970)
Z-Tyr(Z)-OSu	Zahn and Fölsche (1968); Schechter *et al.* (1971)
Z-Tyr(Z)-OTcp	Pless and Boissonnas (1963)
Moz-Tyr-N_2H_3	Fujii and Yajima (1975)
Moz-Tyr(Bzl)-OH	Weygand and Nintz (1965); Klieger (1969); Vandesande (1969)
Moz-Tyr(Bzl)-ONp	Vandesande (1969)
Serine	
Ser(Tms)-NCA[h]	Hirschmann *et al.* (1971); Wies and Pfaender (1973)
Boc-Ser-N_2H_3	Iselin and Schwyzer (1961); Otsuka *et al.* (1966); Storey *et al.* (1972)
Boc-Ser-OPcp	Johnson (1973)
Boc-Ser-OPfp	Kisfaludy *et al.* (1973)
Boc-Ser(Boc)-OH · DCHA	Schnabel *et al.* (1971)
Boc-Ser(*t*Bu)-OH	Schnabel (1967); Schnabel *et al.* (1968)
Z-Ser-N_2H_3	Fruton (1942)
Z-Ser-N_2H_2-Boc	Tomatis *et. al.* (1976b)
Z-Ser-ODnp	Marchiori *et al.* (1963)
Z-Ser-OOBt[c]	König and Geiger (1970b)
Z-Ser-OMor[i]	Okamoto and Shimamura (1973)
Z-Ser-OPcp	Kovacs *et al.* (1967)
Z-Ser-OPfp	Kisfaludy *et al.* (1970)
Z-Ser-OTcp	Pless and Boissonnas (1963)
Z-Ser(Bzl)-ONo	Bodanszky *et al.* (1974a)
Z-Ser(Bzl)-ONp	Ondetti (1963)
Z-Ser(Z)-OH	Oki *et al.* (1970)
Z-DSer(Z)-OH	Oki *et al.* (1970)
Moz-Ser-N_2H_3	Ohno *et al.* (1967)
Moz-Ser-ODnp	Klieger (1969)
Moz-Ser-OMor	Okamoto *et al.* (1979)
Moz-Ser-OPcp	Klieger (1969)
Moz-Ser(Bzl)-OH	Weygand and Nintz (1965); Klieger (1969); Nagasawa *et al.* (1973)
Moz-Ser(Bzl)-OTcp	Klieger (1969)
Threonine	
Thr(Tms)-NCA[h]	Hirschmann *et al.* (1971); Wies and Pfaender (1973)
Boc-Thr-N_2H_3	Schröder and Gibian (1962)
Boc-Thr-ODnp	Bodanszky *et al.* (1967)

Table II *(Continued)*

Derivative	References[b]
Boc-Thr-ONp	Meienhofer *et al.* (1970)
Boc-Thr-OPcp	Johnson (1973)
Boc-Thr-OPfp	Kisfaludy *et al.* (1973)
Boc-Thr-OSu	Ondetti *et al.* (1968a); Storey *et al.* (1972)
Boc-Thr-OTcp	Bodanszky *et al.* (1977b)
Boc-Thr(Boc)-OH	Schnabel *et al.* (1971)
Boc-Thr(*t*Bu)-OH	Schnabel (1967)
Boc-Thr(*t*Bu)-N_2H_3	Schröder (1963)
Nps-Thr-OSu	Moroder *et al.* (1973b)
Z-Thr-N_2H_3	Schröder and Gibian (1962)
Z-Thr-N_2H_2-Boc	Wolman and Klausner (1971); Moroder *et al.* (1973d)
Z-Thr-ODnp	Rocchi *et al.* (1963)
Z-Thr-OObt[c]	König and Geiger (1970b)
Z-Thr-ONp	Bodanszky and Ondetti (1966); Meienhofer *et al.* (1970)
Z-Thr-OPcp	Fauszt and Bajusz (1974); Yajima and Kiso (1974a)
Moz-Thr-N_2H_3	Ohno *et al.* (1967)
Moz-Thr-OPcp	Yajima *et al.* (1974c)
Histidine	
His-NTA	Dewey *et al.* (1971)
Adc-His(Adc)-OH[e]	Haas *et al.* (1966); Moroder *et al.* (1976)
Adc-His(Adc)-OSu	Wünsch *et al.* (1968)
Boc-His-N_2H_3	Schröder and Gibian (1962); Wünsch and Zwick (1964)
Boc-His(Boc)-OH	Schnabel *et al.* (1968, 1971); Fridkin and Goren (1971); Schaich *et al.* (1973)
Boc-His(Boc)-OH · DCHA	Schnabel *et al.* (1971); Yamashiro *et al.* (1972)
Boc-His(Boc)-ONp	Schanbel *et al.* (1968); Schaich *et al.* (1973)
Z-His-N_2H_3	Holley and Sondheimer (1954)
Z-DHis-N_2H_3	Yajima and Kubo (1965)
Z-His-N_2H_2-Boc	Moroder *et al.* (1973b)
Z-His(Z)-OH	Patchornik *et al.* (1957); Sakiyama *et al.* (1958); Inouye and Otsuka (1962)
Z-His(Z)-ONp	Inouye (1965)
Z-His(Z)-OTcp	Bajusz and Fauszt (1973)
Moz-His-N_2H_3	Yajima *et al.* (1973b)
Moz-His(Moz)-OH	Schaich *et al.* (1973)
Moz-His(Moz)-ONp	Schaich *et al.* (1973)

[a] This table comprises a fairly comprehensive list of derivatives of the amino acids which are "pivotal" in defining minimal protection synthesis (groups 2 and 3 of Table I) *and* which are suitable for stage 0 or stage 1 tactics as defined in footnote c of Table III.

[b] See reference list at end of chapter.

[c] OObt = 3-hydroxy-4-oxo-3,4-dihydro-1,2,3-benzotriazine ester.

[d] Ibc = isobornyloxycarbonyl.

[e] Adc = adamantyloxycarbonyl.

[f] Aoc = *tert*-amyloxycarbonyl.

[g] ON < Pht = *N*-hydroxyphthalimide ester.

[h] Tms = trimethylsilyl.

[i] OMor = 4-hydroxymorpholine ester.

were included: Almost any method of activation can then be used; these derivatives simplify activation procedures and yield generally more stable, activated reagents. Advantages related to the presence of a free side chain can be gained in subsequent steps.

B. Aspartic Acid and Glutamic Acid

The tactics used for synthesis of peptides with unprotected aspartic and glutamic acid side chains were discussed in Section I,B,2. Only the use of *N*-carboxyanhydrides allows direct, stepwise introduction of these amino acids in α-peptide linkages with no side-chain protection. In contrast, the aminolysis of N-protected aspartic and glutamic acid α,ω-anhydrides will yield mixtures of α- and ω-peptide bonds. Therefore, in strategies where ω-carboxyl groups are to be left unprotected as much as possible, important use can be made of the alternative approach whereby α-amino and side-chain protecting groups are removed simultaneously.

The potential advantages of free carboxyl side chains are improved solubility (in some cases), improved separations (via ion-exchange chromatography or electrophoresis), decreased tendency toward formation of cyclic imides (especially with β-esters of aspartic acid; Ondetti *et al.*, 1968a; Suzuki *et al.*, 1976; Bodanszky and Kwei, 1978; Martinez and Bodanszky, 1978), and fewer problems in final deprotection where acid-sensitive amino acids are present or where hydrogenation is not practical (Bodanszky and Natarajan, 1975).

While side reactions involving aspartic and glutamic acids during synthesis are fewer when these residues are unprotected, certain problems may arise in the case of aspartic acid peptides in particular. When synthesizing the pentapeptide amide, Boc-Gly-Trp-Met-Asp-Phe-NH_2, by a classical stepwise approach, Bodanszky and Natarajan (1975) noted several less polar by-products. After the final step, one impurity exceeded the desired product in abundance in some trials. This by-product was isolated, characterized and found to be an aspartimide derivative formed by attack on the aspartic acid β-carbonyl by the phenylalanine nitrogen atom. Formation of the aspartimide occurred mainly during reaction of the penultimate tetrapeptide with Boc-Gly-ONp and was particularly serious in cases where the amount of base (diisopropylethylamine) or active ester was in appreciable excess. In conclusion, it was determined that excess base (more than required to free the tetrapeptide amine from its trifluoroacetate salt) brought about ionization of the aspartic acid side chain. The more nucleophilic carboxylate ion was attacked by Boc-Gly-ONp yielding a mixed anhydride, Boc-Gly-Trp-Met-Asp(Boc-Gly-O)-Phe-NH_2, which then mediated an intramolecular acylation forming the aspartimide derivative.

A second by-product was isolated by Natarajan and Bodanszky (1976) from the above reaction. It was characterized and shown to be a nonapeptide resulting from an *inter*molecular acylation of the free tetrapeptide by the proposed β-aspartyl-Boc-glycyl mixed anhydride. The resulting α,β-bis-aspartyl nonapeptide was shown to have the structure, Boc-Gly-Trp-Met-Asp(Trp-Met-Asp-Phe-NH_2)-Phe-NH_2, which further supported the mixed anhydride mechanism. In the previous and subsequent steps of this synthesis, which employed active esters of other amino acids, these side reactions occurred only to a minor extent suggesting that glycine, having no side-chain hindrance, formed the mixed anhydride intermediate with particular ease. Satisfactory results were obtained when the troublesome step was carried out with equimolar amounts of base, amine component, and Boc-Gly-ONp, with gradual addition of the latter reagent (Bodanszky and Natarajan, 1975).

C. Tyrosine, Serine, and Threonine

1. *Tactics*

The hydroxyamino acids often have been introduced into peptides as acyl azides whenever the side chains were left unprotected. Occasionally, active esters have been used for this purpose. A number of investigators have employed the 2,4-dinitrophenyl esters, Z-Ser-ODnp and Z-Thr-ODnp, first described by Marchiori *et al.* (1963) and Rocchi *et al.* (1963), respectively. Other active esters with free side chains have been described and are listed in Table II. The preparation of active esters and other activated forms of O-unprotected hydroxyamino acids can lead to difficulties as mentioned below in the case of Z-Ser-ONp and Z-Thr-ONp, although Meienhofer *et al.* (1970) reported a successful synthesis of the latter compound and its analogue, Boc-Thr-ONp. A very interesting scheme for circumventing the usual side reactions in this kind of synthesis was described by Bodanszky *et al.* (1977b) who prepared Boc-Thr(Bzl)-OTcp (2,4,5-trichlorophenyl ester) and subsequently converted it to Boc-Thr-OTcp by catalytic hydrogenation. Of course, this method cannot be used for making corresponding nitrophenyl esters.

In a surprising number of cases the hydroxyamino acids have been incorporated successfully into small peptides by rather strong activating agents such as chloroformates and carbodiimides. Z-Ser-Ala-N_2H_2-Boc and the corresponding dipeptide methyl ester were prepared from a mixed anhydride of Z-Ser-OH with 76% yield in both instances (Fauszt and Bajusz, 1974). Subsequently, two more amino acids were added to this dipeptide as their mixed anhydrides. Threonine active esters have been prepared via mixed anhydrides: for example, Z-Thr-OPcp (pentachlorophenyl; Fauszt and

Bajusz, 1974) and Boc-Thr-OSu (hydroxysuccinimido; Storey *et al.*, 1972). Threonyl peptides have commonly been prepared via mixed anhydrides. Examples are Z-Thr-Tyr-OMe and Z-Thr-Phe-N_2H_2-Boc (Yanaihara *et al.*, 1977). Also, in the course of this work on the synthesis of secretin, unprotected serine, threonine, and tyrosine residues were present in amine components which were reacted with mixed anhydrides. Camble and Cotton (1975) coupled a benzyloxycarbonylthreonine mixed anhydride to H-Leu-NH_2 and applied benzyloxycarbonylisoleucine mixed anhydride in a later elongation step. Kasafirek *et al.* (1977) prepared Z-Tyr-Gly-OEt by this method.

A more elegant means for generating mixed anhydrides, in the presence of amine components, involves use of the 1,2-dihydroquinoline derivative, EEDQ (Belleau and Malek, 1968) or IIDQ (Kiso and Yajima, 1972). With these reagents the formation of a mixed anhydride is relatively slower than concomitant aminolysis; therefore, the highly reactive anhydride does not accumulate. Peptides with unprotected serine, threonine, and tyrosine residues were prepared satisfactorily with EEDQ or IIDQ by Kiso *et al.* (1973), Klausner *et al.* (1973), Tamura *et al.* (1975), and Vicar *et al.* (1976).

Direct coupling with dicyclohexylcarbodiimide (DCC) has been applied to the synthesis of small peptides containing the hydroxyamino acids. Examples of peptides formed by DCC are Z-Tyr-Ser-OMe (Costopanagiotis *et al.*, 1968), Z-Ser-Glu-Gly-OMe (Handford *et al.*, 1968), Moz-Thr-Phe-OMe and Moz-Tyr-Ser-OMe (Kubota *et al.*, 1976), and Boc-Ser-Met-OMe (Otsuka *et al.*, 1966). The latter workers did not successfully prepare Boc-Ser-Tyr-OMe with DCC, however. In more recent years it has been customary to use additives ("coupling catalysts") such as HOBt and HOSu when employing DCC with components bearing unprotected side chains. The rapid formation of intermediate active esters reduces the chances of side reaction between third functions and the more reactive *O*-acylisourea derivatives which are first formed. It should be cautioned that primary hydroxyl groups may to some extent form ether bonds with HOBt in the presence of DCC (Hemmasi and Bayer, 1977).

2. *Problems*

A very large number of peptides have been synthesized by azide coupling where one or both components possessed unprotected hydroxylic side chains and where no problems with O-acylation of these groups were reported. Either the azides seldom attack hydroxylic side chains or else the modest occurrence of these side reactions has been masked by the usually rich assortment of minor by-products. One report suggests that such a side reaction may have occurred. From the azide coupling mixture of two large segments of the ribonuclease A sequence, Strachan *et al.* (1969) separated a small amount of by-product which appeared to result from condensation of two parts of the azide-bearing segment with one part of the amine component. The latter contained one histidine, one threonine, and two serine residues. In this instance O-acylation may have taken place, possibly catalyzed by the imidazole functionality of histidine; this topic will be discussed below.

Cases involving O-acylation of side chains have been most frequently reported where couplings have been carried out with active esters. Active esters have been widely and successfully employed in synthesis with minimal side-chain protection. Therefore, these reports should be thoughtfully studied since they may have caused many investigators to abandon minmal protection approaches prematurely.

Ramachandran and Li (1963) studied the reaction between equimolar amounts of Z-Val-ONp and HCl·H-Tyr-OMe in acetonitrile at room temperature. By-products were identified on paper chromatograms with specific group reagents. In an experiment where two moles of triethylamine (Et_3N) were added, significant yields of O-(Z-Val)-Tyr-OMe and N,O-$(Z\text{-Val})_2$-Tyr-OMe were obtained. The yield of the former by-product was 40–50%; none of this compound appeared when Et_3N was omitted, and only traces were found when exactly one equivalent of base was added, enough simply to release (and completely solubilize) H-Tyr-OMe from its HCl salt. Excess Et_3N apparently removes a proton from the tyrosine hydroxyl group leaving a highly nucleophilic phenoxide ion open to attack by the *p*-nitrophenyl ester. Nitrophenyl esters of other amino acids gave analogous O-substituted products when allowed to react with H-Tyr-OMe in the presence of excess Et_3N. Likewise, Camble *et al.* (1972) observed a by-product, apparently caused by O-acylation of tyrosine, when Z-Tyr(Z)-OTcp was reacted with a nonapeptide bearing an NH_2-terminal, unprotected tyrosine residue. The extent of this side reaction was enough to lead to serious separation problems when excess Et_3N had been used. In the absence of base the nonapeptide acetate was acylated to give the desired product which was isolated in satisfactory yield. These authors caution, however, that the above and other types of side products accumulate in stepwise active ester syntheses of fairly large, minimally protected segments, and purifications become progressively more difficult as size increases, especially beyond 9 or 10 residues.

Storey *et al.* (1972) attributed by-products to O-acylations of tyrosine residues in the stepwise synthesis of several subsegments of ribonuclease T_1, using 2,4,5-trichlorophenyl and *N*-hydroxysuccinimido esters. In the use of *N*-carboxyanhydrides in peptide synthesis by Hirschmann and Veber (1973), unprotected tyrosine created significant problems in aqueous medium but not in anhydrous solvents. In part, such problems could be dealt with by subsequently breaking unwanted ester bonds with hydroxyamino acids by means of hydrazinolysis.

The primary hydroxyl groups of serine have also been implicated in side reactions with active esters. Zahn *et al.* (1965) carried out the reaction between Z-Cys(Bzl)-ONp and an equimolar amount of H-Ser-OMe, liberated from the HCl salt with Et_3N. Excess base was neutralized with acetic acid (HOAc). No side reactions occurred in dimethylformamide (DMF) where the mixture was in solution. However, when the reaction was done in ethyl acetate (EtOAc) or tetrahydrofuran (THF), HCl · H-Ser-OMe was very slowly dissolved and converted to the free amine. Thus, the active ester was in effective excess much of the time; the result was formation of *N*,*O*-[Z-Cys(Bzl)]$_2$-Ser-OMe in yields up to 25%.

If the nucleophilic component is insufficiently reactive, reactions at hydroxylic side chains will be promoted. Bodanszky and Ondetti (1966) were unable to prepare satisfactorily Z-Ser-ONp and Z-Thr-ONp with dicyclohexylcarbodiimide by the usual method (Bodanszky and du Vigneaud, 1962). Extremely small yields of desired products were obtained; the main products were O-acylated compounds, Z-Ser(Z-Ser)-ONp and Z-Thr(Z-Thr)-ONp, respectively. On the other hand, the corresponding actives esters, Z-Ser-ODnp and Z-Thr-ODnp, were prepared in the normal way from the more nucleophilic 2,4-dinitrophenol (Rocchi *et al.*, 1963; Marchiori *et al.*, 1963; Bodanszky and Ondetti, 1966). The rate of O-acylation of serine in a peptide can best be studied with the NH_2-terminus blocked. Thus, Schafer *et al.* (1971) found that an equimolar mixture of Z-Gly-OTcp and Z-Gly-Ser-OMe in dioxane at room temperature resulted in formation of a detectable (ca 1%) amount of Z-Gly-Ser(Z-Gly)-OMe in several hours. When 2 moles/mole each of Z-Gly-OTcp and Et_3N were added to the dipeptide derivative, O-acylation was complete within 19 hours. As in the case of tyrosine-containing peptides, an excess of base considerably enhanced the rate of this reaction. What is of principal importance is the *relative* rates of N- and O-acylation. Schafer *et al.* (1971) also studied the reaction of Z-Gly-OTcp with H-Ser-OMe in the molar ratio of 0.8 to 1.0 and in the absence of excess base. The effect of different solvents were compared: after 24 hours at room temperature, when the reaction was effectively completed, the yield of Z-Gly-Ser(Z-Gly)-OMe was approximately 1–5% in dioxane, 10–15% in $CHCl_3$ and essentially zero in DMF or DMF–H_2O (95:5). The use of 1.25 moles/mole of the active esters, Z-Gly-OTcp or-ONp, in DMF gave at least 1% of the O-acyl derivative in 16 hours. The authors felt that these modest levels of by-product justified the use of side-chain protection for serine in their projected application. Yet these data suggest that reasonable attention to the avoidance of excess base and the proper choice of solvent could lead to quite satisfactory results without protection of serine residues in most active ester couplings. Even slight excesses of active esters might be satisfactorily employed if reaction times were not prolonged unnecessarily.

Sieber *et al.* (1972) in the stepwise synthesis of human ACTH(25–39), largely with ONp esters in the presence of 0.7 equivalents of base (*N*-methylmorpholine), found that serine-31 was O-acylated in part. Brief exposure to hydrazine hydrate (5% v/v in DMF or MeOH for 15 min at 25°C followed by H_2O precipitation) after each step sufficed to reverse this side reaction. In this special case such treatment could not be employed after introducing asparagine-25 because of the base-induced isomerization of the labile Asn-Gly(25–26) bond giving predominantly the β-linked peptide via the imide. Again, this case may illustrate the problems of unprotected side chains in long stepwise syntheses where excess acylating components must be used to complete effectively the desired steps. During the synthesis of bradykinin, Sakakibara and co-workers (Sakakibara, 1977) noted an impurity arising from the coupling of Aoc-Pro-Pro-Gly-ONp to H-Phe-Ser-Pro-Phe-Arg(NO_2)-OBzl. Following attachment of a terminal arginine residue and deprotection with HF–anisole, the impurity was isolated and shown likely to be Arg-Pro-Pro-Gly-Phe-Ser(Arg-Pro-Pro-Gly)-Pro-Phe-Arg. This composition would not have resulted from an intramolecular acyl migration during deblocking but could be the consequence of an O-acylation of the serine residue in the above coupling reaction.

It appears that the problem of O-acylation can be markedly exacerbated by a specific catalytic effect of the imidazole ring which promotes acyl transfer to oxygen nucleophiles. Tyrosine residues are preferentially O-acetylated over amine groups (at least in relation with most other acetylating reagents) by acetylimidazole (Riordan and Vallee, 1972). Bodanszky and Fagan (1977) used imidazole-catalyzed transesterification of protected amino acid active esters in forming ester linkages with hydroxymethylated polystyrene supports for application in solid-phase peptide synthesis. The reactions went smoothly and efficiently and could be employed with various ONp, ONo, OTcp, OPcp, and OSu amino acid esters. In the preparation of Z- or Boc-Val-Tyr-OEt, Paul (1963) reported that the largest amount of O-acyl by-product and lowest product yield occurred when *N,N′*-carbonyldiimidazole was employed as coupling agent as compared with the use of ONp ester and mixed anhydride methods. Ondetti *et al.* (1968b) used *N,N′*-carbonyldiimidazole to join the NH_2-terminal tetrapeptide segment of secretin to the rest of the molecule which is rich in serine and threonine but contains no tyrosine residues. Partial O-acylation of either serine or threonine side chains was observed.

The message is clear that one should avoid the use or addition of imidazolic catalysts or reagents where ester bond formation with unprotected hydroxylic side chains is not desired. The problem that has been discussed above is not so easily avoided with minimal side-chain protection in the synthesis of peptides containing both histidine and hydroxyamino acids, especially serine or tyrosine. Bodanszky *et al.* (1977a) in a careful study found that acylation of the partially protected hexapeptide H-Pro-Ser(Bzl)-His-Arg(NO_2)-Ile-Ser-OMe with Boc-Asp(OBzl)-OSu in the presence of 1-hydroxybenzotriazole (HOBt) produced, in addition to the desired heptapeptide derivative, significant amounts of a by-product which was identified as the N,O-diacylated peptide, Boc-Asp(OBzl)-Pro-Ser(Bzl)-His-Arg(NO_2)-Ile-Ser[Boc-Asp(OBzl]-OMe. Model experiments showed that extensive O-acylation was due to the imidazole ring of the unprotected histidine residue. The effect could be imitated by using a shorter amine component, lacking histidine, in reaction mixtures to which imidazole had been added. The effects of several active esters and the presence or absence of HOBt or imidazole were studied. The ONp ester tended to produce more diacyl derivative than the OSu compound. Imidazole markedly increased bis-acylation except in the particular case of the ONp ester plus HOBt. Thus, unexpected and perhaps useful variations in outcome can be found by such trial experiments. Geiger *et al.* (1974) also noted O-acylation of a serine residue in a histidine-containing peptide by an OTcp ester. Acylation of serine, threonine, or tyrosine side chains during couplings with active esters can be largely prevented, even in peptides containing histidine, by the use of additives such as 2,4-dinitrophenol and pentachlorophenol (Martinez *et al.*, 1979).

D. Arginine

The guanidino group of arginine under special circumstances can undergo a variety of side reactions in peptide synthesis. However, one is principally concerned with the possibility of undesired acylations at unprotected guanidino nitrogen atoms. Because of the strongly basic character of this side chain (p*K* 12.5), desired couplings can be satisfactorily performed with unprotected ariginine in either COOH- or NH_2-components provided the medium is not highly alkaline. Most workers, in fact, protect arginine side chains in order to reduce water solubility and perhaps to change other physical properties that render isolation of relatively small peptides difficult. The introduction of unprotected arginine residues into large peptides may confer properties advantageous for isolation and can reduce problems arising in final deprotection, especially of nitroarginine residues.

Unprotected arginine can be introduced directly into a growing peptide as its *N*-carboxyanhydride (Hirschmann *et al.*, 1971). Otherwise, arginine with unmasked side chain (i.e., simply protected by protonation) can be introduced with suitable α-amine protection by mixed anhydride, carbodiimide, or azide methods (see Schröder and Lübke, 1965, p. 175 for references to examples). Many syntheses of small arginine peptides or segments have been subsequently carried out in this simple manner. For illustrative examples see Riniker *et al.* (1969), Kamber *et al.* (1976), and Moroder *et al.* (1973c).

Stable, N^{α}-protected, N^{G}-unprotected arginine active esters have been difficult to prepare. Activation of the arginine carboxyl group in absence of a sufficiently strong nucleophile can result mainly in lactam formation involving reaction of the carboxyl carbonyl with the proximal guanidine nitrogen. Nevertheless, Glass and Pelzig (1978) reported the synthesis of stable, characterizable nitrate and picrate salts of Z-Arg-ONp which, however, reacted effectively with amine components only in the presence of an additive such as 1-hydroxybenzotriazole. The attempted synthesis of Z-Arg(NO_2)-ONp was difficult in spite of partial masking by the nitro group (Bodanszky and Sheehan, 1960; Paul *et al.*, 1961). The corresponding ester of the more nucleophilic 2,4-dinitrophenol, Z-Arg(NO_2)-ODnp, was prepared by Bodanszky and Ondetti (1966), and the crude product could be used in peptide bond formation within several days; but the compound was not stable enough to isolate and store. Syntheses of stable N^{α}-protected Arg(NO_2) active esters have been reported (see Table II).

Active esters of arginine with a disubstituted side chain are listed in Table II. Such side-chain protection may be removed along with the α-amine blocking group. A very different way of avoiding the special problems in synthesizing arginine peptides consists of preparing ornithine analogues and finally converting the ornithine residues to arginine by guanidination (Bodanszky *et al.*, 1966; Borin *et al.*, 1971, 1975; Marchiori *et al.*, 1979).

A side reaction which can be troublesome in syntheses with minimal side-chain protection, where azide segment condensations are commonly used, is hydrazinolysis of arginine side chains (to ornithine) whenever COOH-terminal esters of segments are converted to hydrazides. Strachan *et al.* (1969) avoided this problem by carrying out hydrazinolysis with a large excess of hydrazine (in MeOH or DMF) for short periods of time (as brief as 2 min). This problem is also encountered with nitroarginine peptides. Of course, it can be circumvented by starting the synthesis of a segment with a protected hydrazide.

Synthesis of peptides with unprotected arginine residues can lead to by-products resulting from an initial acylation of the guanidino group. Photaki *et al.* (1975) and Photaki and Yiotakis (1976) studied the formation and properties of N^{G}-peptides. The acylating step could be effected with nitropheny esters after the guanidino group had been deprotonated with NaOH and excess Et_3N had been added. They found that the resulting compounds were not stable, especially in alkaline medium, but split into ornithine and the corresponding 2-iminoimidazolidin-4-one (glycocyamidine) derivative (**2**). Thus, N^{G}-acylation by-products (**1**) can lead to formation of peptides (**3**) with ornithine replacing arginine [Eq. (1)].

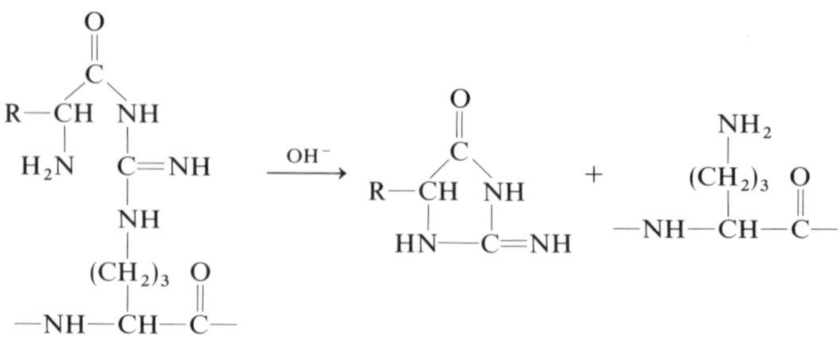 (1)

(1) (2) (3)

E. Histidine

Stable *N*-acyl derivatives of the unprotected imidazole ring of histidine residues do not normally form during synthesis, so that protection of these side chains is not usually required for this reason. On the other hand, acylimidazoles are acylating agents and imidazoles can catalyze ester and amide bond hydrolysis. Therefore, if such difficulties arise, it may be necessary to use a protecting group. As with small arginine peptides purification problems may be improved by protection. In other instances separations have been aided by the polar, ionizable character of the unmodified imidazole ring (Zahn *et al.*, 1964). Schafer and Carlsson (1972) employed the protonated form of histidine in the synthesis of small peptides as an effective cationic "handle" in purifications following each step: the peptide was held in place in an aqueous liquid phase or on an ion-exchange resin while excess reagents and by-products were solvent extracted.

In the majority of cases where histidine has been introduced with its side-chain unsubstituted, workers have started with HCl·H-His-OMe or protected His-N_3 (prepared from stable hydrazides, Table II; see also Schröder and Lübke, 1965, pp. 177–178). The COOH-terminal methyl ester has usually been converted to a hydrazide and azide after completion of a dipeptide or segment. Alternatively, segments have been prepared from COOH-terminal H-His-N_2H_2-Boc (Moroder *et al.*, 1973b,c).

The use of a carbodiimide for activating histidine carboxyl groups can lead to lactam formation with the proximal nitrogen atom of the imidazole ring (Sheehan *et al.*, 1959). N^{α}-Protected histidine and COOH-terminal histidine peptides have been activated and successfully coupled in the presence of amine components with carbodiimide (DCC) plus additives such as HOBt (for example, see Voskuyl-Holtkamp and Schattenkerk, 1977; Schoelkens *et al.*, 1976) or HONb (*N*-hydroxy-5-norbornene-2,3-dicarboximide; Fujino *et al.*, 1972, 1974b). The coupling agent, diphenyl phosphorazidate (DPPA), was used to add Z-His-OH to a tetrapeptide (Hamada *et al.*, 1977). Direct introduction of histidine was accomplished with its *N*-thiocarboxyanhydride or NTA (Dewey *et al.*, 1971; the corresponding NCA was not stable).

Stable active esters of N^{α}-protected histidine have not been reported so far. In Table II are listed active esters which have been prepared from disubstituted N^{α},N^{im}-protected histidine with groups which can be removed simultaneously. For an example of the use of this tactic see Bajusz and Fauszt (1973). H-His(Z)-OMe was prepared from Z-His(Z)-OMe by treatment with HBr in HOAc (Inouye, 1965) for the synthesis of a dipeptide methyl ester; the side-chain benzyloxycarbonyl group was then removed with hydrazine during hydrazinolysis of the ester.

Racemization of histidine appears to be an important problem in carboxyl activation of N^{α}-protected histidine or COOH-terminal histidine peptides (for example, see Rahn *et al.*, 1976). These observations are mainly connected with the widely employed azide method. Increased racemization may occur with aminoacyl or peptidyl azides terminating either in histidine or other amino acids when certain bases (e.g., triethylamine) are used in excess (Sieber *et al.*, 1971). There appear to be additional mechanisms which augment racemization of COOH-terminal histidine that depend upon the unprotected imidazole ring and may include the possibilities of intramolecular base catalysis of azlactone formation or base catalysis of α-proton abstraction. In addition, and perhaps more importantly, Veber and co-workers (Veber, 1975) demonstrated that an equilibrium [see Eq. (2)] exists between a histidyl azide (**4**) and the corresponding bicyclic lactam (**5**) where azide ion acts both as an effective leaving group and as a powerful

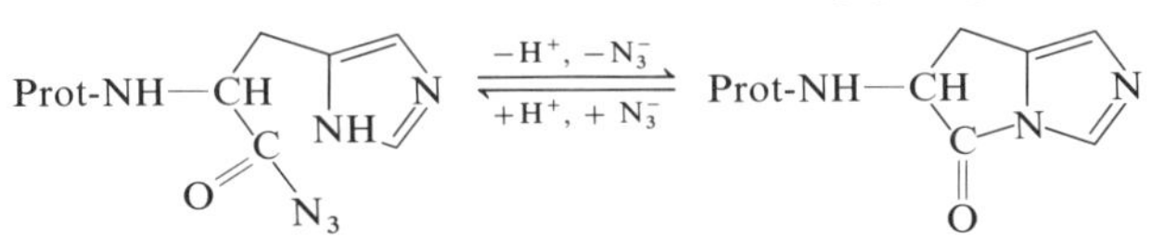

(2)

(**4**) (**5**)

nucleophile. The lactam (**5**) was noted above as being formed in the presence of a carbodiimide and in absence of an amine component (Sheehan *et al.*, 1959). Veber (1975) found that, with slowly reacting amine components such as methyl L-pipecolate or without an amine, compound **5** is rapidly and *reversibly* formed under neutral and alkaline conditions. The slow racemization at the chiral center of histidine which follows takes place more rapidly in compound **5** than in **4**. Racemization is reduced at lower pH values which favor formation of acyl azide (**4**); the latter form reacts substantially more rapidly with amines. Derivatives protected with im-benzyl are still prone to racemize, but electron-withdrawing groups (im-Boc or im-tosyl) do suppress optical inversion. Merrifield and Woolley (1956) proposed the above lactam formation in order to account for the observed self-reaction of Z-Ser-His-N_3 in alkaline medium. Schneider (1960) prepared the same dipeptidyl azide and also isolated a product which gave a negative Pauly reaction.

An addition reaction can occur between DCC and an unprotected side chain of histidine. The histidine residue may reside anywhere in either coupling component. Rink and Riniker (1974) showed that DCC slowly reacts with a nitrogen in imidazole or with either nitrogen in the side chains of Boc-His-OMe or Z-Val-His-OMe. In the latter cases formation of the N^{im}-(N-N'-dicyclohexyl)amidino isomers (**6**) occurs, particularly with excess

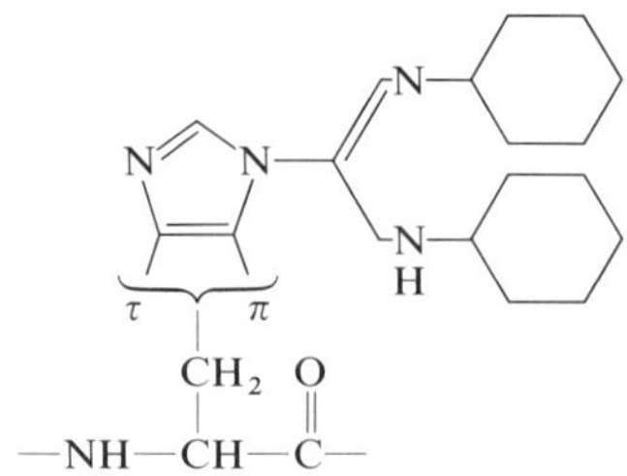

(**6**)

DCC at elevated temperatures. Most notably, the reaction is markedly catalyzed by HOBt in DMF at room temperature. Thus, it can lead to significant amounts of the histidine-substituted by-products during segment condensations with DCC + HOBt, especially where segments are large. Andreatta *et al.* (1973) observed this side reaction (to the extent of 13%) when coupling segments 18–24 and 25–34 of human parathyroid hormone (as cited by Rink and Riniker, 1974). Yajima *et al.* (1976a), in the synthesis of porcine ACTH, obtained an essentially inactive final product after a histidine-containing 6-residue segment had been joined to a 29-residue peptide via DCC + HOBt. Treatment of the final product with MeOH–2 *N* HOAc (4:1) at 60°C for 4 hr did not fully regenerate the active peptide as would have been expected from the dissociation experiments of Rink and Riniker (1974) with smaller amidine derivatives under similar conditions. Thus, removal of attached DCC from a large histidine peptide may proceed somewhat more slowly. The same synthesis was repeated with the above-mentioned segment condensation being carried out by means of pentachlorophenyl trichloroacetate, and a fully active product was obtained (Yajima *et al.*, 1976a). In the syntheses of bovine and human ACTH Yajima's group did, however, obtain the active hormones after treating the final products with MeOH–HOAc mixtures (Koyama *et al.*, 1976a,b). They had applied DCC + HOBt to the segment condensation involving the histidine-containing component.

F. Tryptophan and Methionine

All of the commonly employed coupling methods have been used in syntheses of peptides containing unprotected tryptophan or methionine residues. Side reactions with the indole and thioether functions have often proved troublesome in both coupling and deprotection steps. Furthermore, the presence of methionine has greatly restricted the use of catalytic hydrogenolysis for removing benzylic-type protecting groups; catalysts are poisoned and methionine residues are partially desulfurized. Recently, Felix *et al.* (1977) have reported a successful hydrogenolysis procedure with methionine peptides in liquid ammonia. Hydrogenations can lead to difficulties in the case of tryptophan peptides if atmospheric oxygen is present (D. Brandenburg, as cited by Borin *et al.*, 1973b).

The indole structure of tryptophan is particularly prone to oxidation and rearrangements under the acidic conditions used in deprotections (e.g., with HF, TFA, HCl, or HBr). Therefore, it is best to conduct these steps under nitrogen and to add scavengers—such as indole, skatole, or anisole—and also antioxidants such as ethanedithiol (Yajima *et al.*, 1977; Mori and Yajima, 1976). The other major problem occurring during acidic deprotection of tryptophan peptides has been *tert*-butylation of the indole ring where Boc groups, *tert*-butyl esters, or *tert*-butyl ethers are present (Alakhov *et al.*, 1970). These alkylation reactions can occur to a marked degree in HF or TFA and to a lesser extent in HCl–EtOAc or HCl–HOAc; N^{in}-*tert*-butyltryptophan is a major product, although various other ring substitutions can occur (Jaeger *et al.*, 1978; Löw *et al.*, 1978). The degree of butylation is affected by the position of tryptophan in the peptide (Alakhov *et al.*, 1970; Wünsch *et al.*, 1977). When Nps groups are removed in the presence of HCl in polar solvents, they are partly converted to *o*-nitrophenylsulfenyl chloride which in turn may react with the indole nucleus of tryptophan residues. Other means for removing this protecting group should be employed (Juillerat and Bargetzi, 1976).

It was suspected that tryptophan residues may be partly N-nitrosylated during conversion of peptidyl hydrazides to the corresponding azides with cold nitrous acid (Zahn and Brandenburg, 1966). That this side reaction can be important in azide condensations was corroborated in studies by Agarwal *et al.* (1969) and Bonnett and Holleyhead (1974). The former authors observed the formation of the N^{in}-nitroso derivative when azide formation was carried out either with acidified sodium nitrite or by the anhydrous procedure of Honzl and Rudinger (1961). The problem may be avoided by redesigning the synthesis so that tryptophan-containing carboxyl segments are not activated by the azide method. Alternatively, the indole nitrogen may be protected with a formyl group (Previero *et al.*, 1967; Yamashiro and Li,

1973). The N^{in}-formyl group also confers some protection against oxidation or *tert*-butylation in acidic media that is not altogether checked by the presence of scavengers (Ohno *et al.*, 1973; Löw and Kisfaludy, 1979).

The acidolytic removal of benzyloxycarbonyl (Z) groups results in formation of benzyl carbonium ions and benzyl halides which attack the thioether sulfur in methionine residues yielding sulfonium derivatives. Use of HBr in glacial HOAc may lead to cleavage of the thioether bond and subsequent formation of S-benzylhomocysteine. Demethylation of methionine residues may occur in ethanolic HCl or with excess sodium in liquid ammonia (see Schröder and Lübke, 1965, pp. 241–242). For many reasons the latter method is a risky way to remove benzyloxycarbonyl groups in stepwise procedures. Side reactions from acidolytic cleavages of benzyloxycarbonyl groups can be reduced but not completely eliminated by the use of scavengers (e.g., diethyl phosphite, methylethyl sulfide, or methionine).

Acidolysis of *tert*-butyl-type protecting groups on peptides can lead to *tert*-butylsulfonium derivatives of methionine side chains through the formation of *tert*-butylium ions (Veber *et al.*, 1969; Sieber *et al.*, 1970). When anisole is used as a scavenger in acidolytic deprotection with methanesulfonic acid, there is a danger that its methyl groups may be transferred to the methionine sulfur atom (Yajima *et al.*, 1977). These alkylation reactions may be prevented by prior conversion of methionine residues to the sulfoxide form by oxidation with hydrogen peroxide (Iselin and Schwyzer, 1961) or perborate (Yajima *et al.*, 1977). Synthesis of methionine peptides can begin with L-methionine *d*- or *dl*-sulfoxide (Iselin and Schwyzer, 1961; Hofmann *et al.*, 1965b; Fujii *et al.*, 1978).

Methionine in peptides will readily oxidize partially to sulfoxide during synthetic and purification steps (Hofmann *et al.*, 1965a, 1969). On thin-layer plates the sulfoxides migrate more slowly than corresponding unmodified peptides and complicate interpretation of the chromatograms. It is also for this reason that methionine peptides have been synthesized in the sulfoxide form. The native peptide can be recovered in the end by a mild reduction; for a study of this final step see Houghten and Li (1977).

G. Asparagine and Glutamine

Early reports on the synthesis of asparagine and glutamine peptides have been reviewed by Schröder and Lübke (1965, pp. 197 and 202–206). Carboxyl terminal asparagine and glutamine have been introduced as the free amino acids or as esters. Attempts to remove an ester group by alkaline hydrolysis have yielded primarily α-aminoaspartimide or α-aminoglutarimide derivatives; the latter are especially unstable toward hydrolysis and form mixtures of peptides with COOH-terminal α- and ω-amide residues (e.g., of glutamine and isoglutamine; Sondheimer and Holley, 1954). Ester blocking groups [e.g., OBzl, ONb, O*t*Bu] that can be removed by other means are to be preferred. Peptide segments which are COOH-terminal in asparagine or glutamine may be prepared from blocked hydrazides, Z-Asn-N_2H_2-Boc (Moroder *et al.*, 1973d) and Z-Gln-N_2H_2-Boc (Beacham *et al.*, 1971), following removal of the benzyloxycarbonyl group (Z) by hydrogenolysis. COOH-terminal esters of segments may be hydrazinolyzed without appreciable involvement of the ω-amide group. Segment condensations with COOH-terminal glutamine azides have generally gone smoothly, but one should be cautious about the formation of aspartimide derivatives in analogous situations with asparagine. Many useful derivatives that can be used in the synthesis of asparagine and glutamine peptides (with and without side-chain protection) have been tabulated by Fletcher and Jones (1972, 1975). Still more recent supplementary lists are found in appendices to reviews by Schafer (1976) and Atherton *et al.* (1976, 1978). Also, one may consult the four-volume compilation of Pettit (1970, 1971, 1975, 1976).

Special problems can arise when carboxyl groups of amino-protected asparagine and glutamine are activated while the side chains are unprotected. For example, carbodiimide activation, even in the presence of a nucleophile, may lead to formation of the ω-nitrile (by dehydration and rearrangement) and/or the α,ω-diacylimide (directly from attack of the activated carbonyl on the amide nitrogen). In the first instance, resulting ω-cyano acids will then be activated and incorporated into the growing peptide. It is for this reason that active esters of asparagine and glutamine have been mainly employed in stepwise syntheses, since the above by-products can be removed in purification of the active esters prior to their use. Thus, ω-cyano acids are not incorporated into peptides; although they are produced from active esters during storage and in coupling media (Kisfaludy *et al.*, 1975), they are not reactivated. Diacylimides which also form from active esters may be aminolyzed, in principle, to yield mixtures of α- and ω-peptides. In the case of glutamine this reaction is slower by many orders of magnitude than the desired aminolysis of the intact active ester and therefore is not a problem (Dewey *et al.*, 1969). Kinetics of the above rearrangement reactions of benzyloxycarbonylasparagine and benzyloxycarbonylglutamine active esters in DMF were studied by Kisfaludy *et al.* (1975). The nature of the ester leaving group was shown to have a pronounced effect on stability of the ester. Rearrangement was much more rapid with asparagine esters than with glutamine esters. The *p*-nitrophenyl ester was the most stable one studied, a fact which correlates with the extensive and successful use of these esters in stepwise introductions of asparagine and glutamine residues into peptides. In the preparation of Z-Gln-OSu (Zahn and Fölsche, 1969) and Nps-Gln-OSu (Meyers *et al.*, 1969), the esters rapidly cyclized to the glutarimide derivative when solvent polarity and temperature were raised during workup. Curiously, Nps-Asn-OSu was not similarly labile (Meyers *et al.*, 1969). Finally, it should be mentioned that unprotected asparagine and glutamine have been successfully joined to growing peptides via their *N*-carboxyanhydrides (Hirschmann and Veber, 1973).

Once incorporated into a peptide, asparagine and glutamine side chains are not especially prone to dehydration to nitriles. In the presence of alkali, however, a cyclization reaction can occur resulting from attack of the ω-carbonyl on the α-nitrogen atom of the carboxyl-neighboring amino acid residue with elimination of the ω-amide nitrogen as NH_3. This reaction is particularly facile in the case of the -Asn-Gly- sequence which occurs, for example, in ACTH (Sieber *et al.*, 1972) under mildly alkaline conditions (as with hydrazine in DMF at 25°C). Aspartimide formation was observed in the case of peptides with COOH-terminal Asn-O*t*Bu during alkaline treatment to remove the N-terminal 2-(methylsulfonyl)ethoxycarbonyl protecting group (König and Volk, 1977). Theodoropoulos and Souchleris (1965) observed the apparent formation of a glutarimide derivative from heating Z-LAla-Gln-ONb in EtOH–DMF (5 : 1); nucleophilic attack by the amide nitrogen was probably assisted by the electron-withdrawing effect of the *p*-nitro group.

It has long been known that peptides with glutamine in the N-terminal position are easily converted to pyroglutamyl derivatives with elimination of amide NH_3 (for example, see Pietta *et al.*, 1968; Schnabel *et al.*, 1967). Aside from low yields, this side reaction is undesirable because its products are often difficult to separate from the desired ones. Thus, some workers tend to place glutamine at or near the COOH-terminus of a peptide or segment in order to facilitate such a separation (e.g., Greven and Tax, 1971).

Peptides containing asparagine and glutamine residues should not be treated excessively with hydrazine or with acids and alkalies in order to avoid replacement of amide with hydrazide groups or deamidation, respectively. Strong aqueous acids are especially apt to promote loss of primary amide nitrogen.

Occasionally, masking of asparagine and/or glutamine side chains is employed even in some minimal protection strategies. The usual reason for doing this is to increase the solubility of such peptides in organic solvents required for coupling. Additionally, the more important side reactions discussed above can be prevented by protection. For example, dehydration reactions (nitrile and imide formation) are avoided during direct coupling via mixed anhydrides (Borin *et al.*, 1973a,b) and carbodiimides; pyroglutamyl formation is prevented (Pietta *et al.*, 1968). The most commonly used protecting functions for primary amides are the acidolytically cleavable groups, bis-2,4-dimethoxybenzyl (Weygand *et al.*, 1968) and 4,4′-dimethoxybenzhydryl (Ddm; König and Geiger, 1970c).

III. STRATEGIES AND EXAMPLES OF SYNTHESES WITH MINIMAL SIDE-CHAIN PROTECTION

A. Strategy Considerations

Few if any *a priori* rules can be made that will dictate the degree of side-chain protection needed for synthesizing a given peptide. Peptides possessing a wide variety of features have been prepared successfully with minimal protection approaches. However, once such an approach is chosen in order to realize certain advantages (see Section I,C), the selection of appropriate strategy and tactics may be more crucial for success than is the case with maximal protection methods. There are some possible guidelines (but no hard rules) in making these selections.

The use of many segment condensations and relatively few long stretches of stepwise buildup has been general practice. Smaller segments are easier to purify in order to remove any by-products involving unprotected side chains or unused reactants. In principle, the latter are easily removable after segment condensations because of the relatively larger sizes of products and the significant differences in their properties from the reactants. Use of excess amine component (rather than carboxyl component) and/or not pushing coupling reactions too hard to completion should aid in minimizing side reactions involving third functions and result in easier separation problems.

In all approaches to peptide synthesis involving segment condensations a major problem is racemization at the COOH-terminal residue of the carboxyl component. Division of the sequence into segments must take this problem into consideration first by arranging glycine or proline at these sensitive positions wherever feasible. In a minimal protection synthesis it may be useful to consider placing potentially troublesome residues at or near NH_2-termini of segments so that they will be exposed to fewer steps. An especially reactive side chain (perhaps influenced by its surrounding sequence) may also suggest a particular order for joining segments that will reduce the coupling and deprotection steps to which it is exposed. Segments have often been joined sequentially starting from the final COOH-terminus, a strategy which is sensible when carboxyl components are used in excess. Thus, the growing peptide which comprises the amine component is conserved. The same economy is realized when the amine component is used in excess and the segments are joined sequentially starting from the final NH_2-terminus. In the latter case, as mentioned previously, an opportunity exists for generally reducing over-acylation which can involve unblocked side chains. However, this strategy, which promotes more complete consumption of the carboxyl component, may lead to a higher incorporation of Curtius rearrangement products (isocyanates) during azide condensations since these by-products may couple more slowly than the desired peptidyl azides.

Various methods are used for segment condensations in minimal protection syntheses (see Table III). The azide method (Honzl and Rudinger, 1961; Klausner and Bodanszky, 1974; Inouye *et al.*, 1977b; Inouye and Watanabe, 1977b) has been used most frequently because of its compatibility with unprotected side chains, especially free carboxyl groups. It has been widely used in all approaches involving segment couplings because of the presumed freedom from racemization where COOH-terminal residues are optically active. A number of investigators believe that this assumption is not generally true and cite evidence that racemization can occur in peptide bond formation with peptidyl azides. Accordingly, other methods for segment coupling could be given more serious consideration in planning a minimal protection synthesis provided that attention is given to seeking conditions which are most appropriate for each method and/or peptide.

All commonly used patterns of obligatory and conditionally obligatory protection have been employed with minimal side-chain protection approaches. Since a major objective in these syntheses is avoidance of side reactions resulting from deprotection of third functions, it is wise to consider carefully the methods to be used for removing necessary blocking groups. The mildness of such steps should be an important factor in minimal protection strategies. Fewer protecting groups are used and their removal should be scheduled as early as possible: Removal of any resulting by-products is generally easier from smaller peptide units, and amino acids which are prone toward deprotection side reactions (e.g., tryptophan or methionine) can then sometimes be introduced at a later stage after which only a few groups are left for the final deprotection steps. Purification opportunities may also be enhanced by early, partial removal of side-chain blocking groups. In some instances, obligatory protection can be restored following temporary removal and completion of intermediate purification steps (Moroder *et al.*, 1975a; Schwyzer and Sieber, 1966).

B. Examples of Syntheses

Some representative examples of conventional peptide syntheses using minimal side-chain protection are listed in Table III. They are arranged roughly with regard to degree of minimal usage of blocking groups on the amino acid side chains. The degree of usage depends not only on which amino acids are protected but also on how long any side-chain protection is allowed to remain. The numbers (not in parentheses) under each amino acid subheading represent the stages of a synthesis during which side-chain groups are removed. These stages are defined by a scale of 0 to 5 (see Table III, footnote c) where 0 indicates that the side chains are never protected and 5 signifies that blocking groups are removed with final deprotection after all peptide bonds have been formed. Segment condensation methods apply to minor as well as major couplings involving carboxyl components which are tripeptide or larger in size (see footnote e).

The very minimum degree of protection was employed in the ribonuclease S-protein synthesis by the Merck group as discussed above in Section I,B,2 (also see item 1, Table III). The only known means for introducing aspartic and glutamic acids with side chains unblocked is through their *N*-carboxyanhydrides (NCAs and possibly also the thio analogs or NTAs); therefore, the "very minimum of protection" could also encompass use of stage 1 tactics, such as (1) removing the side-chain group along with the N^{α}-protection after adding a single aspartic or glutamic residue or (2) forming a dipeptide ester, e.g., Boc-Glu(OBzl)-X-OMe, then removing the benzyl group with H_2—Pd and converting the methyl ester to an azide. These useful tactics have not been frequently employed with aspartic and glutamic acids. Among the Table III examples, they were so utilized in the synthesis of a secretin analogue (item 4) and a cytochrome *c* segment (item 8). However, since these syntheses involved blocking groups for arginine and tyrosine side chains, they do not represent the very minimum of protection. Arginine is infrequently introduced unprotected (see items 1 and 11 in Table III) but is sometimes handled with a stage 1 tactic (items 3, 4, 5, 8 and 9) and often deprotected at an early stage (e.g., stages 2 or 3 as in examples 6, 11, 12, 13 and 19).

For the ribonuclease S-protein synthesis (Denkewalter *et al.*, 1969; Strachan *et al.*, 1969; Jenkins *et al.*, 1969; Veber *et al.*, 1969; Hirschmann *et al.*, 1969) a total of 19 segments were prepared. NCA and NTA derivatives, including Arginine-, aspartic acid-, and glutamic acid-NCAs and histidine-NTA, were used in forming about 40% of the peptide bonds in the segments; the remaining bonds were established with Boc-amino acid *N*-hydroxysuccinimido esters which included all asparagine, serine, threonine and NH_2-terminal residues. Segments were joined solely by the azide method on account of the free side-chain carboxyl groups. Azides were prepared from hydrazides derived from COOH-terminal esters. The ε-amino functions of lysine residues were masked with benzyloxycarbonyl groups which were removed with HF at 0°C after final assembly of the 104-residue sequence. Acetamidomethyl groups were subsequently removed from cysteine residues by the action of

Table III Examples of Peptide Syntheses Using Minimal Side-Chain Protection

Laboratory[a]	Peptide	(No. residues)[b]	When deprotected[c] (No. res.)[d]						Methods for segment condensations[e]	Final deprotection means
			Asp + Glu	Arg	Tyr	Ser	Thr	His		
Hirschmann/Veber (1)	Ribonuclease S-protein	(104)	0(7)	0(3)	0(6)	0(12)	0(8)	0(3)	N_3, N_3^*	HF, Hg(II)
Hofmann (2)	Ribonuclease T_1, Segment ABCD	(47)	3(5) 4(1)	−(0)	0(6)	0(7)	0(3)	0(2)	N_3, N_3^*	TFA, Hg(II)
Hofmann (3)	Ribonuclease T_1, Segment EF	(33)	3(3) 4(1)	1(1)	0(1) 1(2)	0(7)	−(0)	−(0)	N_3	TFA, H_2[f]
Yanaihara (4)	N^α-Tyrosylsecretin (porcine)	(28)	1(3)	1(2) 4(2)	0(1)	0(4)	0(2)	0(1)	N_3^*	H_2
Geiger/König (5)	LHRH	(10)	−(0)	1(1)	0(1)	0(1)	−(0)	0(1)	N_3^*, DCC + HOBt	None
Götz (6)	LHRH	(10)	−(0)	2(1)	1(1)	0(1)	−(0)	0(1)	N_3, N_3^*	None
Kasafirek (7)	LHRH	(10)	−(0)	5(1)	0(1)	0(1)	−(0)	0(1)	N_3	Electrolysis
Moroder/Scoffone (8)	Cytochrome *c*, Segment FGH iso-1 (yeast)	(42)	1(1) 3(4)	1(1)	1(2) 3(1)	0(1)	0(4)	−(0)	N_3, N_3^*	TFA, H_2, OH^-
Moroder/Scoffone (9)	Cytochrome *c*, Segment C iso-1 (yeast)	(21)	3(1)	1(1)	−(0)	−(0)	0(1)	0(3)	N_3, N_3^*	TFA, H_2, OH^-
Moroder/Scoffone (10)	Cytochrome *c*, Segment D iso-1 (yeast)	(12)	3(1)	−(0)	1(2)	0(2)	0(1)	−(0)	N_3, N_3^*	TFA, H_2
Schwyzer/Sieber (11)	Corticotropin (ACTH-human)	(39)	5(6)	3(1) 0(2)	0(2)	0(3)	−(0)	0(1)	N_3, DCC + HOBt	TFA
Inouye (12)	Corticotropin (ACTH-human)	(39)	5(6)	3(3)	0(2)	0(3)	−(0)	1(1)	N_3^*, -OSu	TFA
Fujino (13)	Corticotropin (α^{1-24}) (*via* route B)	(24)	5(1)	3(1) 5(2)	0(2)	0(2)	−(0)	0(1)	TCA-OPcp, N_3^* -OTcp	HF
Yajima (14)	Corticotropin (ACTH-human)	(39)	5(6)	5(3)	0(2)	0(3)	−(0)	0(1)	N_3^*, DCC + HOBt	HF

Yajima (15)	Somatostatin N^1-Tyr analogue	(15)	–(0)	–(0)	0(1)	0(1)	0(2)	–(0)	N_3^*, DCC + HOBt	HF
Yajima (16)	Glucagon (duck)	(29)	5(3)	5(2)	0(2)	0(4)	0(4)	0(1)	N_3, N_3^* (*via* -NHNH-Tcc)	MSA[g]
Yajima (17)	Basic Trypsin Inhibitor (Kunitz and Northrup)	(58)	5(4)	5(6)	0(3) 1(1)	0(1)	0(3)	–(0)	TCA-OPcp, DCC + HOSu on solid phase	HF
Greven (18)	Calcitonin (human)	(32)	5(1)	–(0)	0(1)	0(1)	0(5)	0(1)	N_3, N_3^*, DCC + HOSu	HCl
Guttmann (19)	Thyrocalcitonin (porcine)	(32)	4(1)	3(1) 4(1)	0(1)	0(4)	0(2)	4(1)	DCC, N_3^*	TFA

[a] Literature references: (1) Hirschmann *et al.* (1969); (2) Beacham *et al.* (1971), Storey *et al.* (1972); (3) Camble *et al.* (1972); (4) Yanaihara *et al.* (1977); (5) Geiger *et al.* (1971); (6) Immer *et al.* (1974); (7) Kasafirek *et al.* (1977); (8) Moroder *et al.* (1975a,b), Moroder *et al.* (1973d,e,f); (9) Moroder *et al.* (1973c); (10) Borin *et al.* (1973a); (11) Sieber *et al.* (1972); (12) Inouye *et al.* (1977a), Inouye and Watanabe (1977a), Watanabe and Inouye (1977), Otsuka *et al.* (1966), Inouye (1965); (13) Hatanaka *et al.* (1975); (14) Koyama *et al.* (1976b); (15) Fujii and Yajima (1975); (16) Yajima *et al.* (1977); (17) Yajima and Kiso (1974a,b), Yajima *et al.* (1974a,b,c); (18) Greven and Tax (1971); (19) Guttmann *et al.* (1968).

[b] Total number of residues in the peptide as synthesized.

[c] Numbers which are not in parentheses re resent stages of the synthesis where side-chain protection is removed: Side chains are never blocked, 0; removed after introducing the residue in question and before forming still another peptide bond, 1; removed during segment buildup, 2; removed when the given segment is completed and before it is joined with another one, 3; removed sometime after the first segment condensation involving the given segment, 4; removed after the final sequence is completed, 5.

[d] Number of residues of the given amino acid deprotected at the stage of synthesis indicated to the left. When there are no residues of a given amino acid in the peptide, then the corresponding space is marked, –(0).

[e] These methods apply to segment condensations as defined in Section I,B,1. An asterisk (*) signifies that the COOH-terminal amino acid of the carboxyl segment was neither Gly nor Pro. Azide condensations are denoted by N_3; -OSu and -OTcp mean that the corresponding active ester of the COOH-segment was prepared prior to condensation; TCA-OPcp = pentachlorophenyl trichloroacetate; Tcc = β,β,β-trichloroethoxycarbonyl group which is removable with Zn.

[f] Catalytic hydrogenation, usually with palladium-on-charcoal or palladium black catalyst.

[g] Methanesulfonic acid.

$Hg(OAc)_2$ in aqueous acetic acid. Removal of benzyloxycarbonyl groups with HF presented no problems, but alkylation of methionine residues with *tert*-butylium ions generated from cleavage of the single NH_2-terminal Boc group caused significant reduction of enzymatic activity, a problem circumvented by addition of a large excess of methionine to the HF mixture. Segment condensations were generally carried out with substantial excesses of the carboxyl component. Limited solubility in dimethylformamide (DMF) was corrected by admixture with hexamethylphosphoramide. Purifications were carried out in various solvents including 50% aqueous acetic acid.

The synthesis of segments and large portions of the ribonuclease T_1 sequence by Hofmann and co-workers is a leading example of the use of minimal protection (Yanaihara *et al.*, 1969; Beacham *et al.*, 1971; Camble *et al.*, 1972; Storey *et al.*, 1972; Kawasaki *et al.*, 1973; Romovacek *et al.*, 1974, 1979). Several completed portions of the synthesis are summarized briefly in Table III. Seven major segments, denoted A (NH_2-terminal) through G, covering the entire sequence of 104 residues, were prepared. Most of these segments were successfully joined by the azide method using a modification of the Honzl and Rudinger (1961) procedure. Peptide bonds within segments were formed primarily by means of benzyloxycarbonyl (Z) amino acid 2,4,5-trichlorophenyl (OTcp), *p*-nitrophenyl (ONp), and *N*-hydroxysuccinimido (OSu) esters, and Z-amino acid or dipeptidyl azides. Segments B through F were prepared as Z-peptidyl Boc hydrazides with aspartic and glutamic acid side chains blocked as *tert*-butyl esters. Segment A bore a COOH-terminal Z-hydrazide and was masked at the amino end with a formyl group; a single lysine residue, occurring in Segment D was similarly blocked on its side chain. The side-chain O*t*Bu groups were removed with the Boc group on the COOH-terminal hydrazide when each segment was activated. The carboxyl groups liberated played an essential role in subsequent purification steps on ion-exchange columns. Problems with low solubility in DMF were encountered in coupling steps with some major fragments. Satisfactory solubility could be attained in mixtures of DMF and dimethyl sulfoxide (DMSO) in such cases. A single arginine residue was introduced as a mixed anhydride of Z-Arg(NO_2)-OH; the NO_2 function was subsequently removed along with the benzyloxycarbonyl group by catalytic hydrogenation. In several instances tyrosine residues were added in stepwise manner as Z-Tyr(Z)-OTcp and both protecting groups were removed simultaneously. Histidine was always added as Z-His-N_3 and serine either as Z-Ser-N_3 or in Z-dipeptidyl azides.

The use of blocked hydrazides for protecting α-carboxyl functions of segments, first used extensively by K. Hofmann and co-workers, has allowed important flexibility in minimal protection strategies. Such segments can be used either as carboxyl or amine components; the order of segment condensations can be changed without having to restart at the very beginning should severe solubility or other problems arise in a trial synthesis. Hofmann *et al.* (1969) thus prepared pancreatic ribonuclease S-peptide (20 residues) by two pathways using the same segments.

Many aspects of the minimal protection approaches discussed above have been employed subsequently by investigators who had worked in Hofmann's laboratory. An interesting synthesis of a secretin analogue by Yanaihara *et al.* (1977), listed in Table III, began with the preparation of 5 segments. Z^{α}-Peptidyl hydrazides were used frequently, though not exclusively, and segments were joined sequentially, via azides, in the COOH- to NH_2-terminal direction. Except for the use of Z-Ser-N_3 and Z-His-N_3, all peptide bonds within segments were formed with dipeptidyl azides, OSu esters, or mixed anhydrides. An unprotected threonine residue was introduced as a mixed anhydride. Boc-Glu(OBzl)-OH, Z-Asp(OBzl)-OH, and several residues of Z-Arg(NO_2)-OH were added as mixed anhydrides and completely deblocked at the next step (use of stage 1 tactics). The removal of NO_2 from two other arginine residues was delayed by cleaving N^{α}-benzyloxycarbonyl with HBr–HOAc. No *t*Bu or O*t*Bu side-chain protection was used, and final deprotection simply involved the removal of an NH_2-terminal benzyloxycarbonyl group with H_2–Pd. Purifications were readily carried out by gel filtration in 3 *N* HOAc and by ion-exchange chromatography in aqueous buffers.

The group who worked with Luis Moroder and the late E. Scoffone used a minimal protection approach along the lines of Hofmann *et al.* in their synthesis of segments and portions of bakers' yeast iso-l-cytochrome *c* (Moroder *et al.*, 1972, 1973a–f, 1974, 1975a,b; Borin *et al.*, 1973a,b). The entire sequence of 108 residues was divided into eight major segments, designated A (NH_2-terminal) through H (COOH-terminal). They were prepared by combining stepwise elongations with minor segment condensations to yield principally Z-peptidyl Boc-hydrazides using benzyloxycarbonyl groups for N^{α} protection during buildup. Segment B, containing two cysteine residues protected with the acetamidomethyl group (Veber *et al.*, 1968), was built stepwise with N^{α}-Nps-amino acid derivatives.

All segment condensations were made by the azide method of Honzl and Rudinger (1961). Lysine residues were protected with trifluoroacetyl (Tfa) groups; aspartic and glutamic acid side chains were blocked as O*t*Bu esters which were removed with TFA during the conversion of Boc-hydrazides to azides for minor or major segment couplings. Major segments had optically active amino acids as COOH-termini.

Several portions of the cytochrome c synthesis are summarized in Table III (items 8–10). Arginine was introduced either as Z-Arg(NO_2)-OPcp, a mixed anhydride with Z-Arg(NO_2)-OH, or Z-Arg$(Adc)_2$-OSu. In the first two cases side-chain and N^{α}-blocking functions were removed together; the adamantyloxycarbonyl (Adc) groups were cleaved during deprotection of the COOH-terminal hydrazide of a subsegment. In segment D two tyrosine residues were added by preparing dipeptides of the type, Z-Tyr(Z)-X-OMe and converting them to Z-Tyr-X-N_2H_3 with hydrazine which selectively cleaved the side-chain benzyloxycarbonyl group. Another tyrosine residue was added as Z-Tyr(Z)-OSu and immediately deprotected. Other bonds were formed through azides, mixed anhydrides, and a variety of active esters. Histidine residues were introduced as H-His-OMe or H-His-N_2H_2-Boc. A few residues of asparagine and glutamine were initially protected with 4,4′-dimethoxybenzhydryl (Ddm) groups and deprotected with TFA before or after completion of major fragments. NH_2-Terminal benzyloxycarbonyl groups from methionine-containing segments were cleaved with HF in the presence of free methionine as scavenger; otherwise, all final deprotection steps were carried out with H_2–Pd, TFA, and 1 *M* aqueous piperidine (the latter for ε-Tfa groups). Purifications were performed on gel-filtration and ion-exchange columns in H_2O–HOAc mixtures or aqueous buffers.

A number of workers have used minimal protection strategies in synthesizing the decapeptide releasing factor for luteotropic and follicle-stimulating hormones, LH–FSH-RH (for example, see Table III, items 5–7, and Yanaihara *et al.*, 1973). Geiger *et al.* (1971) and Immer *et al.* (1974) used segment condensation strategies. The former investigators carried out all but one segment coupling with DCC + HOBt and added arginine as the mixed anhydride Z-Arg$(Z)_2$-OH; Immer and co-workers joined segments by the azide method alone, introduced Boc-Arg(NO_2)-OH with DCC and removed the NO_2 function with H_2–Pd following formation of a second subsequent peptide bond. Kasafirek *et al.* (1977) and Yanaihara *et al.* (1973) employed stepwise elongation throughout; the former used dipeptide units except for Z-Leu-OH and Z-Arg(Tos)-OH (removing the tosyl group electrolytically in the final step); the latter team added mainly single residues employing a variety of tactics.

Four syntheses of corticotropin (1–39) or (1–24) sequences are compared in Table III (items 11–14). Sieber *et al.* (1972) concluded the preparation of the correct, entire sequence (man) covering work done in part over many years in collaboration with R. Schwyzer, H. Kappeler, B. Iselin, and others. Temporary protection of α-amine functions was afforded by benzyloxycarbonyl groups except for an NH_2-terminal Boc, and side chains of aspartic acid, glutamic acid, and lysine residues were blocked with O*t*Bu and Boc, respectively. Segments were joined via the azide, pivaloyl mixed anhydride, and DCC + HOBt methods. Final deprotection was achieved simply by TFA treatment. Serine-31 was introduced with Z-Ser-OPcp; the side chains of serine, tyrosine, and histidine residues were never protected. Inouye and co-workers (1977a) (see Table III, item 12) applied a similar approach with acidolytically cleavable side-chain protection for aspartic acid, glutamic acid, and lysine; the latter residues were blocked with 1-methylcyclohexyloxycarbonyl (Mch) groups. Nitro groups were removed from arginine residues with H_2–Pd at intermediate stages, and final deprotection was also carried out with TFA. Histidine was introduced as Z-Glu(O*t*Bu)-His(Z)-OMe and deblocked with hydrazine used in converting the methyl ester to hydrazide for azide coupling. Hatanaka *et al.* (1975) and Koyama *et al.* (1976b) (see Table III, items 13 and 14, respectively) blocked lysine side chains with benzyloxycarbonyl groups, N^{α}-amino functions with acid-labile Boc or *p*-methoxybenzyloxycarbonyl Moz groups, and Asp/Glu (in the latter case) with O*t*Bu. Hydrofluoric acid was used for final deprotection which removed NO_2 or tosyl groups from arginine residues. Nishimura *et al.* (1975) of Fujino's group also reported the complete synthesis of human adrenocorticotropic hormone; the sequence 25–39 was prepared mainly by a stepwise elongation employing *N*-hydroxy-5-norbornene-2,3,-dicarboximido esters of benzyloxycarbonyl amino acids. Serine-31 was added as Z-Ser-ODnp. Koyama *et al.* (1976a) and Yajima *et al.* (1976a) reported syntheses of bovine and porcine corticotropins, respectively, which were similar to their approach to preparing the human hormone (Koyama *et al.*, 1976b). More recently, Okamoto *et al.* (1979) reported the synthesis of dogfish ACTH.

Yajima's group has reported a number of other syntheses following the principles of minimal side-chain protection. Three additional examples appear in Table III as items 15–17. Other examples include the gastric inhibitory peptide (Yajima *et al.*, 1975; Ogawa *et al.*, 1976a,b; Kubota *et al.*, 1976), cholecystokinin–pancreozymin (Mori and Yajima, 1976; Mori *et al.*, 1976; Yajima *et al.*, 1976b), and *β*-melanocyte-stimulating hormones from the monkey (Yajima *et al.*, 1969), dogfish (Watanabe *et al.*, 1974), pig (Watanabe *et al.*, 1975), horse (Tamura *et al.*, 1975), camel (Tamura *et al.*, 1977), and mammalian/avian glucagons (Fujino *et al.*, 1978; Wakimasu and Fujino, 1978; Ogawa *et al.*, 1978).

Calcitonin and thyrocalcitonin, peptides rich in hydroxylic amino acids, were synthesized with the OH functions unprotected (see Table III) by Greven and Tax (1971) and Guttmann *et al.* (1968). The latter workers used DCC for condensing segments containing Ser and Thr. In both examples *S*-benzyl protecting groups were removed from two cysteine residues during the preparation of a segment, and the sulfhydryl functions freed were oxidized to form an internal disulfide link which protected these residues during subsequent condensations. A variant sequence of human calcitonin was synthesized by Nozaki (1978) using a minimal protection approach.

Finally, some additional interesting illustrations of minimal protection are found in syntheses of RNase(13–20) peptide (Marchiori *et al.*, 1967), trypsin inhibitor II of Kazel (Rocchi *et al.*, 1974, 1975; Tomatis *et al.*, 1978, 1979), β-lipotropin segments (Medzihradszky *et al.*, 1969), porcine growth hormone releasing hormone (Fauszt and Bajusz, 1974), human ACTH (earlier sequence, Bajusz *et al.*, 1967), salmon calcitonin (Guttmann *et al.*, 1969), vasoactive intestinal peptide (Bondanszky *et al.*, 1972, 1974b; Klausner *et al.*, 1972, 1973; Klausner and Bodanszky, 1973), LHRH, 8-homoarginine analogue (Geiger *et al.*, 1974), another LHRH analogue (Fujino *et al.*, 1974a), ACTH analogues (VanNispen and Tesser, 1975), bovine and human intermediate lobe peptides (CLIP; Kawatani and Yajima, 1974; Kawatani *et al.*, 1974), and a β-corticotropin(1–18) analogue (Riniker and Rittel, 1970).

C. Protein Semisynthesis

Amino acids or peptides may be covalently joined to purified fragments of natural peptides or proteins for the preparation of reconstructed molecules or molecules with modified sequence. Natural sequences, altered by some appropriate semisynthetic method, may be used for studying almost any question concerning structure–function relationships, stability or dynamics of proteins. In addition, structurally nonperturbing probes for isotopic tracer work or nuclear magnetic resonance spectroscopy may be inserted into a primary structure by semisynthesis. In the case of large peptides or proteins, products probably can be obtained with less error and effort by semisynthesis than through total synthesis from constituent amino acids. Although a semisynthesis requires many special techniques and may seem tedious, the difficulty in preparing large, natural fragments for synthetic operations does not increase sharply with size as does the preparation of all-synthetic segments (Rees and Offord, 1976a). The problem of racemization should clearly be less in the former case.

The need for obligatory and conditionally obligatory protection of functional groups as well as many tactical and strategy considerations are the same for both semisynthesis and conventional synthesis with minimal side-chain protection. Thus, these fields of endeavor have, and will continue to have, many areas of overlap and frequent interchange of ideas and techniques. Semisynthesis involves stepwise elongation and segment condensation strategies in the same sense as employed in total synthesis.

Segment condensations have been used less often than stepwise elongations in semisynthesis, yet the former type of approach has been studied carefully by Rees and Offord at Oxford University (Rees and Offord, 1976a,b; Offord *et al.*, 1976, and their earlier papers cited therein). These authors list the stages necessary for producing suitably protected natural fragments which can be activated for coupling reactions. The steps are as follows: (1) reduction of disulfide bridges of the parent protein, and protection of the sulfhydryl functions (preferably as S-sulfonato derivatives), (2) blocking all the protein's amino groups (e.g., by reaction with maleic anhydride), (3) endolytic cleavage of the protein derivative (e.g., by trypsin, specificially at the arginine residues), (4) separation of the resulting peptides, (5) reversible protection of the newly liberated α-amino groups of the peptides (e.g., with Boc groups), (6) reversible protection of the imidazolyl side chains of any histidine residues, (7) reversible protection of all the carboxyl groups of the peptides (by esterification with a diazoalkane), and (8) selective deprotection of the α-carboxyl groups (e.g., with trypsin).

The above stages in some equivalent form will generally be required in any semisynthetic scheme where the natural fragment is to be the carboxyl component. If the fragment is to be an amine component, only the first 4 steps are needed. It may be feasible in very special cases to eliminate steps 6, 7, and 8 for a carboxyl component provided the side-chain COOH groups are significantly less reactive than the COOH-terminal one, and a satisfactory purification method can be applied for removing the expected by-products. An illustration of this shortcut was reported by Obermeier and Geiger (1976) in their conversion of porcine to human insulin. Amino acid residues (or peptide segments), of course, may be added or inserted at some appropriate stage (e.g., after step 4 above) to/into one of the peptides with or without prior sequential degradation (e.g., by the Edman procedure) or further endolytic cleavage. Rees and Offord (1976b) studied various methods for joining protected fragments (including azide coupling) and chose pretreatment of the carboxyl component with DCC + HOBt in DMF as being most compatible with a reasonable efficiency and minimal degree of racemization. Semisynthetic operations with proteins themselves must not cause irreversible denaturation, and, for that reason, all chemical steps with most proteins will have to be carried out in highly polar (usually semiaqueous) media.

In general it has not been found necessary or helpful to protect the side chains of serine, threonine, tyrosine, arginine, asparagine, glutamine, methionine, or tryptophan in the natural fragments. Experiments by Rees and Offord (1976a) demonstrated that protection of the delicate tryptophan residues amounted to a trade-off in problems and was usually not worthwhile. Histidine residues were protected because of the strong alkylating action of the diazoalkanes used subsequently for the esterification of carboxyls.

A number of investigators have been involved with semisynthesis of proteins. At present any appropriate discussion of their work would require a full-length review (cf., Offord and DiBello, 1978; Offord, 1980). For further orientation, the reader is referred to a brief discussion by Smyth (1975), the papers cited above, a recent study of the use of acetimidyl groups for reversible amino protection (DiMarchi *et a.*, 1978), and the excellent review discussions in the yearly Specialist Periodical Reports of the (British) Chemical Society

(Jones, 1971; Jones and Ridge 1972, 1974; Ridge and Jones, 1975; Schafer, 1976, pp. 281–283; Atherton *et al.*, 1976, pp. 281–285; Atherton *et al.*, 1978, pp. 349–354).

IV. CONCLUDING REMARKS

Any definition of minimal usage of side-chain protection, such as the statements given in Section I,B, will have to be somewhat flexible. The various degrees and modes of protection which may be regarded as "minimal" are practiced in a substantial fraction of conventional peptide syntheses; therefore, one is obliged to be highly selective and also a bit arbitrary in choosing from the voluminous peptide literature the pertinent experiences for a reasonably limited and critical discussion of this topic. Hopefully, not too many of the more illustrative facts or syntheses were left unmentioned.

Whether one chooses to minimize the use of side-chain protecting groups or to employ essentially full protection is a decision which must be made early on in planning a synthesis. Lengthy stepwise elongations with infrequent intermediate purification of the growing product (repetitive methods) require rather full protection of side chains throughout. Segment condensation strategies with careful purification of intermediate peptides are compatible with the full range of possibilities regarding the use of blocking groups. Unforseen problems involving unprotected side chains can readily arise. Revision of tactics or strategy and thus repetition of at least a portion of the synthesis may be required. If one wishes to prepare for research purposes one peptide of a particular type only once, it would seem prudent to use full protection. However, it was pointed out in previous Sections that the protection of side chains will also engender problems. In our present state of knowledge, side reactions occurring with the use and removal of protecting groups may be slightly better understood and more controllable than are difficulties which arise from leaving side chains unblocked. Nonetheless, full protection problems tend to be delayed to the end of a synthesis, involve greater risk in the final deprotection step, and may cause increased difficulties in final purification and characterization. A failure at this stage is likely to be very time consuming.

Minimal protection approaches allow greater opportunities for highly discriminatory separations at intermediate stages and involve less risk at final deprotection. The point should be emphasized that paying more direct attention to the mechanisms of side reactions with third functions during synthetic steps will aid significantly in improving the success of minimal protection syntheses. Relatively little useful information has been reported in this area, but the data which were analyzed in Section II consistently point to some practical rules for coupling steps: (1) avoid large excesses of carboxyl components (or else limit reaction time), (2) avoid any excess of base for binding protons, (3) choose a solvent mixture which completely dissolves the amine component at the outset of coupling. For the present, success or failure in using minimal protection may rest on specific empirical knowledge including results from trial syntheses. This information, however, may be applied efficiently to the synthesis of a *series* of peptide analogs, or to an individual synthesis that must be repeated many times as in a commercial application. Preliminary studies for a synthesis should lead to particular selections of solvents, reagents and bases for coupling reactions. Recent advances in analytical techniques (e.g., reverse-phase, high-performance liquid chromatography) should appreciably assist in characterizing the "dirt" which has been customarily "swept under the rug" and aid the methodological research that is perhaps long overdue.

In the earlier days of peptide synthesis, side chains were left unprotected for lack of satisfactory protecting groups and deprotection conditions. Newer methods of protection allowed older approaches to peptide bond formation to be used more effectively and larger peptides to be synthesized. On the other hand, more recent improvements in coupling tactics have permitted large peptides to be prepared with many side chains left unprotected. Further improvements in peptide bond-forming methods should allow still greater use of minimal protection and constitute a trend that is in keeping with the eventual goal in chemical synthesis of peptides: a mild, facile, low-error process that encompasses the virtues of nature's own method which protects none of the side chains.

REFERENCES

Agarwal, K. L., Kenner, G. W., and Sheppard, R. C. (1969). *J. Chem. Soc. C* pp. 954–958.

Alakhov, Y. B., Kiryushkin, A. A., and Lipkin, V. M. (1970). *J. Chem. Soc., Chem. Commun.* pp. 406–407.

Anderson, G. W., Zimmerman, J. E., and Callahan, F. M. (1964). *J. Am. Chem. Soc.* **86**, 1839–1842.

Andreatta, R. H., Hartmann, A., Jöhl, A., Kamber, B., Maier, R., Riniker, B., Rittel, W., and Sieber, P. (1973). *Helv. Chim. Acta* **56**, 470–473.

Atherton, E., Sheppard, R. C., Fletcher, G. A., and Jones, J. H. (1976). *Amino-Acids, Pept., Proteins* **8**, 248–309.

Atherton, E., Sheppard, R. C., and Stachulski, A. V. (1978). *Amino-Acids, Pept., Proteins* **9**, 315–394.

Bajusz, S., and Fauszt, I. (1973). *Acta Chim. Acad. Sci. Hung.* **75**, 419–422.

Bajusz, S., Medzihradszky, K., Paulay, Z., and Lang, Zs. (1967). *Acta Chim. Acad. Sci. Hung.* **52**, 335–341.

Bayer, E., and Mutter, M. (1972). *Nature* (*London*) **237**, 512–513.

Beacham, J., Dupuis, G., Finn, F. M., Storey, H. T., Yanaihara, C., Yanaihara, N., and Hofmann, K. (1971). *J. Am. Chem. Soc.* **93**, 5526–5539.

Belleau, B., and Malek, G. (1968). *J. Am. Chem. Soc.* **90**, 1651–1652.

Benoiton, L. (1962). *Can. J. Chem.* **40**, 570–572.
Beyerman, H. C. (1972). *In* "Chemistry and Biology of Peptides" (J. Meienhofer, ed.), pp. 351–357. Ann Arbor Sci. Publ., Ann Arbor, Michigan.
Bodanszky, M., and du Vigneaud, V. (1962). *Biochem. Prep.* **9**, 110–112.
Bodanszky, M., and Fagan, D. T. (1977). *Int. J. Pept. Protein Res.* **10**, 375–379.
Bodanszky, M., and Kwei, J. Z. (1978). *Int. J. Pept. Protein Res.* **12**, 69–74.
Bodanszky, M., and Natarajan, S. (1975). *J. Org. Chem.* **40**, 2495–2499.
Bodanszky, M., and Ondetti, M. A. (1966). *Chem. Ind.* (*London*) pp. 26–27.
Bodanszky, M., and Sheehan, J. T. (1960). *Chem. Ind.* (*London*) pp. 1268–1269.
Bodanszky, M., Birkhimer, C. A., Lande, S., Ondetti, M. A., Sheenan, J. T., and Williams, N. J. (1966). *In* "Peptides 1963" (L. Zervas, ed.), pp. 3–10. Pergamon, Oxford.
Bodanszky, M., Ondetti, M. A., Levine, S. D., and Williams, N. J. (1967). *J. Am. Chem. Soc.* **89**, 6753–6757.
Bodanszky, M., Klausner, Y. S., and Mutt, V. (1972). *Bioorg. Chem.* **2**, 30–38.
Bodanszky, M., Kondo, M., Lin, C. Y., and Sigler, G. F. (1974a). *J. Org. Chem.* **39**, 444–447.
Bodanszky, M., Klausner, Y. S., Lin, C. Y., Mutt, V., and Said, S. I. (1974b). *J. Am. Chem. Soc.* **96**, 4973–4978.
Bodanszky, M., Klausner, Y. S., and Ondetti, M. A. (1976). "Peptide Synthesis," 2nd ed. Wiley, New York.
Bodanszky, M., Fink, M. L., Klausner, Y. S., Natarajan, S., Tatemoto, K., Yiotakis, A. E., and Bodanszky, A. (1977a). *J. Org. Chem.* **42**, 149–152.
Bodanszky, M., Henes, J. B., Yiotakis, A. E., and Said, S. I. (1977b). *J. Med. Chem.* **20**, 1461–1464.
Bodanszky, M., Tolle, J. C., Deshmane, S. S., and Bodanszky, A. (1978). *Int. J. Pept. Protein Res.* **12**, 57–68.
Bonnett, R., and Holleyhead, R. (1974). *J. Chem. Soc., Perkin Trans. 1* pp. 962–964.
Borin, G., Marchiori, F., Rocchi, R., Moroder, L., and Scoffone, E. (1971). *Gazz. Chim. Ital.* **101**, 426–439.
Borin, G., Moroder, L., Marchiori, F., and Scoffone, E. (1973a). *Biopolymers* **12**, 521–534.
Borin, G., Moroder, L., Marchiori, F., and Scoffone, E. (1973b). *Biopolymers* **12**, 693–700.
Borin, G., Marchiori, F., Moroder, L., and Scoffone, E. (1975). *Gazz. Chim. Ital.* **105**, 137–147.
Brenner, M. (1967). *In* "Peptides 1966" (H. C. Beyerman, A. van de Linde, and W. M. van den Brink, eds.), pp. 1–7. North-Holland Publ., Amsterdam.
Broadbent, W., Morley, J. S., and Stone, B. E. (1967). *J. Chem. Soc. C* pp. 2632–2636.
Bryant, P. M., Moore, R. H., Pimlott, P. J., and Young, G. T. (1959). *J. Chem. Soc.* pp. 3868–3873.
Camble, R., and Cotton, R. (1975). *In* "Peptides, 1974" (Y. Wolman, ed.), pp. 17–26. Wiley, New York.
Camble, R., Dupuis, G., Kawasaki, K., Romovacek, H., Yanaihara, N., and Hofmann, K. (1972). *J. Am. Chem. Soc.* **94**, 2091–2100.
Costopanagiotis, A. A., Handford, B. O., and Weinstein, B. (1968). *J. Org. Chem.* **33**, 1261–1264.
Coy, D. H., and Branyas, N. (1979). *Int. J. Pept. Protein Res.* **14**, 339–343.
Coyle, S., Keller, O., and Young, G. T. (1975). *J. Chem. Soc., Chem. Commun.* pp. 939–940.
Davey, J. M., Laird, A. H., and Morley, J. S. (1966). *J. Chem. Soc. C* pp. 555–566.
Denkewalter, R. G., Schwam, H., Strachan, R. G., Beesley, T. E., Veber, D. F., Schoenewaldt, E. F., Barkemeyer, H., Paleveda, W. J., Jr., Jacob, T. A., and Hirschmann, R. (1966). *J. Am. Chem. Soc.* **88**, 3163–3164.
Denkewalter, R. G., Veber, D. F., Holly, F. W., and Hirschmann, R. (1969). *J. Am. Chem. Soc.* **91**, 502–503.
Determann, H., and Wieland, T. (1961). *Makromol. Chem.* **44–46**, 312–323.

Dewey, R. S., Barkemeyer, H., and Hirschmann, R. (1969). *Chem. Ind.* (*London*) p. 1632 only.
Dewey, R. S., Schoenenwaldt, E. F., Joshua, H., Paleveda, W. J., Jr., Schwam, H., Barkemeyer, H., Arison, B. H., Veber, D. F., Strachan, R. G., Milkowski, J., Denkewalter, R. G., and Hirschmann, R. (1971). *J. Org. Chem.* **36**, 49–59.
DiMarchi, R. D., Garner, W. H., Wang, C.-C., Hanania, G. I. H., and Gurd, F. R. N. (1978). *Biochemistry* **17**, 2822–2829.
Engelhard, M., and Merrifield, R. B. (1978). *J. Am. Chem. Soc.* **100**, 3559–3563.
Fauszt, I., and Bajusz, S. (1974). *Acta Chim. Acad. Sci. Hung.* **82**, 471–480.
Felix, A. M., Jimenez, M. H., and Meienhofer, J. (1977). *In* "Peptides, Proceedings of the Fifth American Peptide Symposium" (M. Goodman and J. Meienhofer, eds.), pp. 532–535. Wiley, New York.
Feurer, M. (1977). *In* "Peptides, Proceedings of the Fifth American Peptide Symposium" (M. Goodman and J. Meienhofer, eds.), pp. 448–457. Wiley, New York.
Fletcher, G. A., and Jones, J. H. (1972). *Int. J. Pept. Protein Res.* **4**, 347–371.
Fletcher, G. A., and Jones, J. H. (1975). *Int. J. Pept. Protein Res.* **7**, 91–102.
Fridkin, M., and Goren, H. J. (1971). *Can. J. Chem.* **49**, 1578–1581.
Fruton, J. S. (1942). *J. Biol. Chem.* **146**, 463–470.
Fujii, N., and Yajima, H. (1975). *Chem. Pharm. Bull.* **23**, 2446–2449.
Fujii, N., Sasaki, T., Funakoshi, S., Irie, H., and Yajima, H. (1978). *Chem. Pharm. Bull.* **26**, 650–653.
Fujino, M., Kobayashi, S., Obayashi, M., Fukuda, T., Shinagawa, S., Yamazaki, I., Nakayama, R., White, W. F., and Rippel, R. H. (1972). *Biochem. Biophys. Res. Commun.* **49**, 698–705.
Fujino, M., Yamazaki, I., Kobayashi, S., Fukuda, T., Shinagawa, S., Nakayama, R., White, W. F., and Rippel, R. H. (1974a). *Biochem. Biophys. Res. Commun.* **57**, 1248–1256.
Fujino, M., Kobayashi, S., Obayashi, M., Fukuda, T., Shinagawa, S., and Nishimura, O. (1974b). *Chem. Pharm. Bull.* **22**, 1857–1863.
Fujino, M., Wakimasu, M., Shinagawa, S., Kitada, C., and Yajima, H. (1978). *Chem. Pharm. Bull.* **26**, 539–548.
Geiger, R., König, W., Wissmann, H., Geisen, K., and Enzmann, F. (1971). *Biochem. Biophys. Res. Commun.* **45**, 767–773.
Geiger, R., König, W., Sandow, J., and Schally, A. V. (1974). *Hoppe-Seyler's Z. Physiol. Chem.* **355**, 1526–1534.
Gibian, H., and Schröder, E. (1961). *Justus Liebigs Ann. Chem.* **642**, 145–162.
Glass, J. D., and Pelzig, M. (1978). *Int. J. Pept. Protein Res.* **12**, 75–80.
Goodman, M., and Stueben, K. C. (1959). *J. Am. Chem. Soc.* **81**, 3980–3983.
Greven, H. M., and Tax, L. J. W. M. (1971). *In* "Peptides 1969" (E. Scoffone, ed.), pp. 38–43. North-Holland Publ., Amsterdam.
Guttmann, St. (1961). *Helv. Chim. Acta* **44**, 721–744.
Guttmann, St., Pless, J., Sandrin, E., Jaquenoud, P.-A., Bossert, H., and Willems, H. (1968). *Helv. Chim. Acta* **51**, 1155–1158.
Guttmann, St., Pless, J., Huguenin, R. L., Sandrin, E., Bossert, H., and Zehnder, K. (1969). *Helv. Chim. Acta* **52**, 1789–1795.
Haas, W. L., Krumkalns, E. V., and Gerzon, K. (1966). *J. Am. Chem. Soc.* **88**, 1988–1992.
Hamada, Y., Shioiri, T., and Yamada, S. (1977). *Chem. Pharm. Bull.* **25**, 221–223.
Hanby, W. E., Waley, S. G., and Watson, J. (1950). *J. Chem. Soc.* pp. 3239–3246.
Handford, B. O., Hylton, T. A., Wang, K.-T., and Weinstein, B. (1968). *J. Org. Chem.* **33**, 4251–4255.
Hardy, P. M., Haylock, J. C., and Rydon, H. N. (1972). *J. Chem. Soc., Perkin Trans. 1* pp. 605–611.
Harington, C. R., and Pitt-Rivers, R. V. (1944). *Biochem. J.* **38**, 417–423.
Hase, S., and Walter, R. (1973). *Int. J. Pept. Protein Res.* **5**, 283–288.

Hase, S., Sakakibara, S., Wahrenburg, M., Kirchberger, M., Schwartz, I. L., and Walter, R. (1972). *J. Am. Chem. Soc.* **94**, 3590–3600.

Hatanaka, C., Nishimura, O., and Fujino, M. (1975). *Chem. Pharm. Bull.* **23**, 1017–1024.

Hayakawa, T., Kondo, Y., Yamamoto, H., and Murakami, Y. (1969). *Bull. Chem. Soc. Jpn.* **42**, 479–482.

Hemmasi, B., and Bayer, E. (1977). *Tetrahedron Lett.* No. 19, pp. 1599–1602.

Hirschmann, R., and Veber, D. F. (1973). *In* "The Chemistry of Polypeptides" (P. G. Katsoyannis, ed.), pp. 125–142. Plenum, New York.

Hirschmann, R., Nutt, R. F., Veber, D. F., Vitali, R. A., Varga, S. L., Jacob, T. A., Holly, F. W., and Denkewalter, R. G. (1969). *J. Am. Chem. Soc.* **91**, 507–508.

Hirschmann, R., Schwam, H., Strachan, R. G., Schoenewaldt, E. F., Barkemeyer, H., Miller, S. M., Conn, J. B., Garsky, V., Veber, D. F., and Denkewalter, R. G. (1971). *J. Am. Chem. Soc.* **93**, 2746–2754.

Hofmann, K., Peckham, W. D., and Rheiner, A. (1956). *J. Am. Chem. Soc.* **78**, 238–242.

Hofmann, K., Haas, W., Smithers, M. J., Wells, R. D., Wolman, Y., Yanaihara, N., and Zanetti, G. (1965a). *J. Am. Chem. Soc.* **87**, 620–631.

Hofmann, K., Haas, W., Smithers, M. J., and Zanetti, G. (1965b). *J. Am. Chem. Soc.* **87**, 631–639.

Hofmann, K., Visser, J. P., and Finn, F. M. (1969). *J. Am. Chem. Soc.* **91**, 4883–4887.

Hofmann, K., Andreatta, R., Finn, F. M., Montibeller, J. Porcelli, G., and Quattrone, A. J. (1971). *Bioorg. Chem.* **1**, 66–83.

Holley, R. W., and Sondheimer, E. (1954). *J. Am. Chem. Soc.* **76**, 1326–1328.

Honzl, J., and Rudinger, J. (1961). *Collect. Czech. Chem. Commun.* **26**, 2333–2344.

Houghten, R. A., and Li, C. H. (1977). *In* "Peptides, Proceedings of the Fifth American Peptide Symposium" (M. Goodman and J. Meienhofer, eds.), pp. 458–460. Wiley, New York.

Immer, H., Nelson, V. R., Revesz, C., Sestanj, K., and Götz, M. (1974). *J. Med. Chem.* **17**, 1060–1065.

Inouye, K. (1965). *Bull. Chem. Soc. Jpn.* **38**, 1148–1157.

Inouye, K., and Otsuka, H. (1962). *J. Org. Chem.* **27**, 4236–4243.

Inouye, K., and Watanabe, K, (1977a). *Bull. Chem. Soc. Jpn.* **50**, 207–210.

Inouye, K., and Watanabe, K. (1977b). *J. Chem. Soc., Perkin Trans. 1*, pp. 1911–1915.

Inouye, K., Watanabe, K., and Otsuka, H. (1977a). *Bull. Chem. Soc. Jpn.* **50**, 211–219.

Inouye, E., Watanabe, K., and Shin, M. (1977b). *J. Chem. Soc., Perkin Trans. 1* pp. 1905–1910.

Iselin, B., and Schwyzer, R. (1961). *Helv. Chim. Acta* **44**, 169–173.

Itoh, M. (1970). *Chem. Pharm. Bull.* **18**, 784–788.

Jaeger, E., Thamm, P., Knof, S., and Wünsch, E. (1978). *Hoppe-Seyler's Z. Physiol. Chem.* **359**, 1629–1636.

Jäger, G. (1973). *Chem. Ber.* **106**, 206–210.

Jäger, G., and Geiger, R. (1970). *Chem. Ber.* **103**, 1727–1747.

Jäger, G., and Geiger, R. (1973). *Justus. Liebigs. Ann. Chem.* pp. 1535–1544.

Jäger, G., König, W., Wissmann, H., and Geiger, R. (1974). *Chem. Ber.* **107**, 215–231.

Jenkins, S. R., Nutt, R. F., Dewey, R. S., Veber, D. F., Holly, F. W., Paleveda, W. J., Jr., Lanza, T., Jr., Strachan, R. G., Schoenewaldt, E. F., Barkemeyer, H., Dickinson, M. J., Sondey, J., Hirschmann, R., and Walton, E. (1969). *J. Am. Chem. Soc.* **91**, 505–506.

Johnson, B. J. (1973). *J. Pharm. Sci.* **62**, 1019–1021.

Johnson, B. J., and Ruettinger, T. A. (1970). *J. Org. Chem.* **35**, 255–257.

Jones, D. A., Jr., Mikulec, R. A., and Mazur, R. H. (1973). *J. Org. Chem.* **38**, 2865–2869.

Jones, J. H. (1971). *Amino-Acids, Pept., Proteins* **3**, 250–252.

Jones, J. H., and Ridge, B. (1972). *Amino-Acids, Pept., Proteins* **4**, 361–363.

Jones, J. H., and Ridge, B. (1974). *Amino-Acids, Pept., Proteins* **5**, 324 and 330–332.

Jost, K. (1971). *Collect. Czech. Chem. Commun.* **36**, 218–233.

Juillerat, M., and Bargetzi, J. P. (1976). *Helv. Chim. Acta* **59**, 855–866.
Kamber, B. (1973). *Helv. Chim. Acta* **56**, 1370–1381.
Kamber, B., Riniker, B., Sieber, P., and Rittel, W. (1976). *Helv. Chim. Acta* **59**, 2830–2840.
Kasafirek, E., Semonsky, M., Hruska, K., and Veznik, Z. (1977). *Collect. Czech. Chem. Commun.* **42**, 2018–2026.
Katchalski, E., and Sela, M. (1953). *J. Am. Chem. Soc.* **75**, 5284–5292.
Kaurov, O. A., Martynov, V. F., and Morozov, V. B. (1970). *J. Gen. Chem. USSR* (*Engl. Transl.*) **40**, 885–888.
Kawasaki, K., Camble, R., Dupuis, G., Romovacek, H., Storey, H. T., Yanaihara, C., and Hofmann, K. (1973). *J. Am. Chem. Soc.* **95**, 6815–6824.
Kawatani, H., and Yajima, H. (1974). *Chem. Pharm. Bull.* **22**, 1872–1878.
Kawatani, H., Tamura, F., and Yajima, H. (1974). *Chem. Pharm. Bull.* **22**, 1879–1888.
Kisfaludy, L., Roberts, J. E., Johnson, R. H., Mayers, G. L., and Kovacs, J. (1970). *J. Org. Chem.* **35**, 3563–3565.
Kisfaludy, L., Low, M., Nyeki, O., Szirtes, T., and Schön, I. (1973). *Justus Liebigs Ann. Chem.* pp. 1421–1429.
Kisfaludy, L., Schön, I., Renyei, M., and Görög, S. (1975). *J. Am. Chem. Soc.* **97**, 5588–5589.
Kiso, Y., and Yajima, H. (1972). *J. Chem. Soc., Chem. Commun.* pp. 942–943.
Kiso, Y., Kai, Y., and Yajima, H. (1973). *Chem. Pharm. Bull.* **21**, 2507–2510.
Klausner, Y. S., and Bodanszky, M. (1973). *Bioorg. Chem.* **2**, 354–362.
Klausner, Y. S., and Bodanszky, M. (1974). *Synthesis* pp. 549–559.
Klausner, Y. S., Mutt, V., and Bodanszky, M. (1972). *Bioorg. Chem.* **2**, 87–94.
Klausner, Y. S., Lin, C. Y., Mutt, V., and Bodanszky, M. (1973). *Bioorg. Chem.* **2**, 345–353.
Klieger, E. (1969). *Justus Liebigs Ann. Chem.* **724**, 204–207.
Klieger, E., and Gibian, H. (1962). *Justus Liebigs Ann. Chem.* **655**, 195–210.
Klieger, E., Schröder, E., and Gibian, H. (1961). *Justus Liebigs Ann. Chem.* **640**, 157–167.
König, W., and Geiger, R. (1970a). *Chem. Ber.* **103**, 788–798.
König, W., and Geiger, R. (1970b). *Chem. Ber.* **103**, 2034–2040.
König, W., and Geiger, R. (1970c). *Chem. Ber.* **103**, 2041–2051.
König, W., and Volk, A. (1977). *Chem. Ber.* **110**, 1–11.
Kovacs, J., Ceprini, M. Q., Dupraz, C. A., and Schmit, G. N. (1967). *J. Org. Chem.* **32**, 3696–3698.
Koyama, K., Kawatani, H., Yajima, H., Fujino, M., and Nishimura, O. (1976a). *Chem. Pharm. Bull.* **24**, 2106–2111.
Koyama, K., Watanabe, H., Kawatani, H., Iwai, J., and Yajima, H. (1976b). *Chem. Pharm. Bull.* **24**, 2558–2563.
Kubota, M., Ogawa, H., and Yajima, H. (1976). *Chem. Pharm. Bull.* **24**, 2435–2446.
Losse, G., Jeschkeit, H., and Knopf, D. (1964). *Chem. Ber.* **97**, 1789–1798.
Löw, M., and Kisfaludy, L. (1979). *Hoppe-Seyler's Z. Physiol. Chem.* **360**, 13–18.
Löw, M., Kisfaludy, L., and Sohar, P. (1978). *Hoppe-Seyler's Z. Physiol. Chem.* **359**, 1643–1651.
Lundt, B. F., Johansen, N. L., Vølund, A., and Markussen, J. (1978). *Int. J. Pept. Protein Res.* **12**, 258–268.
Lundt, B. F., Johansen, N. L., and Markussen, J. (1979). *Int. J. Pept. Protein Res.* **14**, 344–346.
Marchiori, F., Rocchi, R., and Scoffne, E. (1963). *Gazz. Chim. Ital.* **93**, 834–840.
Marchiori, F., Rocchi, R., Moroder, L., Vidali, G., and Scoffne, E. (1967). *J. Chem. Soc.* pp. 89–91.
Marchiori, F., Borin, G., Filippi, B., Moretto, V., Bonora, G. M., and Toniolo, C. (1979). *Int. J. Pept. Protein Res.* **14**, 143–152.
Martinez, J., and Bodanszky, M. (1978). *Int. J. Pept. Protein Res.* **12**, 277–283.
Martinez, J., Tolle, J. C., and Bodanszky, M. (1979). *Int. J. Pept. Protein Res.* **13**, 22–27.
Mazur, R. H., and Schlatter, J. M. (1963). *J. Org. Chem.* **28**, 1025–1029.

Mazur, R. H., Schlatter, J. M., and Goldkamp, A. H. (1969). *J. Am. Chem. Soc.* **91**, 2684–2691.
Medzihradszky, K., Bajusz, S., Vargha, H. S., and Lang, Z. (1969). *Acta Chim. Acad. Sci. Hung.* **59**, 165–168.
Mehlis, B., Apelt, H., Bergmann, J., and Niedrich, H. (1972). *J. Prakt. Chem.* **314**, 390–400.
Meienhofer, J., Sano, Y., and Patel, R. P. (1970). *In* "Peptides: Chemistry and Biochemistry" (B. Weinstein and S. Lande, eds.), pp. 419–434. Dekker, New York.
Merrifield, R. B. (1963). *J. Am. Chem. Soc.* **85**, 2149–2154.
Merrifield, R. B., and Woolley, D. W. (1956). *J. Am. Chem. Soc.* **78**, 4646–4649.
Meyers, C., Havran, R. T., Schwartz, I. L., and Walter, R. (1969). *Chem. Ind.* (*London*) pp. 136–137.
Mori, Y., and Yajima, H. (1976). *Chem. Pharm. Bull.* **24**, 2781–2787.
Mori, Y., Koyama, K., Kiso, Y., and Yajima, H. (1976). *Chem. Pharm. Bull.* **24**, 2788–2793.
Morley, J. S. (1967). *J. Chem. Soc. C* pp. 2410–2421.
Moroder, L., Borin, G., Marchiori, F., and Scoffone, E. (1972). *Biopolymers* **11**, 2191–2194.
Moroder, L., Borin, G., Marchiori, F., and Scoffone, E. (1973a). *Biopolymers* **12**, 477–492.
Moroder, L., Marchiori, F., Borin, G., and Scoffone, E. (1973b). *Biopolymers* **12**, 493–505.
Moroder, L., Borin, G., Marchiori, F., and Scoffone, E. (1973c). *Biopolymers* **12**, 507–520.
Moroder, L., Marchiori, F., Borin, G., and Scoffone, E. (1973d). *Biopolymers* **12**, 701–720.
Moroder, L., Borin, G., Marchiori, F., and Scoffone, E. (1973e). *Biopolymers* **12**, 721–728.
Moroder, L., Marchiori, F., Borin, G., and Scoffone, E. (1973f). *Biopolymers* **12**, 729–750.
Moroder, L., Borin, G., Marchiori, F., and Scoffone, E. (1974). *Justus Liebigs Ann. Chem.* pp. 213–224.
Moroder, L., Borin, G., Filippi, B., and Marchiori, F. (1975a). *In* "Peptides 1974" (Y. Wolman, ed.), pp. 1–8. Wiley, New York.
Moroder, L., Filippi, B., Borin, G., and Marchiori, F. (1975b). *Biopolymers* **14**, 2061–2074.
Moroder, L., Wackerle, L., and Wünsch, E. (1976). *Hoppe-Seyler's Z. Physiol. Chem.* **357**, 1647–1650.
Nagasawa, T., Kuroiwa, K., Narita, K., and Isowa, Y. (1973). *Bull. Chem. Soc. Jpn.* **46**, 1269–1272.
Natarajan, S., and Bodanszky, M. (1976). *J. Org. Chem.* **41**, 1269–1272.
Nishimura, O., Hatanaka, C., and Fujino, M. (1975). *Chem. Pharm. Bull.* **23**, 1212–1220.
Nosaki, S. (1978). *Bull. Chem. Soc. Jpn.* **51**, 2995–3003.
Obermeier, R., and Geiger, R. (1976). *Hoppe-Seyler's Z. Physiol. Chem.* **357**, 759–767.
Offord, R. E. (1980). "Semisynthetic Proteins." Wiley, New York.
Offord, R. E., and DiBello, C., Eds. (1978). "Semisynthetic Peptides and Proteins." Academic Press, New York.
Offord, R. E., Storey, H. T., Rees, A. R., Hayward, C. F., Johnson, W. H., Pheasey, M. H., and Wightman, D. A. (1976). *Biochem. J.* **159**, 480–486.
Ogawa, H., Kubota, M., and Yajima, H. (1976a). *Chem. Pharm. Bull.* **24**, 2428–2434.
Ogawa, H., Kubota, M., Yajima, H., Tobe, T., Fujimura, M., Henmi, K., Torizuka, K., Adachi, H., Imura, H., and Taminato, T. (1976b). *Chem. Pharm. Bull.* **24**, 2447–2456.
Ogawa, H., Sugiura, M., Yajima, H., Sakurai, H., and Tsuda, K. (1978). *Chem. Pharm. Bull.* **26**, 1549–1557.
Ohno, M., Kato, T., Mitsuyasu, N., Waki, M., Makisuma, S., and Izumiya, N. (1967). *Bull. Chem. Soc. Jpn.* **40**, 204–208.
Ohno, M., Tsukamoto, S., Sato, S., and Izumiya, N. (1973). *Bull. Chem. Soc. Jpn.* **46**, 3280–3285.
Okamoto, K., and Shimamura, S. (1973). *Yakugaku Zasshi* **93**, 333–337.
Okamoto, K., Yasumura, K., Shimamura, S., Nakamura, M., Tanaka, A., and Yajima, H. (1979). *Chem. Pharm. Bull.* **27**, 499–507.
Oki, K., Suzuki, K., Tuchida, S., Saito, T., and Kotake, H. (1970). *Bull. Chem. Soc. Jpn.* **43**, 2554–2558.

Ondetti, M. A. (1963). *J. Med. Chem.* **6**, 10–16.
Ondetti, M. A., Deer, A., Sheehan, J. T., Pluscec, J., and Kocy, O. (1968a). *Biochemistry* **7**, 4069–4075.
Ondetti, M. A., Narayanan, V. L., von Saltza, M., Sheehan, J. T., Sabo, E. F., and Bodanszky, M. (1968b). *J. Am. Chem. Soc.* **90**, 4711–4716.
Otsuka, H., Inouye, K., Shinozaki, F., and Kanayama, M. (1966). *Bull. Chem. Soc. Jpn.* **39**, 1711–1178.
Ozawa, K., Shiori, T., and Yamada, S. (1977). *Chem. Pharm. Bull.* **25**, 122–125.
Patchornik, A., Berger, A., and Katchalski E. (1957). *J. Am. Chem. Soc.* **79**, 6416–6420.
Paul, R. (1963). *J. Org. Chem.* **28**, 236–237.
Paul, R., Anderson, G. W., and Callahan, F. M. (1961). *J. Org. Chem.* **26**, 3347–3350.
Pettit, G. R. (1970). "Synthetic Peptides," Vol. 1. Van Nostrand-Reinhold, Princeton, New Jersey.
Pettit, G. R. (1971). "Synthetic Peptides," Vol. 2. Van Nostrand-Reinhold, Princeton, New Jersey.
Pettit, G. R. (1975). "Synthetic Peptides," Vol. 3. Academic Press, New York.
Pettit, G. R. (1976). "Synthetic Peptides," Vol. 4. Elsevier, Amsterdam.
Photaki, I., and Yiotakis, A. (1976). *J. Chem. Soc., Perkin Trans. 1* pp. 259–264.
Photaki, I., Sakarellos, C., Yiotakis, A., and Zervas, L. (1975). *In* "Peptides 1974" (Y. Wolman, ed.), pp. 65–69. Wiley, New York.
Pietta, P. G. Chillemi, F., and Corbellini, A. (1968). *In* "Peptides 1968" (E. Bricas, ed.), pp. 104–106. North-Holland Publ., Amsterdam.
Pless, J., and Boissonnas, R. A. (1963). *Helv. Chim. Acta* **46**, 1609–1625.
Previero, A., Coletti-Previero, M.-A., and Cavadore, J.-C. (1967). *Biochim. Biophys. Acta* **147**, 453–461.
Rahn, W., Eckstein, H., and König, W. A. (1976). *Hoppe-Seyler's Z. Physiol. Chem.* **357**, 1223–1227.
Ramachandran, J., and Li, C. H. (1963). *J. Org. Chem.* **28**, 173–177.
Rees, A. R., and Offord, R. E. (1976a). *Biochem. J.* **159**, 467–479.
Rees, A. R., and Offord, R. E. (1976b). *Biochem. J.* **159**, 487–493.
Ridge, B., and Jones, J. H. (1975). *Amino-Acids, Pept., Proteins* **6**, 349–352.
Riniker, B., and Rittel, W. (1970) *Helv. Chim. Acta* **53**, 513–519.
Riniker, B., Brugger, M., Kamber, B., Sieber, P., and Rittel, W. (1969). *Helv. Chim. Acta* **52**, 1058–1073.
Rink, H., and Riniker, B. (1974). *Helv. Chim. Acta* **57**, 831–835.
Riordan, J. F., and Vallee, B. L. (1972). *In* "Methods in Enzymology" (C. H. W. Hirs and N. Timasheff, eds.), Vol. 25, pp. 500–506. Academic Press, New York.
Rocchi, R., Marchiori, F., and Scoffone, E. (1963). *Gazz. Chim. Ital.* **93**, 823–833.
Rocchi, R., Benassi, C. A., Guarneri, M., Guggi, A., and Tomatis, R. (1974). *Int. J. Pept. Protein Res.* **6**, 391–395.
Rocchi, R., Benassi, C. A., Guarneri, M., Guggi, A., and Tomatis, R. (1975). *In* "Peptides, 1974" (Y. Wolman, ed.), pp. 9–15. Wiley, New York.
Romovacek, H., Drabarek, S., Kawasaki, K., Dowd, S. R., Obermeier, R., and Hofmann, K. (1974). *Int. J. Pept. Protein Res.* **6**, 435–445.
Romovacek, H., Dowd, S. R., Kawasaki, K., Nishi, N., and Hofmann, K. (1979). *J. Am. Chem. Soc.* **101**, 6081–6091.
Sakakibara, S. (1977). *In* "Peptides, Proceedings of the Fifth American Peptide Symposium" (M. Goodman and J. Meienhofer, eds.), pp. 436–447. Wiley, New York.
Sakiyama, F., Okawa, K., Yamakawa, T., and Akabori, S. (1958). *Bull. Chem. Soc. Jpn.* **31**, 926–932.
Schafer, D. J. (1976). *Amino-Acids, Pept., Proteins* **7**, 247–319.

Schafer, D. J., and Carlsson, L. (1972). *J. Chem. Soc., Chem. Commun.* p. 276 only.
Schafer, D. J., Young, G. T., Elliott, D. F., and Wade, R. (1971). *J. Chem. Soc. C* pp. 46–49.
Schaich, E., Fretzdorff, A.-M., and Schneider, F. (1973). *Hoppe-Seyler's Z. Physiol. Chem.* **354**, 897–902.
Schechter, B., Schechter, I., Ramachandran, J., Conway-Jacobs, A., and Sela, M. (1971). *Eur. J. Biochem.* **20**, 301–308.
Schnabel, E. (1967). *Justus Liebigs Ann. Chem.* **702**, 188–196.
Schnabel, E., Klostermeyer, H., Dahlmans, J., and Zahn, H. (1967). *Justus Leibigs Ann. Chem.* **707**, 227–241.
Schnabel, E., Herzog, H., Hoffman, P., Klauke, E., and Ugi, I. (1968). *Justus Liebigs Ann. Chem.* **716**, 175–185.
Schnabel, E., Stoltefuss, J., Offe, H. A., and Klauke, E. (1971). *Justus Liebigs Ann. Chem.* **743**, 57–68.
Schneider, F. (1960). *Hoppe-Seyler's Z. Physiol. Chem.* **321**, 38–48.
Schoelkens, B. A., Wissmann, H., Lindner, E., and Geiger, R. (1976). *Hoppe-Seyler's Z. Physiol. Chem.* **357**, 825–838.
Schröder, E. (1963). *Justus Liebigs Ann. Chem.* **670**, 127–136.
Schröder, E., and Gibian, H. (1962). *Justus Liebigs Ann. Chem.* **656**, 190–204.
Schröder, E., and Klieger, E. (1964a). *Justus Liebigs Ann. Chem.* **673**, 196–207.
Schröder, E., and Klieger, E. (1964b). *Justus Liebigs Ann. Chem.* **673**, 208–220.
Schröder, E., and Lübke, K. (1965). "The Peptides," Vol. 1. Academic Press, New York.
Schwyzer, R., and Sieber, P. (1966). *Helv. Chim. Acta* **49**, 134–158.
Sheehan, J. C., Hasspacker, K., and Yeh, Y. L. (1959). *J. Am. Chem. Soc.* **81**, 6086 only.
Shioiri, T., and Yamada, S. (1974a). *Chem. Pharm. Bull.* **22**, 849–854.
Shioiri, T., and Yamada, S. (1974b). *Chem. Pharm. Bull.* **22**, 855–858.
Shioiri, T., and Yamada, S. (1974c). *Chem. Pharm. Bull.* **22**, 859–863.
Shioiri, T., Ninomiya, K., and Yamada, S. (1972). *J. Am. Chem. Soc.* **94**, 6203–6205.
Sieber, P., Riniker, B., Brugger, M., Kamber, B., and Rittel, W. (1970). *Helv. Chim. Acta* **53**, 2135–2150.
Sieber, P., Brugger, M., and Rittel, W. (1971). *In* "Peptides 1969" (E. Scoffone, ed.), pp. 60–61. North-Holland Publ., Amsterdam.
Sieber, P., Rittel, W., and Riniker, B. (1972). *Helv. Chim. Acta* **55**, 1243–1248.
Smithwick, E. L., Jr., and Shuman, R. T. (1974). *J. Org. Chem.* **39**, 3441–3442.
Smyth, D. G. (1975). *Nature (London)* **256**, 699–700.
Sondheimer, E., and Holley, R. W. (1954). *J. Am. Chem. Soc.* **76**, 2467–2470.
Storey, H. T., Beacham, J., Cernosek, S. F., Finn, F. M., Yanaihara, C., and Hofmann, K. (1972). *J. Am. Chem. Soc.* **94**, 6170–6178.
Strachan, R. G., Paleveda, W. J., Jr., Nutt, R. F., Vitali, R. A., Veber, D. F., Dickinson, M. J., Garsky, V., Deak, J. E., Walton, E., Jenkins, S. R., Holly, F. W., and Hirschmann, R. (1969). *J. Am. Chem. Soc.* **91**, 503–505.
Suzuki, K., Nitta, K., and Sasaki, Y. (1976). *Chem. Pharm. Bull.* **24**, 3025–3033.
Tamura, F., Kiso, Y., Yajima, H., Tanaka, A., and Nakamura, M. (1975). *Chem. Pharm. Bull.* **23**, 2405–2409.
Tamura, F., Ogawa, H., Fujii, N., Yajima, H., Miyata, K., Nakamura, M., and Tanaka, A. (1977) *Chem. Pharm. Bull.* **25**, 767–774.
Theodoropoulos, D., and Souchleris, I. (1965). *Acta Chim. Acad. Sci. Hung.* **44**, 183–185.
Tilak, M. A. (1970). *Tetrahedron Lett.* pp. 849–854.
Tomasz, J. (1971). *Acta Chim. Acad. Sci. Hung.* **70**, 255–261.
Tomatis, R., Guggi, A., Menegatti, E., and Rocchi, R. (1976a). *Int. J. Pept. Protein Res.* **8**, 107–113.

Tomatis, R., Guggi, A., Benassi, C. A., Salvadori, S., and Rocchi, R. (1976b). *Int. J. Pept. Protein Res.* **8**, 65–77.
Tomatis, R., Guggi, A., Salvadori, S., Periotto, V., and Rocchi, R. (1978). *Int. J. Pept. Protein Res.* **11**, 269–281.
Tomatis, R., Guarneri, M., Guggi, A., Salvadori, S., and Rocchi, R. (1979). *Int. J. Pept. Protein Res.* **14**, 347–355.
Vandesdande, F. (1969). *Bull. Soc. Chim. Belg.* **78**, 395–406.
VanNispen, J. W., and Tesser, G. I. (1975). *Int. J. Pept. Protein Res.* **7**, 57–67.
Veber, D. F. (1975). *In* "Peptides: Chemistry, Structure and Biology" (R. Walter and J. Meienhofer, eds.), pp. 307–316. Ann Arbor Sci. Publ., Ann Arbor, Michigan.
Veber, D. F., Milkowski, J. D., Denkewalter, R. G., and Hirschmann, R. (1968). *Tetrahedron Lett.* pp. 3057–3058.
Veber, D. F., Varga, S. L., Milkowski, J. D., Joshua, H., Conn, J. B., Hirschmann, R., and Denkewalter, R. G. (1969). *J. Am. Chem. Soc.* **91**, 506–507.
Vičar, I., Gut, V., Fric, I. and Blàha, K. (1976). *Collect. Czech. Chem. Commun.* **41**, 3467–3473.
Visser, S., Raap, J., Kerling, K.E.T., and Havinga, E. (1970). *Recl. Trav. Chim. Pays-Bas* **89**, 865–875.
Voelter, W., Zech, K., Grimminger, W., Breitmaier, E., and Jung, G. (1972). *Chem. Ber.* **105**, 3650–3657.
Voskuyl-Holtkamp, I., and Schattenkerkc, (1977). *Int. J. Pept. Protein Res* **10**, 60–70.
Wakimasu, M., and Fujino, M. (1978). *Chem. Pharm. Bull.* **26**, 1522–1526.
Watanabe, H., Kubota, M., Yajima, H., Tanaka, A., Nakamura, M., and Kawabata, T. (1974). *Chem. Pharm. Bull.* **22**, 1889–1894.
Watanabe, H., Ogawa, H., and Yajima, H. (1975). *Chem. Pharm. Bull.* **23**, 375–383.
Watanabe, K., and Inouye K. (1977). *Bull. Chem. Soc. Jpn.* **50**, 201–206.
Weinert, M., Kircher, K., Brandenburg, D., and Zahn, H. (1971). *Hoppe-Seyler's Z. Physiol. Chem.* **352**, 719–724.
Wendel, A., Schaich, E., Weber, U., and Flohe, L. (1972). *Hoppe-Seyler's Z. Physiol. Chem.* **353**, 514–522.
Weygand, F., and Hunger, K. (1962). *Chem. Ber.* **95**, 7–16.
Weygand, F., and Nintz, E. (1965). *Z. Naturforsch., Teil B* **20**, 429–434.
Weygand, F., Hoffmann, D., and Wünsch, E. (1966). *Z. Naturforsch., Teil B* **21**, 426–428.
Weygand, F., Steglich, W., and Bjarnason, J. (1968). *Chem. Ber.* **101**, 3642–3648.
Wies, R., and Pfaender, P. (1973). *Justus Liebigs Ann. Chem.* pp. 1269–1274.
Wolman, Y., and Klausner, Y. S. (1971). *Isr. J. Chem.* **9**, 205–209.
Wolman, Y., Schejter, A., and Solovsky, M. (1972). *J. Am. Chem. Soc.* **94**, 1720–1723.
Wünsch, E. (1967). *Z. Naturforsch., Teil B* **22**, 1269–1276.
Wünsch, E. (1971). *Angew. Chem., Int. Ed. Engl.* **10**, 786–795.
Wünsch, E. (1972). *Naturwissenschaften* **59**, 239–246.
Wünsch, E., ed. (1974). "Houben-Weyl Methoden der Organischen Chemie," Vol. 15, Part 1, pp. 43–45. Thieme, Stuttgart.
Wünsch, E., and Wendlberger, G. (1967). *Chem. Ber.* **100**, 160–172.
Wünsch, E., and Zwick, A. (1963). *Hoppe-Seyler's Z. Physiol. Chem.* **333**, 108–114.
Wünsch, E., and Zwick, A. (1964). *Chem. Ber.* **97**, 2497–2503.
Wünsch, E., Zwick, A., and Jaeger, E. (1968). *Chem. Ber.* **101**, 336–340.
Wünsch, E., Jaeger, E., Kisfaludy, L., and Löw, M. (1977). *Angew. Chem. Int. Ed. Engl.* **16**, 317–318.
Yajima, H., and Kiso, Y. (1974a). *Chem. Pharm. Bull.* **22**, 1061–1066.
Yajima, H., and Kiso, Y. (1974b). *Chem. Pharm. Bull.* **22**, 1087–1094.
Yajima, H., and Kubo, K. (1965). *J. Am. Chem. Soc.* **87**, 2039–2044.

Yajima, H., Okada, Y., Kinomura, Y., Mizokami, N., and Kawatani, H. (1969). *Chem. Pharm. Bull.* **17**, 1237–1245.
Yajima, H., Kitagawa, K., and Kurobe, M. (1973a). *Chem. Pharm. Bull.* **21**, 2566–2567.
Yajima, H., Tamura, F., Kiso, Y., and Kurobe, M. (1973b). *Chem. Pharm. Bull.* **21**, 1380–1382.
Yajima, H., Okada, Y., Watanabe, H., and Kiso, Y. (1974a). *Chem. Pharm. Bull.* **22**, 1067–1074.
Yajima, H., Mizokami, N., Kiso, M., Jinnouchi, T., Kai, Y., and Kiso, Y. (1974b). *Chem. Pharm. Bull.* **22**, 1075–1078.
Yajima, H., Kiso, Y., and Kitagawa, K. (1974c). *Chem. Pharm. Bull.* **22**, 1079–1086.
Yajima, H., Ogawa, H., Kubota, M., Tobe, T., Fujimura, M., Henmi, K., Torizuka, K., Adachi, H., Imura, H., and Taminato, T. (1975). *J. Am. Chem. Soc.* **97**, 5593–5594.
Yajima, H., Koyama, K., Kiso, Y., Tanaka, A., and Nakamura, M. (1976a). *Chem. Pharm. Bull.* **24**, 492–499.
Yajima, H., Mori, Y., Koyama, K., Tobe, T., Setoyama, M., Adachi, H., Kanno, T., and Saito, A. (1976b). *Chem. Pharm. Bull.* **24**, 2794–2802.
Yajima, H., Ogawa, H., and Sakurai, H. (1977). *J. Chem. Soc., Chem. Commun.* pp. 909–910.
Yamashiro, D., and Li, C. H. (1973). *J. Org. Chem.* **38**, 2594–2597.
Yamashiro, D., Blake, J., and Li, C. H. (1972). *J. Am. Chem. Soc.* **94**, 2855–2859.
Yanaihara, N., Yanaihara, C., Dupuis, G., Beacham, J., Camble, R., and Hofmann, K. (1969). *J. Am. Chem. Soc.* **91**, 2184–2185.
Yanaihara, N., Yanaihara, C., Sakagami, M., Tsuji, K., Hashimoto, T., Kaneko, T., Oka, H., Schally, A. V., Arimura, A., and Redding, T. W. (1973). *J. Med. Chem.* **16**, 373–377.
Yanaihara, N., Kubota, M., Sakagami, M., Sato, H., Mochizuki, T., Sakura, N., Hashimoto, T., and Yanaihara, C. (1977). *J. Med. Chem.* **20**, 648–655.
Zahn, H., and Brandenburg, D. (1966). *Justus Liebigs Ann. Chem.* **692**, 220–230.
Zahn, H., and Fölsche, E. T. J. (1968). *Justus Liebigs Ann. Chem.* **716**, 164–174.
Zahn, H., and Fölsche, E. T. J. (1969). *Chem. Ber.* **102**, 2158–2159.
Zahn, H., and Schmidt, G. (1970a). *Justus Liebigs Ann. Chem.* **731**, 91–100.
Zahn, H., and Schmidt, G. (1970b). *Justus Liebigs Ann. Chem.* **731**, 101–112.
Zahn, H., Meienhofer, J., and Klostermeyer, H. (1964). *Z. Naturforsch., Teil B* **19**, 110–113.
Zahn, H., Bremer, H., and Zabel, R. (1965). *Z. Naturforsch., Teil B* **20**, 653–660.
Zervas, L., and Hamalidis, C. (1965). *J. Am. Chem. Soc.* **87**, 99–104.
Zervas, L., Winitz, M., and Greenstein, J. P. (1957). *J. Org. Chem.* **22**, 1515–1521.
Zervas, L., Otani, T. T., Winitz, M., and Greenstein, J. P. (1959). *J. Am. Chem. Soc.* **81**, 2878–2884.

Chapter 7

Dual Function Groups

BRIAN J. JOHNSON

Many of the protecting groups used in peptide chemistry have to be removed selectively at judicious times during synthesis before further manipulations can be made to the peptide. Often this process produces unwanted

THE PEPTIDES, VOLUME 3

ISBN 0-12-304203-8

side reactions as illustrated by the racemization associated with the alkaline hydrolysis of unactivated carboxyl protecting esters (McDermott and Benoiton, 1972; see also Kenner and Seely, 1972).

It occurred to several investigators that peptide chemistry would benefit from having a number of dual function groups that could be used for protective purposes and then, without removal, be converted to moieties that could be used for another purpose, e.g., coupling. Furthermore, this conversion should be easily accomplished without racemization and damage to the attached peptide. Obvious uses for such dual function groups occur in the preparation of activated monomers for the synthesis of sequential polypeptides, cyclic peptides, and peptide intermediates for the synthesis of peptides by segment condensation.

The following dual function groups have been investigated using the solution- and solid-phase techniques of peptide synthesis with varying degrees of success.

I. SUBSTITUTED HYDRAZIDES

A. Deprotection Followed by Activation through Azide Formation

Generally, the hydrazide group is not suitable for carboxyl protection although it has been reported (Cheung and Blout, 1965) that the peptide hydrazide chain can be lengthened by the use of N-protected amino acid activacted esters.

It was first suggested by Hofmann *et al.* (1952) that the use of N-protected hydrazine derivatives as carboxyl protective groups was a feasible method to enable the ready preparation of peptide hydrazides. Using this type of carboxyl protection, the required peptide can be synthesized by additions to its NH_2-terminus. At the appropriate time the hydrazide protecting group can be removed to release the hydrazide which can then be converted to the azide in anticipation for further coupling through the COOH-terminus of the peptide (Scheme 1). An attraction of this method is its use of the azide for the coupling step, a process that can be made racemization-free, provided that careful selection of the conditions to be used in the azide syntheses are chosen (Kemp *et al.*, 1970. Sieber and Riniker, 1973; see also Kemp, Vol. 1 Chapter 7).

$$—CON_2H_2Prot \longrightarrow —CON_2H_3 \longrightarrow —CON_3$$

Scheme 1

The requirement of an additional amine protecting group for the substituted hydrazide can be viewed as a disadvantage of the method. This additional protecting group must be chosen so that its selective removal in the presence of other protecting groups attached to the peptide can be achieved unambiguously. The great advantage of the method is that the peptide hydrazide can be generated without having to go through the hydrazinolysis of the corresponding peptide ester, especially since this latter procedure can be, as shown by Yajima and Kiso (1971), fraught with danger. In their case it was found that the hydrazinolysis of the tripeptide ester (**1**) produced material (**2**) that possessed two hydrazide moieties (Scheme 2). A similar situation occurred in the hydrazinolysis of a protected tetrapeptide methyl ester segment of rubredoxin (Weinstein *et al.*, 1976).

$$\underset{(\mathbf{1})}{\text{Boc-Asp(O}t\text{Bu)-Glu(O}t\text{Bu)-Gly-OMe}} \xrightarrow{N_2H_4} \underset{(\mathbf{2})}{\text{Boc-Asp(}N_2H_3\text{)-Glu(O}t\text{Bu)-Gly-}N_2H_3}$$

Scheme 2

1. *Benzyloxycarbonylhydrazide*

The combination of phthalyl, tosyl, and benzyloxycarbonyl groups for amine protection with benzyloxycarbonylhydrazide for carboxyl protection was originally used by Hofmann *et al.*, (1952). Later work by the same group on the synthesis of S-peptide antagonists (Hofmann *et al.*, 1970) lengthened the list of protecting groups that can be used with this carboxyl protecting moiety to include the *tert*-butyloxycarbonyl and *tert*-butyl ester groups. In the synthesis of a peptide related to scotophobin (Ali *et al.*, 1972), a similar variety of protecting groups were used including the benzyl ester moiety. Hydrogenolysis successfully deprotected the benzyloxycarbonylhydrazide heptapeptide fragment to release the hydrazide. This material was converted to its azide and used for the final coupling to produce the protected 15-peptide possessing the sequence attributed to scotophobin. Subsequently, the list of protecting groups used in combination with the benzyloxycarbonylhydrazide protecting group has been expanded to include the formyl and *S*-ethylcarbamoyl moieties as illustrated in the synthesis of an 11-peptide segment of ribonuclease T_1 (Storey *et al.*, 1972). Furthermore, exposure of this 11-peptide segment to HBr in trifluoroacetic acid yielded the 11-peptide hydrazide hydrobromide, which was successfully converted to the azide and then coupled to a suitably protected 12-peptide to give the protected 23-peptide (Storey *et al.*, 1972). This success indicated the utility of dual function groups in peptide chemistry.

Benzyloxycarbonylhydrazide has been synthesized by the partial hydrazinolysis of dibenzyl carbonate (Hofmann *et al.*, 1952) and by the benzyloxycarbonylation of hydrazine hydrate (Böshagen and Ullrich, 1959; Schröder and Lübke, 1963). The carboxyl protecting group can be attached to the N-protected amino acid by any of the conventional coupling methods.

2. *tert-Butyloxycarbonylhydrazide*

This variation of the substituted hydrazide method for carboxyl protection was introduced simultaneously by Boissonnas *et al.* (1960) and Schwyzer *et al.* (1960). The applicability of this protecting group, in combination with the benzyloxycarbonyl amine protecting group, was verified by its use in the synthesis of bradykinin and a sequence of tyrocidin. Subsequently, this carboxyl protecting group has enjoyed popularity both in the solution and, by appropriate modifications, in the solid-phase techniques of peptide synthesis. The wide use of the *tert*-butyloxycarbonylhydrazide as a carboxyl protecting group in the synthesis of very different and also large peptides is indicative of the reliability and reproducibility of the method.

The *tert*-butyloxycarbonylhydrazide carboxyl protecting group has been used in combination with the benzyloxycarbonyl, *tert*-butyl ester, nitro, formyl, *S*-ethylcarbamoyl, *tert*-butyl ether, trifluoroacetyl, *S*-acetamidomethyl, diadamantyloxycarbonyl, and dimethoxybenzhydryl moieties. Examples of these are illustrated in the syntheses of protected segments of ribonuclease T_1 (Yanaihara *et al.*, 1969; Beacham *et al.*, 1971; Camble *et al.*, 1972; Storey *et al.*, 1972; Kawasaki *et al.*, 1973) and also cytochrome *C* (Moroder *et al.*, 1972).

One of the criticisms of a number of protecting groups is the lack of demonstrable use in the synthesis of large peptides. Such is not the case with the dual function *tert*-butyloxycarbonylhydrazide protecting group. It has been used to great success in the synthesis of a *N*-benzyloxycarbonyl 36-peptide *tert*-butyloxycarbonylhydrazide corresponding to positions 12–47 of ribonuclease T_1 (Scheme 3). This peptide fragment was synthesized from the corresponding 12-peptide, positions 12–23, 11-peptide, positions 24–34, and 13-peptide, positions 35–47, and it entailed the deprotection, using 90% trifluoroacetic acid, of two of the segment *tert*-butyloxycarboxyl hydrazides, positions 24–34 and 35–47, and then subsequent conversion to their respective azides (Beacham *et al.*, 1971). In later work this dual function group has been used in the synthesis of a protected 47-peptide, corresponding to the positions 1–47 of ribonuclease T_1 (Storey *et al.*, 1972). This particular segment was synthesized by coupling peptide (1–23), which had been obtained from the corresponding *tert*-butyloxycarbonylhydrazide segment, with the suitably protected 24-peptide, positions 24–47.

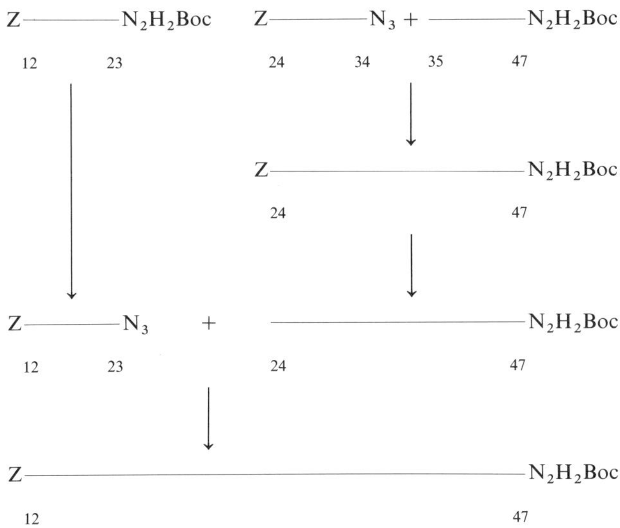

Scheme 3

The same approach was used to synthesize a *N*-benzyloxycarbonyl 33-peptide corresponding to the positions 48–80 of ribonuclease T_1 (Camble *et al.*, 1972). In this example an 18-peptide *tert*-butyloxycarbonylhydrazide, positions 48–65, was converted to its azide and then coupled to the suitably protected 15-peptide segment, positions 66–80.

The protected hydrazide approach was used to synthesize the protected 24-peptide amide, corresponding to positions 81–104 (Kawasaki *et al.*, 1973), and the 39-peptide, corresponding to positions 66–104 of ribonuclease T_1 (Romovacek *et al.*, 1974).

Application of the dual function *tert*-butyloxycarbonylhydrazide group has been made in the solid-phase technique of peptide synthesis. Obviously hydrazinolysis of the peptide–polymer anchor is one way of producing a protected peptide hydrazide intermediate suitable for azide formation. However, such a method has limitations and in order to overcome these Wang and Merrifield (1969) introduced a solid-phase polymer support that incorporated the *tert*-alkyloxycarbonylhydrazide function (Scheme 4). A somewhat simpler method of preparing the solid-phase polymer support has appeared (Wolters *et al.*, 1974). The successful use of the modified resin relies on the very mild conditions required to remove the 2-(4-biphenylyl) propyl (2) oxycarbonyl (Bpoc) amine protecting group of the attached amino acid or peptide without disrupting the anchoring bond to the resin. Subsequent work by these investigators (Wang and Merrifield, 1972) indicated that this

$N_2H_3COOC(Me)_2CH_2CH_2$—Resin

↓ SPS, Bpoc-Leu-OH, Bpoc-Ala-OH, Bpoc-Val-OH, Z-Phe-OH

Z-Phe-Val-Ala-Leu-NHNHCOOC$(Me)_2CH_2CH_2$—Resin

↓ TFA

Z-Phe-Val-Ala-Leu-N_2H_3

Scheme 4

solid-phase technique could be used to synthesize a protected decapeptide hydrazide.

Another resin incorporating the *p*-alkoxybenzyloxycarbonylhydrazide (Wang, 1972) has been used for the synthesis of a protected peptide hydrazide (Wang, 1973) as illustrated in Scheme 5. Subsequently, Wang (1975) has sug-

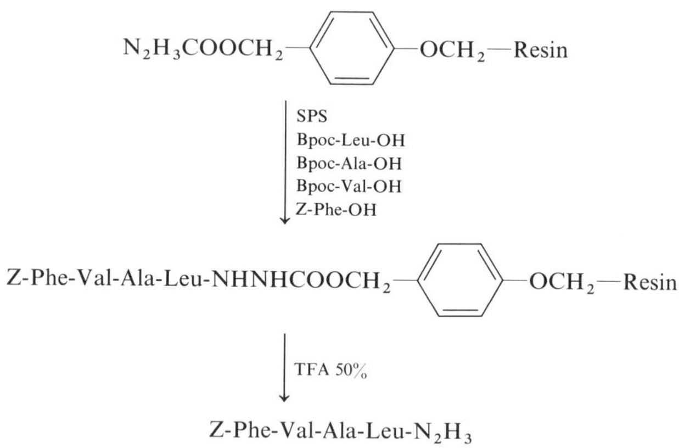

Scheme 5

gested a further modification of the resin to incorporate the 3-(*p*-benzyloxyphenyl)-1,1-dimethylpropyloxycarbonylhydrazide moiety (**3**). Apparently, the last variation is easier to prepare than the previously used anchoring groups.

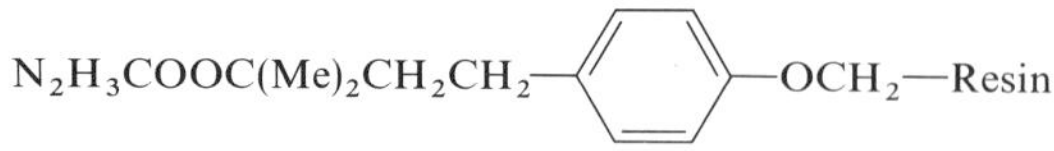

(3)

3. Tritylhydrazide

This carboxyl protecting group has not been employed very widely. However, it has been used in combination with the trifluoroacetyl moiety (Weygand and Steglich, 1959).

4. Trichloroethyloxycarbonylhydrazide

This carboxyl protecting roup was introduced by Yajima and Kiso (1971). The required intermediates are prepared by coupling N-protected amino acids with trichloroethyloxycarbonylhydrazine using either the dicyclohexylcarbodiimide or the mixed anhydride methods. The precursor, trichloroethyloxycarbonylhydrazine, is prepared from either trichloroethyloxycarbonyl chloride and hydrazine or via the hydrazinolysis of trichloroethyl *p*-nitrophenyl carbonate.

The originators of this carboxyl protecting group have used it in combination with the *p*-methoxybenzyloxcarbonyl and benzyl ester protecting groups. Subsequently, Weinstein *et al*., (1976) used it in conjunction with the benzyloxycarbonyl and *tert*-butyl ester moieties; however, these latter investigators experienced difficulties upon removal of the benzyloxycarbonyl groups using hydrogenolysis conditions. Apparently, the trichloroethyloxycarbonyl moiety is also hydrogenated and HCl is released. A similar situation occurs with the hydrogenolysis of benzyloxycarbonyl peptide pentachlorophenyl esters.

The peptide trichloroethyloxycarbonylhydrazides are converted to their corresponding peptide hydrazides by means of zinc dust in acetic acid (Yajima and Kiso, 1971).

5. Trifluoroacetylhydrazide

This carboxyl protecting group was introduced by Preston and Weinstein (1967). It is similar to the other substituted hydrazides possessing the property of masking the hydrazide function in peptides where chain extension by conversion to the azide is anticipated as a later synthetic stage. The trifluoroacetylhydrazide moiety has been used in combination with the *o*-nitrophenylsulfenyl and benzyloxycarbonyl protecting groups. This carboxyl protecting group has not been used to any great extent.

6. Picolyloxycarbonylhydrazide

Macrea and Young (1975) introduced this carboxyl protecting group with the view of synthesizing peptide fragments suitable for further coupling by the azide method. One of the advantages of this protecting group over other N-protected hydrazides is that it possesses a weakly basic residue which aids in the isolation of the product peptide after coupling. The procedure has been illustrated by the synthesis of several picolyloxycarbonylhydrazides (Scheme 6) including the 13-peptide picolyloxycarbonylhydrazide corresponding to residues 1–13 of porcine gastric inhibitory polypeptide. The picolyl group can be removed by catalytic hydrogenation or electrolytic reduction to produce the unprotected hydrazide ready for conversion to the azide in anticipation for further coupling.

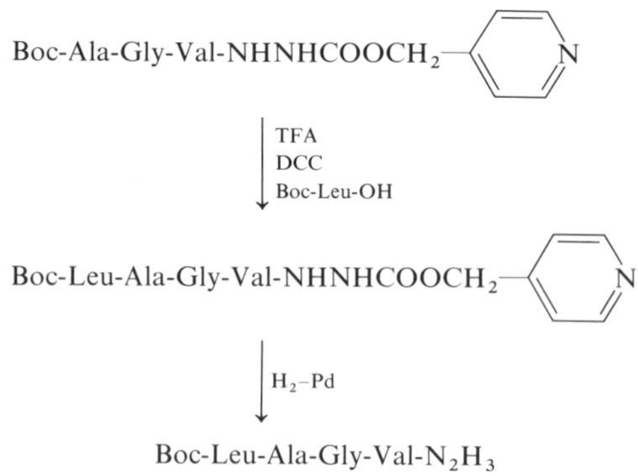

Scheme 6

B. Activation by Oxidation

1. Phenylhydrazides

The phenylhydrazide group is an effective amino acid carboxyl protecting group, being stable to mild acid and base. Derivatives of this moiety have been prepared from the benzyloxycarbonylamino acids and phenylhydrazine, using papain as catalyst, or by the azide method (Milne and Most, 1968). The phenyl group cannot be removed from acylamino acid phenylhydrazides to produce the corresponding acylamino acid hydrazides necessary as precursors for azide formation. However, these phenylhydrazides can be oxidized by either lead tetraacetate or *N*-bromosuccinimide (NBS) to give the unstable acylamino acid phenyldiimides (Milne and Kilday, 1965) which are capable of peptide coupling. Milne and Most (1968) extended this procedure to synthesize benzyloxycarbonyl-glycyl-phenylalanyl-glycyl-phenylalanyl-phenylhydrazide from the corresponding dipeptides (Scheme 7). Subsequently, Wieland *et al.* (1970) used the phenylhydrazide carboxyl protecting group, which was activated by oxidation with *N*-bromosuccinimide, to prepare the decapeptide DAla-antamanide.

These latter investigators also introduced variants of the phenylhydrazide carboxyl protecting group. In these cases, N-protected amino acids and peptides were linked to hydrazobenzene to give the diphenylhydrazides

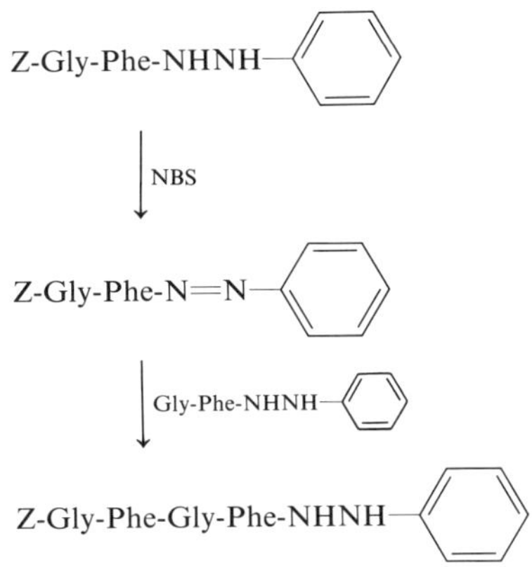

Scheme 7

(**4**) or to cyclic phthalhydrazide to give the *O*-acyl-3-hydroxyphthalazones (**5**) (Scheme 8).

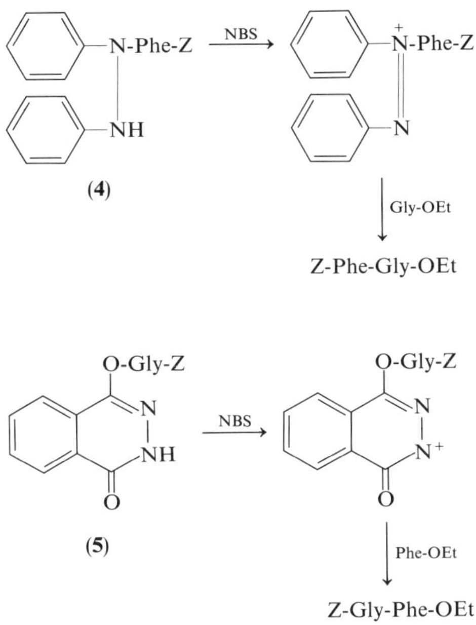

Scheme 8

Furthermore, this method of protection and then activation by oxidation with *N*-bromosuccinimide can be applied to the solid-phase technique of peptide synthesis (Wieland *et al.*, 1970). This is illustrated in Scheme 9.

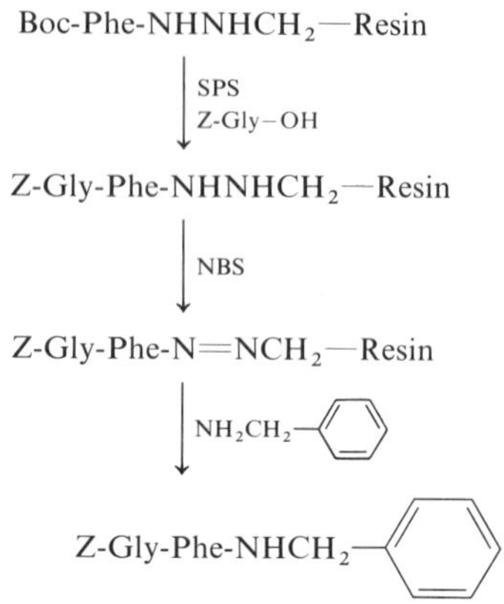

Scheme 9

Another variant on the activation of phenylhydrazides was introduced by Lewalter and Birr (1970). It was shown that such substituted hydrazides can be oxidized at a platinum anode in organic solution and in the presence of tetra-*n*-butylammonium perchlorate to form the corresponding diimide. Such activated materials react with amino acid esters to give peptides. The utility of the method was illustrated by the synthesis of a number of dipeptides.

II. ESTERS

Esterification is the most popular method of protecting the carboxyl group of amino acids and peptides. Although such moieties can be aminolyzed, the ease by which this occurs is dependent upon a number of parameters. These include the steric hindrance encountered by the attacking amine component and the electrophilic character of the carbonyl carbon which is usually influenced by the electronegativity of the ester alkoxy or aryloxy groups. This latter parameter has been used in the design of dual function ester groups for use in peptide chemistry.

A. Protective Esters with the Capability for Conversion to Activated Esters

1. 4-Methylthiophenyl Ester

This carboxyl protecting group was introduced by Johnson and Jacobs (1968a). The dual function 4-methylthiophenyl esters (**6**) are sufficiently unreactive toward aminolysis that they provide carboxyl protection. The attractive feature of this protective ester is its facile conversion without racemization (Johnson and Jacobs, 1968a) to the active 4-methylsulfonylphenyl ester (**7**). This latter active ester had been previc by Schwyzer and Sieber (1958). The conversion of the prote ester is achieved by oxidation using either hydrogen pe acid or *m*-chloroperoxybenzoic acid (Johnson and Tracey,

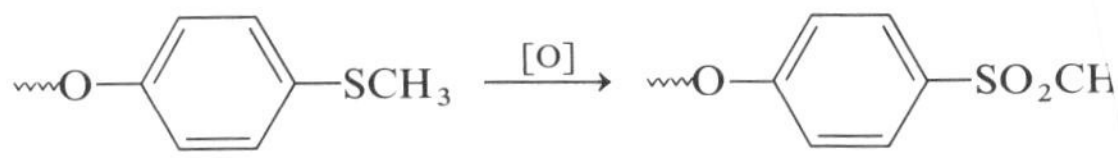

(**6**) (**7**)

The 4-methylthiophenyl esters are easily prepared by the dicyclohexylcarbodiimide method and they have been used in combination with the benzyloxycarbonyl, *tert*-butyloxycarbonyl, *tert*-butyl ether, benzyl ether, and trifluoracetyl protecting groups. No damage was observed to these moieties under the oxidative conditions used to prepare the 4-methylsulfonylphenyl active esters (Johnson and Ruettinger, 1970). Illustrations of the utility of this dual function ester has been reported for peptide (Johnson and Trask, 1968; Johnson 1969), polypeptide (Johnson and Rea, 1970) and *O*-depsipeptide (Johnson, 1969) synthesis by the solution technique.

Subsequently, the protective 4-methylthiophenyl ester has been applied to the solid-phase technique of peptide synthesis. Marshall and Liener (1970) prepared an insoluble polymer (**8**) using the *O*-carbonate ester of *p*-mercaptophenol and a chloromethylated polystyrene resin. The resin was used in the normal manner for solid-phase peptide synthesis by the stepwise addition of amino acid residues. Oxidation converted the anchoring ester linkage to the activated ester (**9**) which can be aminolyzed. By this

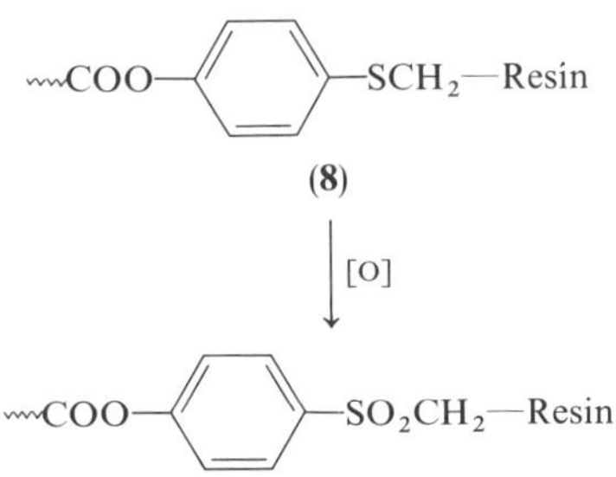

(**8**)

(**9**)

means, a protected peptide can be released from the resin by aminolysis with an appropriate amino acid moiety to give a protected peptide lengthened by one amino acid residue at the COOH-terminus. Racemization could possibly occur during this process. The method has been illustrated by the synthesis of several small peptides and a decapeptide (Flanigan and Marshall, 1972). These investigators have also shown that cyclic peptides can be prepared using this dual function method applied to the solid-phase technique. Removal of the N-terminal protecting group from a peptide attached to the activated form of the resin allows self aminolysis to occur and release of protected cyclic peptides. This procedure is also applicable to the synthesis of protected peptide amides and hydrazides by the use of the appropriate reactants on the peptides attached to the activated form of the resin.

2. *o*-Hydroxyphenyl Ester

The monoesters of catechol have been reported (Hansen, 1963; Biggins *et al.*, 1963) to solvolyze abnormally easily, and this has been atributed to intramolecular general base catalysis (Capon and Ghosh, 1966).

N-Protected amino acid *o*-hydroxyphenyl esters also aminolyze readily; their use for peptide synthesis was first suggested by Young (1967). Furthermore, their attractiveness has been enhanced by the claim that they are highly resistant to racemization (Jones and Young, 1968). This property is attributed to their possession of a neighboring group capable of hydrogen bonding to the incoming amine and accepting its proton, causing acceleration of aminolysis but not oxazolone formation.

The dual function nature of these monoesters of catechol is apparent by their use in their protected forms as *o*-benzyloxyphenyl (**10**) (Jones and Young, 1968) or *o*-phenacyloxyphenyl (**11**) esters (Trudelle, 1971). These moieties behave as normal phenyl esters suitable for COOH-terminal protection. However, at the appropriate time the removal of the hydroxyl protection from these esters exposes the accelerating hydroxyl moiety and thereby creates the active *o*-hydroxyphenyl ester (**12**).

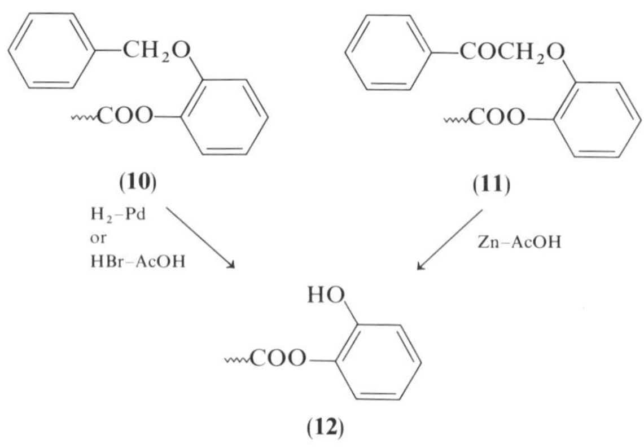

The benzyloxy protecting group can be removed by catalytic hydrogenation or by hydrogen bromide in acetic acid. These latter conditions are rather drastic and the advantage of the phenacyloxy protecting moiety is its removal by the use of zinc dust in aqueous anhydrous acetic acid (Hendrickson and Kandall, 1970). These relatively mild conditions allow the presence of other protecting groups, based on *tert*-butyl or benzyl alcohol, to be used for amine and side-chain protection.

Illustrations of the utility of this dual function group have been afforded by the synthesis of a number of sequential polypeptides (Jones, 1969; Trudelle, 1971; 1973; Cowell and Jones, 1971a,b, 1972; Fairweather and Jones, 1972; Jones and Walker, 1972; Bell *et al.*, 1974).

3. 2,2-Diphenyl-2-hydroxyethyl Ester

This dual function ester was introduced by Wieland *et al.* (1970). The 2,2-diphenyl-2-hydroxyethyl esters (**13**) are used for COOH-terminal protection during chain elongation. In anticipation of further peptide coupling through the COOH-terminus, acid-catalyzed dehydration with trifluoroacetic acid, converts the protective ester to the active 2,2-diphenylvinyl ester (**14**). An illustration of the method has been reported (Wieland *et al.* 1970).

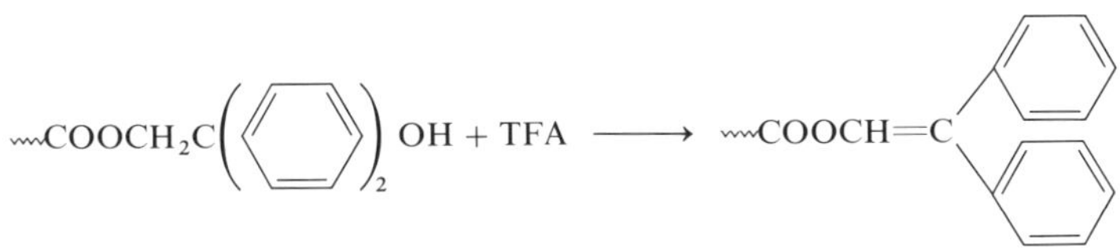

(**13**) (**14**)

The protective 2,2-diphenyl-2-hydroxyethyl esters have been used in combination with the trityl and benzyloxycarbonyl protecting groups. However, the combination of acid-labile groups for side-chain protection is effectively precluded by the dehydration conditions necessary for the generation of the activated ester (**14**).

This dual function group has been applied to the solid-phase technique of peptide synthesis by the attachment of 4′-aminomethyl-2,2-diphenylethanediol (**15**) to a chloromethylated polymer via its aminomethyl function (Wieland *et al.*, 1972). N-Protected amino acids have been esterified to the free hydroxy group and dehydration from the resin-bound ester has generated the reactive enol esters.

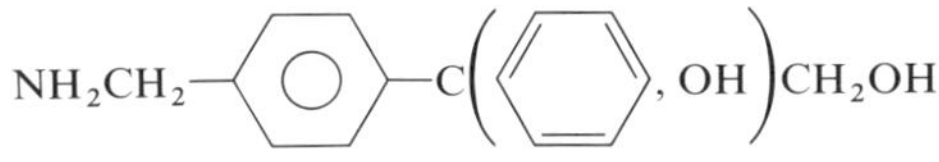

(**15**)

4. Phenacyloxy Ester

The protective phenacyloxy esters were introduced by Stelakatos *et al.* (1966) for carboxyl protection during chain elongation and maniputation. These esters are quite stable to acid hydrolysis (Hendrickson and Kandall, 1970) and acidolysis (Stelakatos *et al.*, 1966). However, Treiber (1973) has found that this protective ester (**16**) can be converted to the active 2-cyano-2-phenylvinyl ester (**17**) by the action of hydrogen cyanide in the presence of base (Scheme 10). This vinyl-type ester has been used to prepare peptides, and it is claimed that no racemization can be expected during its conversion from a protective to an activated ester.

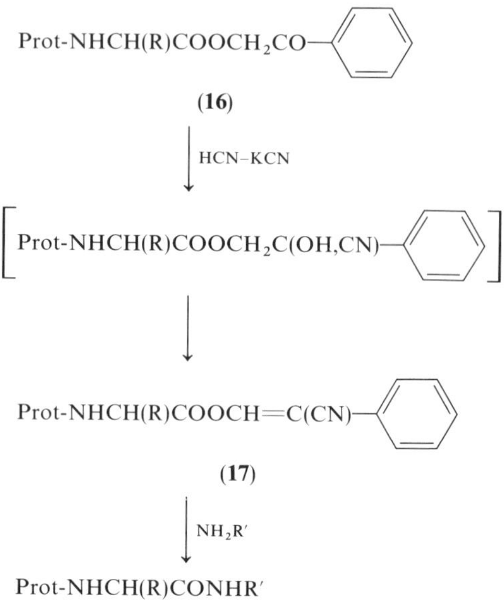

Scheme 10

B. Activated Esters Used for Protection and Subsequent Coupling

1. Pentachlorophenyl, Trichlorophenyl, and p-Nitrophenyl Esters

One of the reasons for the introduction of dual function groups was to overcome some of the problems associated with COOH-terminal activation necessary for peptide segment condensations. An approach to this problem has been to use active esters for carboxyl protection with the objective of synthesizing protected peptide activated esters. The methodology requires the stepwise addition of N-protected amino acid residues (**18**) to an amino acid or peptide active ester (**19**) to give the protected peptide active ester (**20**). The success of the method depends upon the relative rates of formation of the peptide compared to the rates of intramolecular and polycondensations of the peptide active esters. Peptide formation is usually achieved by the use of the mixed anhydride or dicyclohexylcarbodiimide procedures.

Z-Asp(O*t*Bu)OH + Pro-ONp·HBr

(18) **(19)**

↓ MA

Z-Asp(O*t*Bu)-Pro-ONp

(20)

Illustrations of this process are given in the synthesis of a number of activated peptide monomers designed for the preparation of sequential polypeptides. Carboxyl protection has been afforded by pentachlorophenyl esters (Kapoor and Gerencser, 1969; Kapoor *et al.*, 1969; Poroshin *et al.*, 1969; Yusupov *et al.*, 1969). These latter investigators also used trichlorophenyl esters for protective purposes. Similarly, Stevenson *et al.* (1972) have used the *p*-nitrophenyl ester for carboxyl protection during the synthesis of a protected nonapeptide of rubredoxin.

III. MISCELLANEOUS GROUPS

A. Activation of Methyl Esters

Methyl esters can be activated by a simple procedure using reagents derived from monothiols and trimethylaluminum to give the corresponding thioesters (Corey and Beames, 1973) (Scheme 11). The application of this method of activation to peptide synthesis has not been reported.

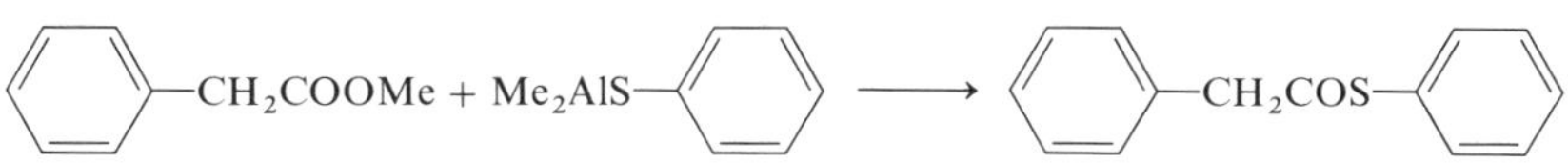

Scheme 11

B. Acylsulfonamides

Acylsulfonamides have been proposed by Kenner *et al.* (1971) as dual function groups. This role was suggested by their resistance to cleavage under a variety of conditions usually encountered in peptide chemistry. However, upon their N-methylation, cleavage can be achieved under mild conditions to produce the protected peptide free acid, amide or hydrazide (Scheme 12).

The carboxyl protecting acylsulfonamides are stable to the acidic conditions used for the removal of the benzyloxycarboxyl and *tert*-butyloxycarbonyl protecting groups. They are therefore highly applicable for the solid-phase technique of peptide synthesis which Kenner *et al.* (1971) have used

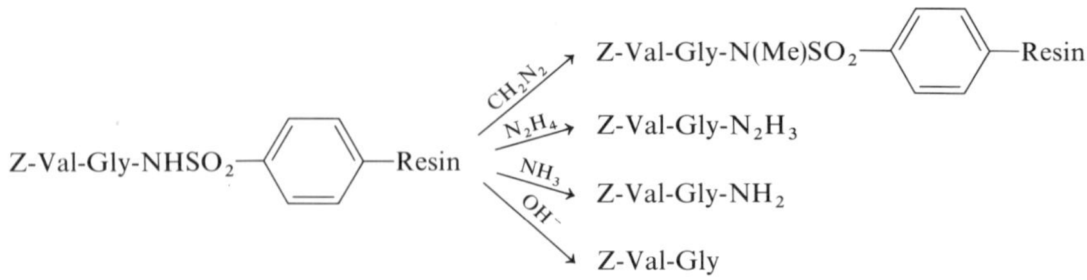

Scheme 12

to synthesize a number of protected peptides. During these studies some limitations, such as transpeptidation and slight racemization of the COOH-terminal residues, were noted.

C. Picolyl Esters

Picolyl esters (**21**) were introduced by Camble *et al.* (1968). Although these esters behave as carboxyl-protecting moieties during peptide synthesis, their inclusion here is based on their additional capacity to aid in peptide

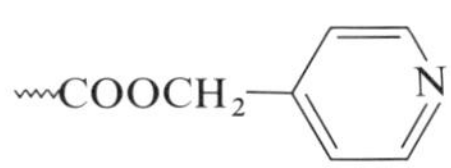

(**21**)

purification (Camble *et al.*, 1969). The basic nature of the picolyl ester allows, after each coupling reaction, the separation of the growing peptide from the various by-products and excess reagents. This separation is achieved by sulfoethyl–Sephadex (Schafer *et al.*, 1971) or Amberlyst-15 ion-exchange resins (Burton *et al.*, 1971) to which the picolyl peptide is bound and from which the various impurities are washed away. The purified peptide picolyl ester can be released from sulfoethyl–Sephadex by extraction into aqueous citric acid or from Amberlyst-15 by a strong tertiary base. Illustrations of the method have been reported in syntheses of bradykinin (Schafer *et al.*, 1971, [Val^5]angiotensin II (Garner and Young, 1971), [Val^5]angiotensin II (Garner and Young, 1971), some sequences of fibrinogen (Schafer, 1972), and some bradykinin analogues (Fletcher and Young, 1972).

REFERENCES

Ali, A., Faesel, J. H. R., Sarantakis, D., Stevenson, D., and Weinstein, B. (1972). *Int. J. Protein Res.* **4**, 395–404.

Beacham, J. Dupuis, G., Finn, F., M. Storey, H. T., Yanaihara, C., Yanaihara, N., and Hofmann, K. (1971). *J. Am. Chem. Soc.* **93**, 5526–5539.

Bell, J. R., Jones, J. H., Regester, D. M., and Webb, T. C. (1974). *J. Chem. Soc., Perkin Trans. 1* pp. 1961–1964.
Biggins, R., Cavins, T., Eglinton, G., Haslam, E., and Haworth, R. D. (1963). *J. Chem. Soc. B* pp. 1750–1756.
Boissonnas, R. A., Guttmann, S., and Jaquenoud, P. A. (1960). *Helv. Chim. Acta* **43**, 1349–1358.
Böshagen, H., and Ullrich, J. (1959). *Chem. Ber.* **92**, 1478–1480.
Burton, J., Fletcher, G. A., and Young, G. T. (1971). *Chem. Commun.* pp. 1057–1058.
Camble, R., Garner, R., and Young, G. T. (1968). *Nature (London)* **217**, 247–248.
Camble, R., Garner R., and Young, G. T. (1969). *J. Chem. Soc. C* pp. 1911–1916.
Camble, R., Dupuis, G., Kowasaki, K., Romovacek, H., Yanaihara, N., and Hofmann, K. (1972). *J. Am. Chem. Soc.* **94**, 2091–2100.
Capon, B., and Ghosh, B. C. (1966). *J. Chem. Soc. B* pp. 472–478.
Cheung, H. T., and Blout, E. R. (1965). *J. Org. Chem.* **30**, 315–316.
Corey, E. J., and Beames, D. J. (1973) *J. Am. Chem. Soc.* **95**, 5829–5831.
Cowell, R. D., and Jones, J. H. (1971a). *J. Chem. Soc. C* pp. 1082–1090.
Cowell, R. D., and Jones, J. H. (1971b). *Chem. Commun.* pp. 1009–1010.
Cowell, R. D., and Jones, J. H. (1972). *J. Chem. Soc., Perkin Trans. 1* pp. 2236–2240.
Fairweather, R., and Jones, J. H. (1972). *J. Chem. Soc., Perkin Trans. 1* pp. 2475–2481.
Flanigan, E., and Marshall, G. R. (1972). *In* "Progress in Peptide Research" (S. Lande, ed.), pp. 7–14. Gordon & Breach, New York.
Fletcher, G. A., and Young, G. T. (1972). *J. Chem. Soc., Perkin Trans. 1* pp. 1867–1874.
Garner, R., and Young, G. T. (1971). *J. Chem. Soc. C* pp. 50–53.
Hansen, B. (1963). *Acta Chem. Scand.* **17**, 1375–1379.
Hendrickson, J. B., and Kandall, C. (1970). *Tetrahedron Lett.* pp. 343–344.
Hofmann, K., Lindenmann, A., Magee, M. Z., and Kahn, N. H. (1952). *J. Am. Chem. Soc.* 74, 470–476.
Hofmann, K., Visser, J. P., and Finn, F. M. (1970). *J. Am. Chem. Soc.* **92**, 2900–2909.
Johnson, B. J., and Jacobs, P. M. (1968a). *J. Org. Chem.* **33**, 4525–4526.
Johnson, B. J., and Jacobs, P. M. (1968b). *Chem. Commun.* pp. 73–75.
Johnson, B. J. (1969). *J. Org. Chem.* **34**, 1178–1179.
Johnson, B. J., and Rea, D. S. (1970). *Can. J. Chem.* **48**, 2509–2511.
Johnson, B. J., and Ruettinger, T. A. (1970) *J. Org. Chem.* **35**, 255–257.
Johnson, B. J., and Tracey, D. E. (1969). *J. Pharm. Sci.* **58**, 1299.
Johnson, B. J., and Trask, E. G. (1968). *J. Org. Chem.* **33**, 4521–4524
Jones, J. H. (1969). *Chem. Commun.* pp. 1436–1437.
Jones, J. H., and Walker, J. (1972). *J. Chem. Soc., Perkin Trans. 1* pp. 2923–2926.
Jones, J. H., and Young, G. T. (1968). *J. Chem. Soc. C* pp. 436–441.
Kapoor, A., and Gereneser, L. W. (1969). *J. Pharm. Sci.* **58**, 976–979.
Kapoor, A., Gereneser, L. W., and Koutnik, W. R. (1969). *J. Pharm. Sci.* **58**, 281–282.
Kawasaki, K., Camble, R., Dupuis, G., Romovacek, H., Storey, H. T., Yanaihara, C., and Hofmann, K. (1973). *J. Am. Chem. Soc.* **95**, 6815–6824.
Kemp, D. S., Bernstein, Z., and Rebek, J. (1970). *J. Am. Chem. Soc.* **92**, 4756–4757.
Kenner, G. W., and Seely, J. H. (1972). *J. Am. Chem. Soc.* **94**, 3259–3260.
Kenner, G. W., McDermott, J. R., and Sheppard, R. C. (1971). *Chem. Commun.* pp. 636–637.
Lewalter, J., and Birr, C. (1970). *Justus Liebigs Ann. Chem.* **740**, 48–51.
Macrae, R., and Young, G. T. (1975). *J. Chem. Soc., Perkin Trans. 1* pp. 1185–1191.
McDermott, J. R., and Benoiton, N. L. (1972). *In* "Chemistry and Biology of Peptides" (J. Meienhofer, ed.), pp. 369–372. Ann Arbor Sci. Publ., Ann Arbor, Michigan.
Marshall, D. L., and Liener, I. E. (1970). *J. Org. Chem.* **35**, 867–868.
Milne, H. B., and Kilday, W. (1965). *J. Org. Chem.* **30**, 64–66.
Milne, H. B., and Most, C. F. (1968). *J. Org. Chem.* **33**, 169–175.

Moroder, L., Borin, G., Marchiori, F., and Scoffone, E. (1972). *Biopolymers* **11**, 2191–2194.
Poroshin, K. T., Zegel'man, A. B., and Yusupov, T. Y. (1969). *Dokl. Akad. Nauk Tadzh SSR* **12**, 20–23.
Preston, J., and Weinstein, B. (1967). *Experientia* **23**, 265–266.
Romovacek, H., Drabarek, S., Kawasaki, K., Dowd, S. R. Obermeier, R., and Hofmann, K. (1974). *Int. Pept. Protein Res.* **6**, 435–445.
Schafer, D. J. (1972). *J. Chem. Soc., Perkin Trans. 1* pp. 1452–1456.
Schafer, D. J., Young, G. T., Elliott, D. F., and Wade, R. (1971). *J. Chem. Soc. C* pp. 46–49.
Schröder, E., and Lübke, K. (1963). *In* "Peptides 1962" (G. T. Young, ed.), pp. 195–206. Pergamon, Oxford.
Schwyzer, R., and Sieber, P. (1958). *Helv. Chim. Acta* **41**, 2190–2199.
Schwyzer, R., Surbeck-Wegmann, E., and Dietrich, H. (1960). *Chimia* **14**, 366–370.
Sieber, P., and Riniker, B. (1973). *In* "Peptides 1971" (H. Nesvadba, ed.), pp. 49–53. North-Holland Publ. Amsterdam.
Stelakatos, G. C., Paganou, A., and Zervas, L. (1966). *J. Chem. Soc. C* pp. 1191–1197.
Stevenson, D., Cook, R. M., and Weinstein, B. (1972). *Int. J. Protein Res.* **4**, 101–108.
Storey, H. T., Beecham, J., Cernosek, S. F., Finn, F. M., Yanaihara, C., and Hofmann, K. (1972). *J. Am. Chem. Soc.* **94**, 6170–6178.
Treiber, L. R. (1973). *Experientia* **29**, 1335–1337.
Trudelle, Y. (1971). *Chem. Commun.* pp. 639–641.
Trudelle, Y. (1973). *J. Chem. Soc., Perkin Trans. 1* pp. 1001–1005.
Wang, S. (1972). *In* "Chemistry and Biology of Peptides" (J. Meienhofer, ed.), pp. 179–182. Ann Arbor Sci. Publ., Ann Arbor, Michigan.
Wang, S. (1973). *J. Am. Chem. Soc.* **95**, 1328–1333.
Wang, S. (1975). *J. Org. Chem.* **40**, 1235–2139.
Wang, S., and Merrifield, R. B. (1969). *J. Am. Chem. Soc.* **91**, 6488–6491.
Wang, S., and Merrifield, R. B. (1972). *Int. J. Protein Res.* **4**, 309–318.
Weinstein, B., Cook, R. M., and Heffernan, J. T. (1976). *Int. J. Pept. Protein Res.* **8**, 21–25.
Weygand, F., and Steglich, W. (1959). *Chem. Ber.* **92**, 313–319.
Wieland, T., Lewalter, J., and Birr, C. (1970). *Justus Liebigs Ann. Chem.* **740**, 31–47.
Wieland, T. Birr, C., and Fleckenstein, P. (1972). *Justus Liebigs Ann. Chem.* **756**, 14–19.
Wolters, E. M. T., Tesser, G. I., and Nivard, R. J. F. (1974). *J. Org. Chem.* **39**, 3388–3392.
Yajima, H., and Kiso, Y. (1971). *Chem. Pharm. Bull.* **19**, 420–423.
Yanaihara, N., Yanaihara, C., Dupuis, G., Beacham, J., Camble, R., and Hofmann, K. (1969). *J. Am. Chem. Soc.* **91**, 2184–2185.
Young, G. T. (1967). *In* "Peptides 1966" (H. C. Beyermann, A. van de Linde, and W. Maassen van den Brink, eds.), pp. 55–66. North-Holland Publ., Amsterdam.
Yusupov, T. Y., Zegel'man, A., and Poroshin, K. T. (1969). *Dokl. Akad. Nauk Tadzh. SSR* **12**, 27–31.

Author Index

Numbers in italics refer to the pages on which the complete references are listed.

D

E

F

G

H

I

S

T

Y

Z

Subject Index

B

E

F

I

K

L

N

Q

R

S

V

W

X,Y,Z